INTERNATIONAL WETLANDS: ECOLOGY, CONSERVATION AND RESTORATION

INTERNATIONAL WETLANDS: ECOLOGY, CONSERVATION AND RESTORATION

JOSE R. HERRERA
EDITOR

Nova Science Publishers, Inc.
New York

NOTICE TO THE READER

The Publisher has taken reasonable care in the preparation of this book, but makes no expressed or implied warranty of any kind and assumes no responsibility for any errors or omissions. No liability is assumed for incidental or consequential damages in connection with or arising out of information contained in this book. The Publisher shall not be liable for any special, consequential, or exemplary damages resulting, in whole or in part, from the readers' use of, or reliance upon, this material. Any parts of this book based on government reports are so indicated and copyright is claimed for those parts to the extent applicable to compilations of such works.

Independent verification should be sought for any data, advice or recommendations contained in this book. In addition, no responsibility is assumed by the publisher for any injury and/or damage to persons or property arising from any methods, products, instructions, ideas or otherwise contained in this publication.

This publication is designed to provide accurate and authoritative information with regard to the subject matter covered herein. It is sold with the clear understanding that the Publisher is not engaged in rendering legal or any other professional services. If legal or any other expert assistance is required, the services of a competent person should be sought. FROM A DECLARATION OF PARTICIPANTS JOINTLY ADOPTED BY A COMMITTEE OF THE AMERICAN BAR ASSOCIATION AND A COMMITTEE OF PUBLISHERS.

LIBRARY OF CONGRESS CATALOGING-IN-PUBLICATION DATA

International wetlands : ecology, conservation, and restoration / Jose R. Herrera, editor.
  p. cm.
 ISBN 978-1-60456-999-5 (hardcover)
 1. Wetland ecology. 2. Wetland conservation. 3. Wetland restoration. I. Herrera, Jose R.
 QH541.5.M3I585 2008
 577.68--dc22
                          2008031095

Published by Nova Science Publishers, Inc. ✦ New York

CONTENTS

PREFACE

Wetlands are lands where saturation with water is the dominant factor determining the nature of soil development and the types of plant and animal communities living in the soil and on its surface. Wetlands vary widely because of regional and local differences in soils, topography, climate, hydrology, water chemistry, vegetation, and other factors, including human disturbance. Indeed, wetlands are found from the tundra to the tropics and on every continent except Antarctica. This book focuses on international wetlands.

Chapter 1 – Despite Spanish wetland richness, many have disappeared in the last century and many others are threatened. Desiccation, eutrophication and pollution, organic matter accumulation, siltation, salinisation and the invasion of exotic species are the most frequent processes affecting Spanish wetlands. We here briefly outline these problems, paying attention to some case studies that illustrate this picture. Miopic attitudes of stakeholders (irrigation farmers most often), struggles among different environmental offices and poor scientific knowledge of environmental managers are responsible for widespread failures in Spanish wetland management and restoration despite enough funding.

Chapter 2 - Freshwater wetlands are highly productive and important features of terrestrial landscapes, yet knowledge of their biotas and understanding of their function has lagged behind that of other ecosystems. Superficially, their communities are known to include bacteria, protists, algae, fungi, higher plants, and invertebrate and vertebrate animals that come together to benefit from the rich resources and oftentimes relatively predator-free space that these habitats provide. The cost of these benefits requires adaptation to a highly variable, though often predictable, hydroperiod. This chapter will focus on wetland habitats that are truly intermittent in their water-balance, and will summarize what is known of the structure of the biota at different trophic levels and how they interact at a dynamic land/water interface. For example, fluctuations in flooding regime are known to cause fluxes in carbon, nitrogen and phosphorus in basin sediments as well as in the water column. Such nutrient pulses can lead to increased productivity followed by anoxia and fluxes of methane in these sediments significantly, microbes that die during the drying period can represent an important source carbon upon re-wetting at the same time that phosphorus and nitrogen enter the system fluxes produce a range of responses from both prokaryote and eukaryote components wetland community. Detritivore and herbivore consumers in wetlands are represented rich variety of protists and invertebrates, especially crustaceans and insects, many appear to be predictably represented, taxonomically, across wetland types. provide the crucial trophic link between producers and the top predators

trophic status of some of the latter is variable. For example, many amphibians that breed in wetlands represent both prey and predator depending on the stage in their life cycle. While salamanders may input large amounts of energy via egg deposition, many larger vertebrate inhabitants of wetlands are seen as net removers of energy from these systems (grazing, predation, etc.), with very little return (chiefly wastes) – although intuitively, they must represent important links between aquatic and terrestrial compartments. However, it must be said that the role of vertebrates in nutrient cycling and energy flow in wetlands is not well understood, and that the consequences of, for example the global decline in amphibian populations, are largely unknown. Coverage of these topics will be global in extent. Future research needs to be focused on piecing together the components of intermittent wetland function through study of: 1) physiological and phenological response to natural (e.g., hydroperiod regime) and human-induced stress by individual taxa; 2) community response to these same factors; 3) trophic function, including bottom-up resources and top-down pressures; 4) interaction with, especially energy flow through, permanent sections of wetland, the riparian zone, adjacent woodlands and grasslands, and beyond.

Chapter 3 - The biomass and productivity of macrophytes on a spatio-temporal scale was studied in the monsoonal wetland system of Keoladeo National Park (KNP), Bharatpur, India during July 2005 through April 2006. Macrophyte samples were collected from 7 blocks of the wetland, which remain inundated for most of the months in the year. Species richness, assemblage of all the macrophytes and aboveground net primary productivity (NPP, 2.71 g m^{-2} day^{-1} on dry weight basis) were examined. 37 species of macrophytes were recorded in the quadrats sampled. Of the macrophytes studied, the order of the major species in terms of their biomass per unit area was *Pseudoraphis spinescens* > *Cynodon dactylon* > *Sporobolus* sp. > *Paspalum distichum* > *Vallisneria natans* > *Alternanthera* sp. > *Utricularia* sp. > *Hemiadelphis polyspermus* > *Najas minor*. The maximum average biomass of macrophytes was recorded during November 2005 (92.21 g m^{-2}) and the minimum during January 2006 (27.52 g m^{-2}). Schluter's test for the macrophyte species showed positive association with each other (W = 147.79, $P < 0.05$). Univariate analysis of variance showed significant variation in biomass among the months, the blocks and the categories of macrophytes. The water depth during the study period varied from 0.5 – 1.15 meter. Water depth seems to be the significant explanatory variable for aquatic vegetation assemblage in our study, conjointly with other factors including location specific history of the plant communities. The biological ...ity often observed in the park is strongly related to hydro period.

...ter 4 - We examined the arrival pattern of fishes into the wetland system of ...ional Park (KNP), Bharatpur, India for a period of three years, i.e. 2003 through ...was undertaken to assess the temporal variation in the arrival pattern of ...n and explore the influence of the processes in the catchment on ...ce of fishes in the feeder canal. A total of 40,766 individuals ...enera and 24 species of fishes were recorded during the study. ...however, suggests that there might be 30 species in the ...vestigation recorded 7 new arrivals to the park, viz. ...pio communis, Clarias gariepinus, Danio dangila, ...and Securicula gora. The highest fish diversity was ...2003 (0.934). The drift density and rate differed among ...and were positively correlated with each other (r = 0.703, P ...ec) was negatively correlated with the number of individuals per catch. Except *B. albolineatus, Catla catla, C. gariepinus* and *Mastacembelus armatus*, all other fish species were distinct and independent with respect to their abundance (χ^2, $P < 0.05$). During hierarchical cluster analysis of all the recorded species *Puntius sophore* evidently formed a distinct and separate cluster. *Chanda ranga* and *Puntius sarana* formed one cluster while the rest of the species together formed another cluster.

The quantity and duration of water release, a function of total rainfall in the catchment, influenced the composition of fishes and number entering the park. The decline in the fish species recruitment in comparison to the previous reports necessitates a review of the ongoing anthropogenic pressures on fishes, and habitat destruction for infrastructure development and urbanization in the catchments.

Chapter 5 - This chapter summarizes the nutrient and metal removal of a free water surface constructed wetland and compares it with a previous small-scale prototype. The wetland was built to treat wastewater containing metals (Cr, Ni, Zn) and nutrients (P and N) from a tool factory in Santo Tomé, Santa Fe, Argentina. Water, sediment and macrophytes were sampled at the inlet and outlet areas of the constructed wetland during three years. Both wetlands showed high efficiencies in contaminant removal. The development of the vegetation showed the same pattern in both the small-scale prototype and the large-scale wetland. Three successive phases of vegetation dominance were developed and three different patterns of contaminant retention were observed. A first phase of floating macrophyte colonization with dominance of *Eichhornia crassipes* was followed by a combined floating-emergent phase ending in an long, on-going stand of emergent macrophyte phase, with dominance of *Typha domingensis*. During the *E. crassipes* dominance, contaminants were retained in the macrophyte biomass; during the *E. crassipes* + *Typha domingensis* stage, contaminants were retained in the sediment and in the *T. domingensis* dominance stage, contaminants were retained in sediment and in the macrophyte biomass. Even though retention mechanisms were different, removal efficiencies did not show significant differences among the three vegetation stages, except for NH_4^+ and soluble reactive phosphorus (SRP). Greenhouse experiments were carried out to explain the early disappearance of floating macrophytes. Thresholds for conductivity, pH and metal damage were determined. High conductivity and pH of the incoming wastewater were the cause of the disappearance of the floating species. Because of its highest tolerance, *T. domingensis* is the best choice to treat wastewater of high pH and conductivity with heavy metals, a common result from many industrial processes.

Chapter 6 - We have analysed the planktonic communities and their trophic interactions in a vegetated wetland from the Lower Paraná River (Otamendi Natural Reserve, Argentina). The studies combined field surveys, and *in situ* experiments at microcosms and mesocosms scales.

Fields studies carried out in the main shallow lake of the wetland and in abandoned meanders of the river (relict oxbow lakes) show that the free-floating plants cover was one of the driving forces in the regulation of the structure of planktonic communities. Later on, the effect of the light attenuation due to floating macrophytes was simulated at microcosms scale, which showed changes in the structure of the microbial planktonic communities in response to the light penetration. In particular we found changes in the relationship autotrophs/heterotrophs + mixotrophs, depending on light conditions inside the enclosures. Further experiments with fluorescent-labeled bacteria (FLB), revealed the presence of mixotrophic phytoplankton species, which are frequent components of the phytoplankton in this wetland.

The zooplankton predation impact on phytoplankton was assessed using experiments at mesocosms scale in terms of abundance, size fraction structure and species composition. Likewise, the cascading responses of other components the microbial chains were also explored (heterotrophic nanoflagellates, ciliates and picoplankton)

On the other hand, the effect of nutrients (bottom up) and predation (top down) on the phytoplankton communities were simultaneously assessed. Regarding the top down effect, the impact of planktivorous fish predation over zooplankton was analyzed, as well as the consecutive cascading effect on phytoplankton. Moreover, the effect of nutrient release from natural lake sediments, was assessed.

In this communication we propose a first model of functioning of the plankton communities and their main interactions for this South American wetland, based on field and experimental data, which have been obtained along a six- years study period.

Chapter 7 - In order to alleviate the water shortage problem in North China, China government has been building the South-to-North Water Diversion Project since 2000. The project included three systems: clean water passages, water using security system and water quality improvement engineering system. No municipal wastewater along the line would allow discharging to the passages directly. Constructed wetlands were adopted for municipal wastewater treatment along the line.

Sihong County is located in the northwest of Jiangsu province, P.R.China. In order to meet the demands of the South-to-North Water Diversion Project, under the national finance support, Sihong County Sewage Water Treatment Plant was built in 1999. Subsurface flow constructed wetland system (SSFCW) was adopted in this plant. It was used for treating municipal wastewater. The design capacity is 50,000 cubic meter per day.

According to the monitoring data, the kinetic equation of COD_{Cr} removal with hydraulic detention time in SSFCW fitted well with the first order reaction. The total removal rates of COD_{Cr} in both spring and autumn were 74.1% and 50.9%, respectively. The change of seasons had a great effect on the purifying effect. In spring and autumn, the removal rates for the total nitrogen were 64.6% and 49.3%, 70.7% and 50.5% for ammonia, respectively.

Tengzhou City is also along the South-to-North Water Diversion Project. It is located in the south of Shandong province, P.R.China. Surface flow constructed wetland system (SFCW) was adopted for treating micro-polluted river water. In 2006, a pilot-scale SFCW was built on coal mine subsidence area.

The results showed that hydrophytes played an important role, and the SFCW had good purification to micro-polluted river water. Comparing the reed system and the cattail system, the former had good removal effect on BOD_5, COD_{Cr}, ammonia and phosphorous. The effluent of two systems can reach the third level of China Surface Water Quality Standard (GB3838-2002).

As an ecological wastewater treatment system, constructed wetland system adapts to municipal wastewater treatment plants in small towns. This technology is in low cost at construction and operation, also the effluent quality is well. And the efficiency of removal of nitrogen and phosphorus is higher than conventional activated sludge. So it has a great value of popularization.

Chapter 8 - The distribution of China's mangrove forests has experienced a considerable loss during the past several decades. Systematic research on current mangrove ecosystems is urgently needed to protect and restore diminishing mangrove forests. The Gaoqiao mangrove forest is the core area of the Zhanjiang Mangrove National Nature Reserve, the largest state mangrove nature reserve in China. We investigated species diversity, population biomass, succession, composition and soil attributes of the major community types of the Gaoqiao mangrove forest. Comparing with the previous controversial studies, we pointed out nine mangrove species within eight genera and four families occurred in Gaoqiao, and there were six main mangrove community types. Community composition indicated that the well-grown *Rhizophora stylosa* + *Bruguiera gymnorrhiza* community perhaps represented the climax community in Gaoqiao. As to species diversity, communities with two canopy layers had relatively richer species diversity, lower ecological dominance and higher evenness. In terms of biomass, big difference existed between different population and different organ. For example, ratios of root biomass to total biomass of pioneer species were relatively larger as they possessed stronger roots to live in large tidal fluxes. Six communities covered early, middle and late succession stages, respectively were distributed at different intertidal sites. Soils under mangrove communities at different succession stages showed that with succession soil acidification, organic matter content and salinity increased, but the total N, P content didn't change along the succession series. Local mangrove succession mode was presented, based on different species' abilities to suit the changing soil attributes and tidal influence at different intertidal sites. This study provides a valuable prerequisite for mangrove conservation and management.

Chapter 9 - Wetlands are widespread on the Southern Kuril Islands. Wetland development is associated with late Pleistocene-Holocene climate warming. There are several types of wetlands. Blanket peat bogs are more ancient and have existed on flattened surfaces since the Postglacial warming (about 12 ka) and the accompanying increase in precipitation. Wetlands in coastal zones were formed in bays after the Holocene transgression maximum and the age of the oldest peat bogs are less than 6-4 ka. The oldest peat bogs are more than 7 ka and are found in coastal zones of volcanic islands; local conditions determined their formation. Wetland environments attained peak expansion in the Late Holocene when isthmus areas arose, coastal plains formed and surfaces turned into swamps. Wetlands in the lower portions of river valleys are also typical. Wetlands were formed in different ecological environments and during the Holocene had different peat buildup rates. Factors modulating wetland development are discussed. Wetland evolution during Holocene climatic changes is considered. Reconstructions are based on pollen, diatom, botanic analysis of peat and radiocarbon data. Sea-level changes played a key role in coastal wetland development. Specific local factors influencing wetland development on the islands are volcanic activity, varied neo-tectonic movements and tsunamis.

Chapter 10 - The large quantity of publications is devoted to the studies of bog ecosystems in the Lower Amur River basin. To identification of regularities of the bogs placement, development and classification in the region under study, the most significant contribution was made by Yu.S. Prozorov [16, 17, 18, 19]. In accordance with the zoning scheme of the USSR peat fund, the Lower Amur River territory was identified as the greatest in area peat basin of the Far East. In the Lower Amur River basin, the typological composition of the bog systems is the most representative as compared with other regions of the Russian Far East [22]. This work presents materials concerning the vegetation, some properties of peat, evolution and stability of the bog ecosystems in the Lower Amur River basin.

Chapter 11 - The emergence of an epistemology of complexity is a contemporary phenomenon strongly based upon technology that has claimed new research needs and revitalize old studied themes. The study of wetlands can be considered in this context. The ability to respond in an organized manner to disturbances caused by seasonal flooding, a characteristic of wetlands ecosystems, define them as complex systems. The spatio-temporal heterogeneity of floodplain river systems is responsible for a diverse array of dynamic aquatic habitats. Thus, natural disturbances, represented by flooding and responsible for intensifying ecosystem heterogeneity, are the main factor in maintaining the ecological integrity. This requires knowledge of long-term patterns of inundation to preserve geomorphic formations, in other words, habitat diversity. Nevertheless, the processes that control water, sediment and nutrients transfer through the floodplain are not well understood. The maintenance of river-floodplain connectivity has been recognized as a central strategy in ecosystem management. In this sense, remote sensing technology for a broad-scale systematic focus is particularly relevant. Research studies in this area provide new approaches for models useful in management on a regional scale. This chapter aims therefore to illustrate this wide research field involving current concepts and methodologies in order to improve the comprehension of the wetlands ecosystems as well as establish conservation criteria.

Chapter 12 - Lake Kasumigaura is the second largest lake in Japan and well known for its eutrophication. Surface-flow wetlands (area: 38,000 m^2; 50 m × 40 m × 19 units) planted with common reed (*Phragmites australis*) were constructed near the estuary of the Seimeigawa River, which is a nutrient polluted river flowing into Lake Kasumigaura, to treat the river water. The hydraulic loading rate, reduction in nutrient contents, plant transition, and components of sediment were investigated after construction and further surveyed periodically from 1999 to 2005. Although nutrient removal efficiencies varied after several years running, particularly from spring to summer, considerable SS, nitrogen, and phosphorus amounts were retained by the constructed wetland in 2005 after 10 years running. The average total sediment accumulation was 3.61 – 4.43 kg m^{-2} yr^{-1}. The total phosphorus content in sediment was similar or slightly more than the amount removed from the inflow river water, whereas the total nitrogen content in sediment was significantly less than that removed, suggesting that phosphorus was removed with particle sedimentation while most nitrogen was removed by denitrification.

On the other hand, the average community diversity index of four representative constructed wetland units was 0.88 in 2004, which was significantly higher than 10 years ago, indicating greater diversity. In addition, compared with the *P. australis* population in native wetland, *P. australis* in the constructed wetland had more gene types in 2005 probably owing to the constructed wetland being in an establishment stage, whereas the native wetland had already entered a stationary stage.

ISBN: 978-1-60456-999-5

Chapter 1

WETLAND MANAGEMENT IN SPAIN: A HISTORY OF CONTROVERSIAL CONSERVATION

Santos Cirujano[1], Miguel Álvarez-Cobelas[2],
Salvador Sánchez-Carrillo[2], David G. Angeler[3,1]
and Pablo García-Murillo[4]

[1]CSIC, Real Jardín Botánico, Madrid, Spain
[2]CSIC, Instituto de Recursos Naturales, Madrid, Spain
[3]Inst. Ciencias Ambientales, Universidad de Castilla-La Mancha, Toledo, Spain
[4]Dept. Botánica, Fac. Farmacia, Univ Sevilla, Sevilla, Spain

ABSTRACT

Despite Spanish wetland richness, many have disappeared in the last century and many others are threatened. Desiccation, eutrophication and pollution, organic matter accumulation, siltation, salinisation and the invasion of exotic species are the most frequent processes affecting Spanish wetlands. We here briefly outline these problems, paying attention to some case studies that illustrate this picture. Miopic attitudes of stakeholders (irrigation farmers most often), struggles among different environmental offices and poor scientific knowledge of environmental managers are responsible for widespread failures in Spanish wetland management and restoration despite enough funding.

INTRODUCTION

Spain's landscape is dotted with wetlands of variable size (< 0.5 ha to ca. 24,000 ha covered by the Doñana marshes) and they show diverse environmental and ecological settings (oligotrophic temporary mountain ponds, permanent karst lagoons, floodplain wetlands, coastal wetlands (so-called albuferas), ephemeral water bodies, endorheic hypersaline areas,

---

[1] Current address: Dept. Environmental Assessment, Swedish Univ. Agricultural Sciences, Uppsala, Sweden

etc; Casado & Montes, 1995). Currently, from the thousands of Spain's wetlands, 63 (281,768 ha) are listed in the Ramsar list (Bernués, 1998) and a large proportion of the scientific, cultural and economic value of this natural patrimony still remains to be explored.

The management and rehabilitation of wetlands is a relatively recent topic in Spain. Although the first attempts to protect wetlands dates back to 1969 (Doñana) and 1973 (Las Tablas de Daimiel), when these sites were declared National Parks, it has not been until the 1980s when sound management and conservation programmes have been launched (Vallès, 1989). Although the need for management based on ecological criteria was advocated in the first scientific wetland meeting in 1987 (VV. AA., 1987), integral management schemes are only starting to be implemented recently.

Until the 1970s wetland management was perceived to be most 'effective if human interventions were reduced to a minimum. However, large-scale socioeconomic changes in this country during the last century soon required artificial management if wetlands and their component species (many of socioeconomic value) are to be conserved appropriately.

In this sense, initial management schemes focused on local habitat with the aim to conserve target waterfowl species. Management aimed chiefly at increasing artificially the flooded area and altering vegetation to increase the habitat use of rare and threatened waterfowl, rather than dealing with overall wetland integrity. Projects were considered to be successful when censuses detected an increase of target bird populations and the appearance of protected species. Other important wetland features, including water quality and vegetation dynamics were widely ignored. While a certain short-term success was achieved reflected by the establishment of threatened birds like the white-headed duck (*Oxyura leucocephala)*, the marbeled duck (*Anas angustirostris)* or the gull-billed tern *(Gelochelidon nilotica*) (Sargatal & Fèlix, 1989; Ambiental S.L., 1992; Jubete et al. 2006, Matamala & Aguilar, 2003) the impact on other structural and functional wetland features became evident only in the long term. Shifts in trophic state conditions (eutrophication) with consequent biodiversity declines, and impoverishment of habitat quality ultimately resulted in a population decrease of target waterbirds, thereby jeopardizing wetland ecology and management.

These failures lead to changes in views with ecosystem dynamics being increasingly considered. Management increasingly embraces a global ecosystem view where physical and chemical aspects, including water quality and hydrology, faunistic and floristic characteristics and wetland history are considered simultaneously. What nowadays matters is the rehabilitation of wetland physiognomy and achieving maximal biodiversity bearing in mind wetland idiosyncrasies. However, limited water availability and often poor quality are too frequently challenging effective management.

In addition to problems associated with water availability, large-scale land-use change poses an additional threat. Intensive agriculture and conversion of forests to croplands contributes to water quality deterioration and disrupts natural hydrological cycles through groundwater overexploitation and altered surface runoff. These anthropogenic stressors are currently limiting the effective management of many of the most valued wetland ecosystems (e.g. the National Parks of Doñana and Las Tablas de Daimiel and Laguna de la Nava Regional Park).

A BRIEF REVISION OF THE ACTUAL WETLAND THREATS IN SPAIN

Until the late 1960s wetlands were consideresd as unproductive land and the home of disease vectors (mosquitoes) which had to be converted to serve better human needs. In fact, there existed a law of 1918 (Ley Cambó; Ceballos Moreno, 2001) that advocated wetland desiccation, however this law was discontinued and substituted by the Spanish Water Regulation in 1985. Facing increased loss of ecological and socioeconomic values of wetlands, including biodiversity, social attitudes toward wetlands began to change, and their enormous biological and environmental value started to form the basis for designing rehabilitation and management plans. However, much of the historical damage inflicted on wetlands that affect their hydrological functioning, their geomorphology, water quality, etc, is irreversible and limit seriously wetland restoration.

Currently, the problems related to wetland management in Spain boil down to five major points: desiccation, eutrophication and contamination, organic matter accumulation, siltation, salinisation and the invasion of exotic species (Fig. 1). In many cases these stressors act synergistically (Alvarez-Cobelas et al., 2001) and contribute to the loss of values which motivate protection programs.

Furthermore, a scientifically-oriented wetland management, when rarely occurring in Spain, is spoiled by societal needs and attitudes towards wetlands. Unfortunately, many people in Spain still consider Nature values as being second to economic values and hence economic profit is searched agonistically despite the harmful effects this might create to wetlands. In this way, irrigation farmers with their miopic practices that impinge dramatically on water availability and quality has been and still is one of the most serious threats to Spanish wetlands (López Sanz, 1999).

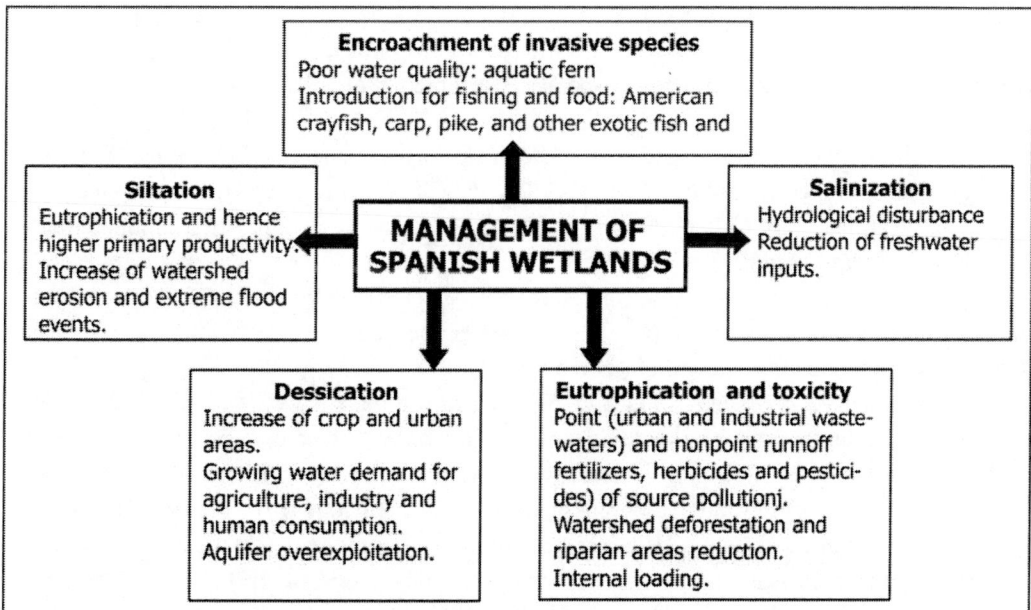

Figure 1. Common problems of Spanish wetland conservation.

And last but not least, administration issues on environmental management of wetlands are also in case. In the last 20 years, Spain has experienced a strong administrative decentralization resulting in a quasi-federal State. This has been desired by most Spaniards, but has had and still poses great inefficiencies to arrive at environmentally-sounded decissions that concern wetland management and restoration. Clashes between different offices at the local, the regional and the State levels (Nieto, 1996) often result in either no or ill-founded environmental decissions.

THE FIRST CASE STUDY: WATER QUALITY DETERIORATION AND THE INVASIVE AQUATIC PLANT *AZOLLA FILICULOIDES* IN DOÑANA NATIONAL PARK (SOUTHERN SPAIN)

Species introductions to wetlands in the Iberian Peninsula are an ancient practice, and aimed to increase productivity and socioeconomic benefit. Many of the intentional introductions, like those of the common carp (*Cyprinus carpio;* Doadrio, 2001), the pike (*Esox lucius*; Pena, 1986) and the American red swamp crayfish (*Procambarus clarkii*; Habsburgo-Lorena, 1979) had negative effects on native communities, which often contain singular and endemic elements (e.g. impoverishment of the autochthonous ichthyofauna as a result of pike introductions), and as a consequence catastrophic ecosystem shifts (loss of submerged vegetation as a result of carp and crayfish activity). Many other introductions are unintentional, and therefore more difficult to control, and arise from human activities or from the incidence which human population have in the environment.

Azolla filiculoides is a floating fern that is native to American water bodies. The eutrophication of surface waters is without doubt the main factor contributing to the expansion of this plant in the Iberian Peninsula, but also in the rest of the World. There exists strong consensus in the scientific literature that dissolved phosphorus controls the growth of *Azolla,* thereby converting it into a pest species (Carrapico et al., 1996; Wagner, 1997; Kar et al., 2001).

First collections of *Azolla* in the Iberian Peninsula were made 1920 in the Rio Mondego delta (Portugal), in the vicinity of rice fields (Ruiz de Clavijo et al., 1984). It was detected in the Doñana National Park (37° N, 6° W; Fig. 2) in 2000, and among the first among aquatic invaders in this area. Doñana is located in a Mediterranean climatic area with an average rainfall of 565 mm and a yearly temperature of 24.3 °C. It covers 53,835 ha, embracing ecosystems such as pasture lands, pine forests, dunes, beaches and freshwater and estuarine marshlands of some 24,000 ha (García-Novo et al., 2007).

In subsequent years, except 2005 and 2006 when the marshes were almost dry, Azolla reached an alarming biomass and formed floating carpets of > 10 cm thickness (Cobo et al., 2004; García-Murillo et al., 2006; Fernández-Zamudio et al., 2006; Fig. 3). The presence of *Azolla filiculoides* has important repercussions on ecosystem integrity, because submerged vegetation, in which charophyte beds are an important component for the waterfowl diet, succumbs to the presence of this invader when dry biomasses of 230 g/m^2 are reached. Moreover, it reduces the concentration of dissolved oxygen (Fig. 5) and can lead to respiratory stress of secondary producers. As a result, biological diversity in the Doñana marshes is tremendously reduced (Fig. 6).

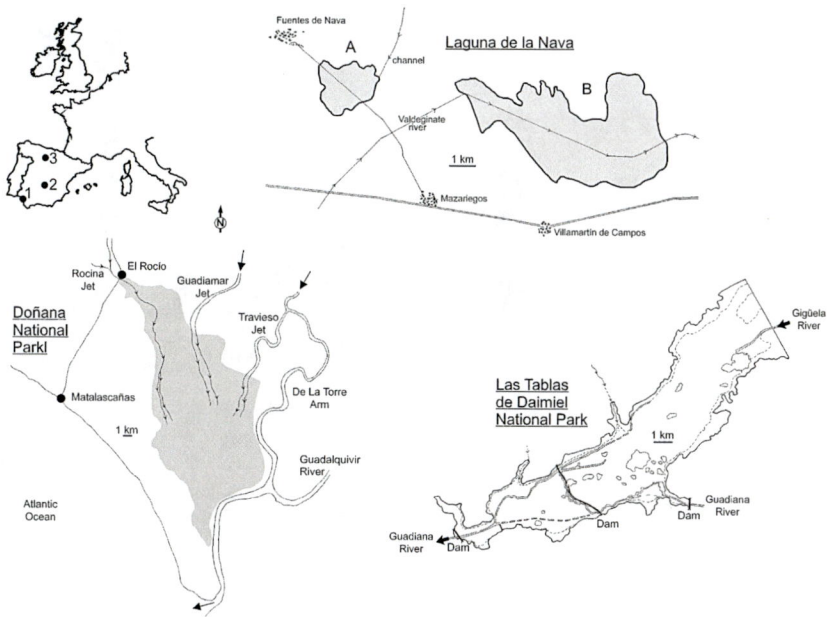

Figure 2. Location of Doñana National Park (1), Las Tablas de Daimiel National Park (2) and Laguna de la Nava (3), three examples of managed wetlands from Spain.

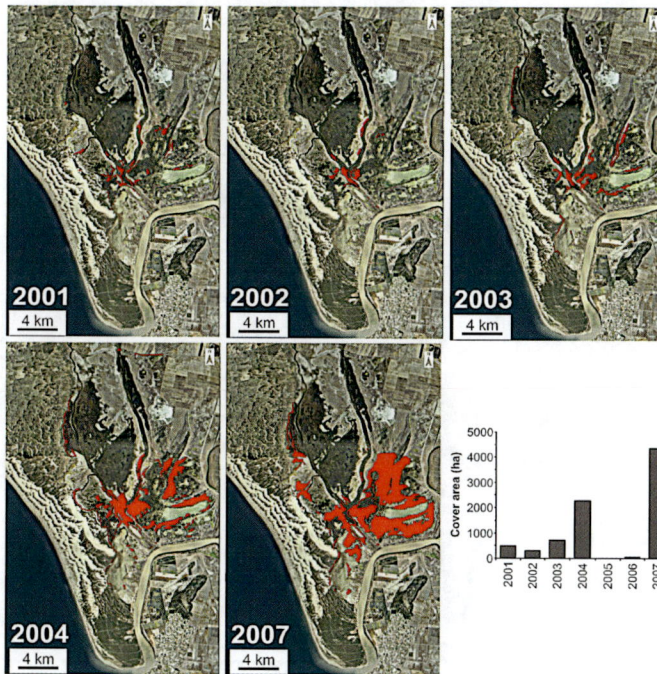

Figure 3. Temporal trend of *Azolla filiculoides* cover (red area) in Doñana National Park for the period 2001-2007. Images were not included for 2005 and 2006 because the marsh remained dry.

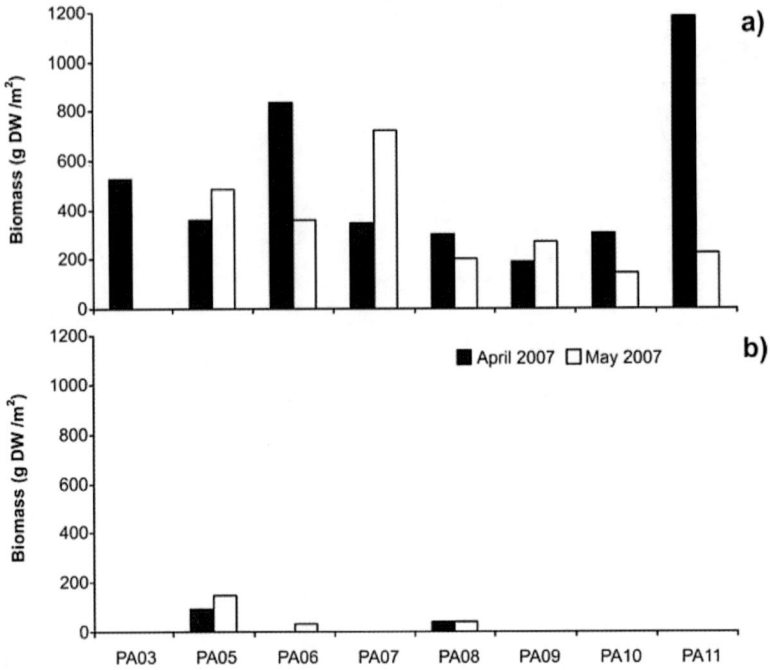

Figure 4. *Azolla filiculoides* effects on biomass of aquatic submerged plants (*Chara galioides, Callitriche truncata, Zannichellia obtusifolia, Ruppia drepanensis, Ranunculus peltatus* subsp. *fucoides*) in Doñana National Park during April and May 2007. Macrophyte biomass where *Azolla* was absent (a) and present (b).

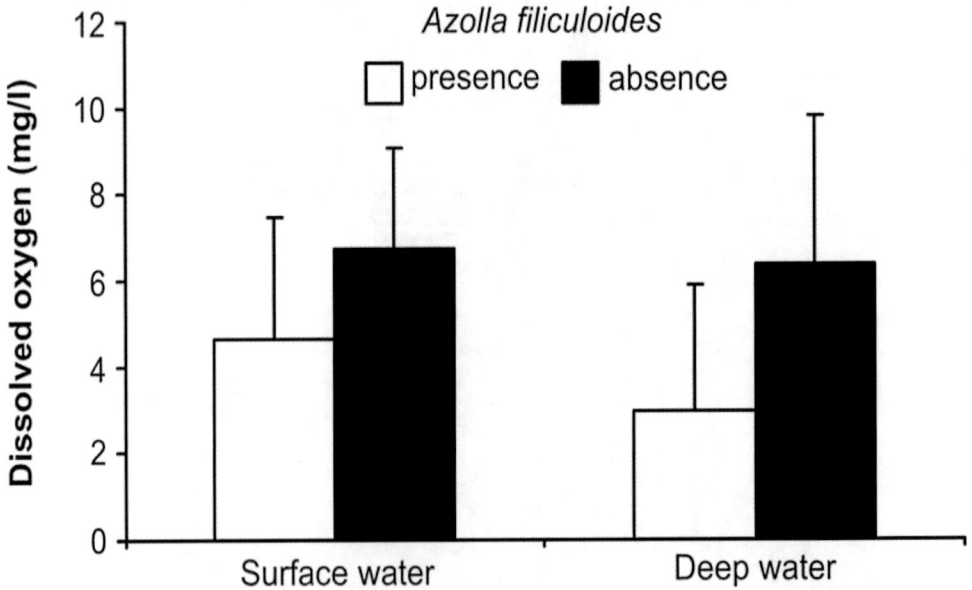

Figure 5. Effects of *Azolla filiculoides* on dissolved oxygen (mg/l) at midday in Doñana National Park (mean depth 0.5 m).

Poor water quality of waters that discharge into the marshes are the main factor that control the spread of *Azolla filiculoides* in Doñana. Phosphorus concentration in tributaries range 0.12 to 2.10 mg P/l, and the biomass production of this plant is especially favoured upon concentrations of 0.27 mg P/l.

Without and adequate control of water quality biological diversity is jeopardized in the Doñana marshes in the short-term (Fig. 6). Integral management schemes to controlling *Azolla* should therefore focus on water quality restoration.

Figure 6. Contrasting aspect of aquatic plants in April 2007 (upper) and May 2007 (lower) at Lucio del Rey area in Doñana National Park. Plant community composition: *Chara galioides, Callitriche truncata, Zannichellia obtusifolia, Ruppia drepanensis, Lemna minor, Ranunculus peltatus* subsp. *fucoides, Azolla filiculoides*. Emergent macrophytes: *Scirpus maritimus, Arthrocnemum macrostachyum, Juncus subulatus*.

THE SECOND CASE STUDY: THE EFFECTS OF HYDROLOGIC DEGRADATION ON LAS TABLAS DE DAIMIEL NATIONAL PARK (CENTRAL SPAIN)

Las Tablas de Daimiel floodplain wetland was declared National Park in 1973. It is situated in central Spain (39° N, 3° W; Fig. 2) at an altitude of 606-603 m. It is shallow (0.5-2 m) and has flat slopes (0.02 %). Its catchment area is 15,000 km². The wetland actually covers 1,928 ha while the wetland area prior to degradation was 20,000 ha that formed as a result of the confluence of the rivers Gigüela and Guadiana. The former contributed saline water on a seasonal basis, while the latter permanently discharged freshwater which was rich in bicarbonates. The climate is dry (439 mm annual rainfall; 14.3° C mean temperature) (Álvarez-Cobelas & Cirujano, 1996).

This continental marsh embodies all types of anthropogenic stress that currently affect wetlands in Spain: hydrological alterations, channel constructions, groundwater overexploitation (which lead to the disappearance of the Guadiana river in 1986, and causing a disconnection of the wetland with the aquifer), contamination episodes, prescribed burnings (1986, 1987) which affected one third of *Cladium mariscus* cover (an emblematic and dominant plant in the Park; Alvarez-Cobelas et al., 2001), increased sedimentation (Sánchez-Carrillo et al., 2001), increased internal loading (Sánchez-Carrillo & Álvarez-Cobelas, 2001; Alvarez-Cobelas et al., 2007) and exotic species introductions (*Procambarus clarkii* and *Cyprinus carpio;* Álvarez-Cobelas & Cirujano, 1996). The combined effects of these stressors profoundly modified wetland ecology resulting in large-scale ecological shifts (Fig. 7) (Cirujano et al., 1996; Camargo et al., 1997; Álvarez-Cobelas et al., 2008).

An emergency remediation plan was instated in 1986 to halt further degradation and ultimately the disappearance of the wetland. This "hydrological regeneration" plan aimed at maintaining inundation levels by constructing two dams in the wetland, which should help in the rehabilitation of the lost natural values of the Park.

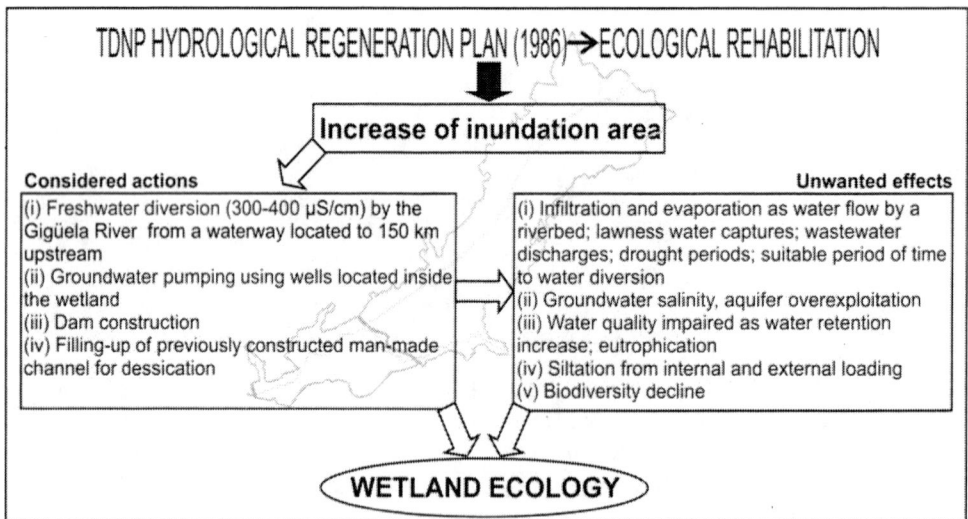

Figure 7. Unwanted indirect effects of the emergency remediation plan (Hydrological Regeneration Plan – 1986) for Las Tablas de Daimiel National Park (TDNP) to achieve an ecological rehabilitation of the wetland.

The efficiency of this plan was conditioned by several factors that were not considered initially. No detailed prior assessment of ecological characteristics has been carried out and recovery was thought to be conditioned chiefly by hydrological features. Although this plan helped maintaining certain aquatic values, the desired recovery has not been achieved because of recurring contamination events and poor efficiency of water diverted inputs (Figs 8, 9).

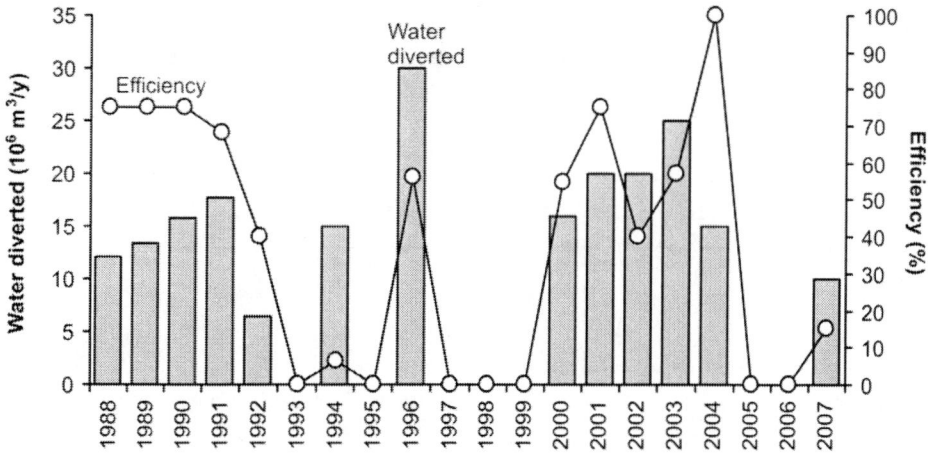

Figure 8. Water diverted to Las Tablas de Daimiel National Park and efficiencies (%) since the implementation of the Hydrological Regeneration Plan (1986).

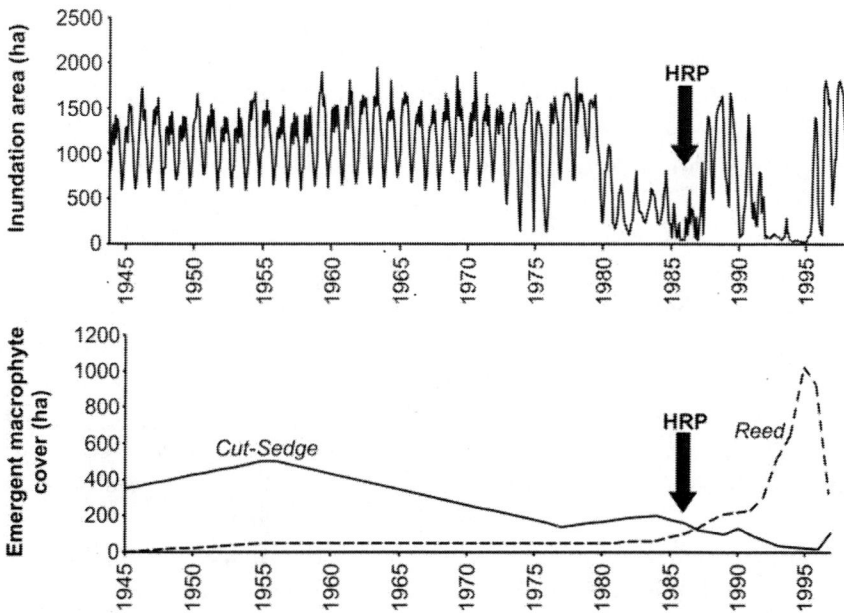

Figure 9. Effects of the Hydrological Regeneration Plan (HRP) implemented in Las Tablas de Daimiel National Park on inundation area (upper panel) and cut-sedge (*Cladium mariscus*) recovery.

Future management faces the problem that many altered features of this wetland are irreversible. While the wetland showed relatively stable and predictable hydroperiods prior to 1956, actual hydroperiods and flood frequencies are highly variable and depend on meteorological conditions. During exceptionally dry years, this wetland can dry up almost completely (Fig. 9).

The use of efficiently treated waste waters in the whole catchment, combined with efficient transport systems and measures to control wetland water quality are key aspects which need to be considered in sound management of this wetland. This requires also that water use is more efficient in the whole territory. These aspects currently form the background in the new action plan of the Alto Guadiana, instated in 2008 by the Spanish

National Government. This plan aims at recovering surface and groundwater water quality and an integral ecological functioning of aquatic ecosystems affected by hydrological overexploitation in the territory, especially in Las Tablas de Daimiel area, before 2027.

THE THIRD CASE STUDY: WETLAND SILTATION IN LAGUNA DE LA NAVA (NORTHERN SPAIN)

Laguna de la Nava (42° N, 4° W; Fig. 2) is a temporary, continental, shallow and semi-endorheic wetland (mean depth 60-100 cm; maximum depth 200 cm). It is situated in 737 m altitude, over impermeable clay substrate. This lagoon has a variable surface area (2,200-3,550 ha) within a small catchment area (864 km^2) in which various temporary stream feed the lagoon especially in winter months when the amount of water carried by these streams can be substantial (25 hm^3). Rainfall (14 hm^3) is an other major contributor of water to the wetland (Dantín-Cereceda, 1929; González-Garrido, 1993).

Although plans to convert the lagoon into croplands date back to the 19th century, the first written testimony is available from March, 1831. La Nava lagoon was definitively dried up in 1965 (Plans, 1970). In 1990 a rehabilitation plan was started where in a first phase 56 ha (period 1990-1995) and in a second phase (354 ha) were recovered. From these, 156 ha function as wetland, while 198 ha are pasture lands dedicated to the cattle industry of nearby villages.

Because the natural catchment hydrology has been irreversibly altered, current hydroperiods are achieved by artificial water discharge. Discharge patterns are similar between years, which ignores interannual variability inherent to these ecosystems. Water quality is generally poor (0.09-0.26 mg P/l) and dominated by bicarbonates, chlorides, and Na, Mg and Ca which lead to electrical conductivity levels of 674-1147 µS/cm.

Wetland rehabilitation took place without prior evaluation of the impacts of artificial hydroperiod management on the structure and dynamics of emergent vegetation. From the initial pastures dominated by grasses (*Lolium perenne, Bromus rubens, Hordeum marinum, Elymus hispidus*, etc.), Fabacea (*Melilotus indica, Trifolium campestre, T. repens, T. resupinatum*) etc.), Compositae (*Scozonera laciniata, Leontodon saxatilis, Picris echioides, Bellis anual*, etc.) and others (*Plantago coronopus, P. lancelolata, Convolvulus arvensis, Carex divisa*, etc) a gradual replacement to compact *Carex divisa* zones took place. In more deeper zones helophytes started to grow (*Typha domingensis, T. latifolia, Scirpus lacustris, Eleocharis palustris*, etc), and in open water zones submerged aquatic macrophytes appeared (*Chara vulgaris, Zannichellia pedunculata, Ranunculus peltatus, R. trichoplyllus, Callitriche brutia*, etc; Table 1).

Over time the helophyte stands became to dominate the vegetation of the entire wetland and open water zones disappeared. Several methods are currently used to clear vegetation to maintain 15-20% of the wetland surface as open water areas. These methods consist of prescribed burnings, sediment removal of surficial layers, and grazing through horses, but no post monitoring is carried out to evaluate the efficiency of each method (Fig. 10).

Table 1. Wetland vegetation changes from 1993 to 2005 in Laguna de la Nava (Northern Spain).

	1993	1996	1999	2005
Aquatic plants:				
Algae (Charophytes)				
Chara aspera		•	•	•
Chara canescens	•	•	•	•
Chara connivens	•	•	•	
Chara fragilis	•	•	•	
Chara vulgaris var. longibracteata	•	•	•	•
Chara vulgaris var. oedophylla		•	•	•
Chara vulgaris var. vulgaris	•	•	•	•
Chara vulgaris var. contraria			•	
Nitella flexilis				•
Tolypella glomerata	•	•		•
Mosses				
Drepanocladus aduncus	•	•	•	•
Vascular plants (submerged)				
Callitriche brutia	•	•	•	•
Hippuris vulgaris				•
Lemna gibba	•	•	•	
Lemna minor	•	•	•	•
Myriophyllum spicatum	•			
Polygonum amphibium	•	•	•	•
Potamogeton crispus	•			
Potamogeton pectinatus	•	•	•	•
Potamogeton pusillus	•	•	•	
Ranunculus peltatus	•	•	•	•
Ranunculus penicillatus			•	
Ranunculus trichophyllus	•	•	•	
Utricularia australis				•
Zannichellia obtusifolia			•	
Zannichellia palustres	•	•	•	•
Zannichellia pedunculata	•	•	•	
Zannichellia peltata		•	•	
Total	**19**	**20**	**22**	**16**
Emergent plants (vascular taxa):				
Alisma lanceolatum	•	•	•	•
Alisma plantago-aquatica				•
Baldellia ranunculoides	•	•	•	•
Damasonium polyspermum	•	•	•	•
Butomus umbellatus	•	•	•	•
Carex divisa	•	•	•	•
Eleocharis palustres	•	•	•	•
Eleocharis uniglumis	•	•	•	•
Scirpus lacustris subsp. Lacustris	•	•	•	•
Scirpus maritimus	•	•	•	•
Glyceria declinata	•	•	•	•
Phragmites australis	•	•	•	•
Verónica anagallis-aquatica	•	•	•	•
Sparganium erectum subsp. neglectum	•	•	•	•
Typha domingensis	•	•	•	•
Typha latifolia	•	•	•	•
Total	**15**	**15**	**15**	**16**

La Nava is actually presenting recovered biodiversity. Especially considering the avifauna several species are cited from this wetland, many of them being considered to have a high extinction risk (e.g. *Acrocephalus paludicota*; Jubete et al. 2006). However, we highlight, that the artificial management of open water zones interferes with the natural succession dynamics of this wetland. If no management will be carried out, this wetland tends to form tight stands dominated by *Scirpus maritimus* and *Eleocharis palustris*. Definitely, these elements would not reflect conditions most similar to those prior to degradation.

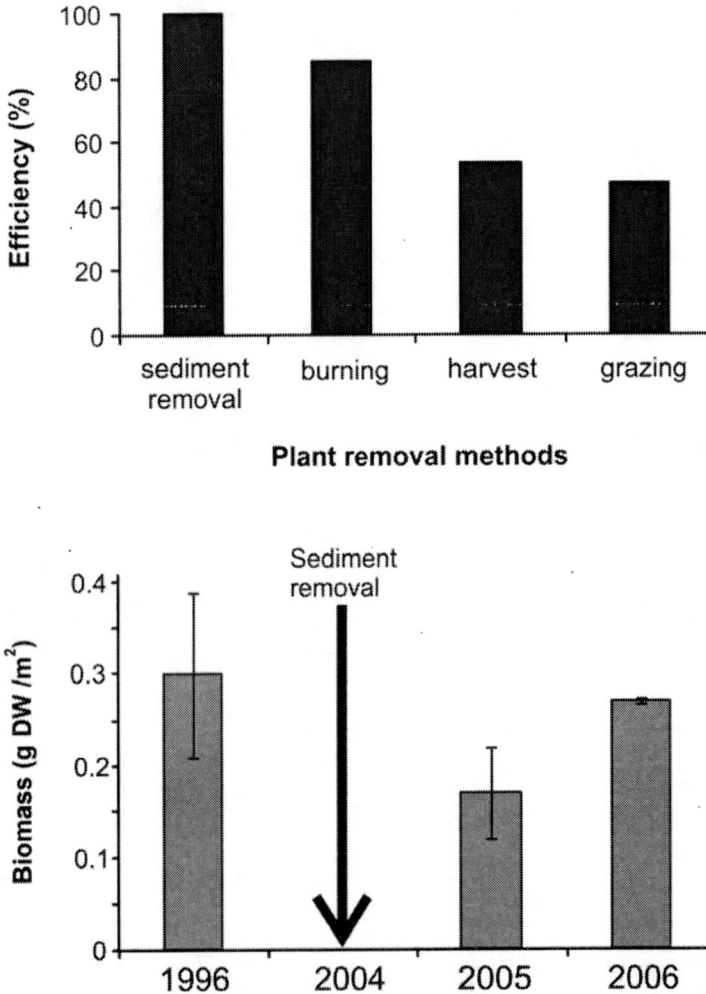

Figure 10. Efficiency of plant removal a year later after different management methods of vegetation in Laguna de la Nava (upper pannel) and measured biomass (g/m^2) of *Zannichellia* and charophyte prairies during 2005-2006 after sediment crapping occurring in 2004 (lower pannel).

CONCLUSIONS

Wetland management in Spain is recent and requires a good basic ecological knowledge if applied issues are to be resolved efficiently. Each wetland has site-specific abiotic and biotic characteristics which make them different from other wetlands. These idiosyncrasies must be considered, suggesting that case-by-case approaches to management and conservation, rather than universally applicable/identical schemes, are needed.

Because many managers have little wetland and scientific knowledge, most studies are contracted and carried out by scientific institutions. Unfortunately, obtained scientific results are often not implemented, because they deviate from those expected by managers. The dichotomy of views related to conservation of selected target organisms (managers view) or overall ecosystem integrity (scientific view) is currently an obstacle to sound wetland management.

Widespread, wetland-uninformed attitudes of stakeholders (irrigation farmers, more often) and conflicting views of environmental officers at different levels of the Spanish Administration make often difficult achieve projected goals of wetland management and restoration.

Also, while currently much effort and resources, often provided by the European Community, are devoted to wetland conservation in Spain, no follow-up research is carried out to evaluate the efficacy of interventions. These are essential for understanding to what degree ecosystem structure and function can be recovered. Only then can the role of wetlands as ecosystem service providers to humans and wildlife be evaluated.

REFERENCES

Álvarez-Cobelas, M. & Cirujano, S. (eds.), 1996. *Las Tablas de Daimiel. Ecología acuática y Sociedad.* Organismo Parques Nacionales, Serie Técnica. Madrid.

Alvarez-Cobelas, M., Cirujano, S. & Sánchez Carrillo, S. 2001. Hydrological and botanical man-made changes in the Spanish wetland of Las Tablas de Daimiel. *Biological Conservation* 97: 89-97.

Álvarez-Cobelas, M., Sánchez-Carrillo, S., Cirujano, S. & Angeler, D.G. 2008. Long-term changes in spatial patterns of emergent vegetation in a Mediterranean floodplain: natural versus anthropogenic constraints. *Plant Ecology* 194: 257-271

Ambiental, S.L. 1992. *Plan de recuperación de la cerceta pardilla (Marmaronetta angustiostris) en la Comunidad Valenciana.* Consejería de Medi Ambient de la Generalitat Valenciana. Internal Report.

Bernués, M. 1998. *Humedales españoles inscritos en la lista del Convenio de Ramsar.* Ministerio de Medio Ambiente. 386 pp.

Camargo, J. A. & Cirujano, S. 1996. Reduction in diversity of aquatic plants in a Spanish Wetland: The effect of the size of inundated area. *Journal of Freshwater Ecology* 12: 539-543.

Carrapico, F., Costa M.H., Texeira, G., Frazao, M., Santos, C. & Baioa, M. 1996. The uncontrolled growth of *Azolla* in the Guadiana River. *Aquaphyte* 16(2): 11.

Casado, S. & Montes, C. 1995. *Guía de los lagos y humedales de España.* J. M. Reyero Editor. Madri. 255 pp.

Ceballos Moreno, M. 2001. La problemática jurídico-administrativa de las zonas húmedas. *Humedales Mediterráneos (SEHUMED)* 1: 155-162.

Cirujano, S. Casado, C., Bernués, M. & Camargo, J. A. 1996. Ecological study of the National Park of Las Tablas de Daimiel (Ciudad Real, Spain): changes in the physico-chemical characteristics of the waters and the vegetation between 1974-1989. *Biological Conservation* 75: 211-215.

Cobo, M. D., Garcia Murillo, P., Sánchez-Gullón, E. & Garrido, H. 2004. Plantas exóticas e invasoras del Parque Nacional de Doñana. *Medio Ambiente (Junta de Andalucía)* 46: 46-53.

Dantín-Cereceda, J. 1929. La cuenca endorreica de la Nava (Palencia). XII *Congreso de la Asociación Española para el progreso de las Ciencias*, 3-4: 97-100.

Doadrio, I. (ed.) 2001. Atlas y libro rojo de los peces continentales de España. CSIC. Madrid. 364 pp.

Fernández-Zamudio, R., Cirujano, S., Espinar, J.M., Rubio, A., López-Bravo, M.I., Cobo, M.D. & García-Murillo, P. 2006. Datos preliminares sobre la biología y ecología de Azolla filiculoides en el Parque Nacional de Doñana. *II Congreso Nacional sobre Especies Exóticas Invasoras*. Universidad de León.

García-Murillo, P., Fernández-Zamudio, R., Cirujano, S. & Sousa, A. 2006. Aquatic macrophytes in Doñana protected area (SW Spain): An overview. *Limnetica* 25: 71-80.

García-Novo, F., Martín-Vicente, A. & Toja, J. 2007. *La frontera de Doñana*. Universidad de Sevilla. 317 pp.

González-Garrido, J. 1993. *La Tierra de Campos. Región Natural.* 2ª edición. Ámbito Ediciones. 462 pp.

Habsburgo-Lorena, A.S. 1979. Present situation of exotic species of crayfish introduced into Spanish continental waters. *Freshwater Crayfish* 4: 175-184.

Jubete, F., Torres, M., Gómez, E. Cirujano, S. & Zuazua, P. 2006. *The aquatic warbler. Manual for managing helophytic vegetation and monitoring populations.* Fundación Global Nature, Palencia. 144 pp.

Kar, P., Mishra, M. & Singh, P. 2001. Influence of different phosphorus management strategies on the sporulation and grow of *Azolla. Experimental Agriculture* 37: 53-64.

López Sanz, G. 1999. Irrigated agriculture in the Guadiana River high basin (Castilla-La Mancha, Spain): environmental and socioeconomic impacts. *Agricultural Water Management* 40: 171-181.

Matamala J.J. & Aguilar, F.J. 2003. Humedales almerienses. In: PARACUELLOS, M. (ed.), *Ecología, manejo y conservación de los humedales.* 221-244. Instituto de Estudios Almerienses, Almería.

Nieto, A. 1996. *La · "nueva" organización del desgobierno.* Editorial Ariel. Barcelona. 240 pp.

Pena, J.C. 1986. Introducción y expansión del lucio (*Esox lucius* L. 1758) en la Península Ibérica: síntesis general y estudio de las poblaciones de la cuenca del Esla. *Limnetica* 2: 241-251.

Plans, P. 1970. *La Tierra de Campos.* Instituto de Geografía Aplicada del Patronato Alonso de Herrera, CSIC. Madrid. 298 pp.

Ruiz De Clavijo, E., Muñoz, J. & Salvo, A.E. 1984. Sobre la presencia de *Azolla filiculoides Lam.* en España. *Acta Botanica Malacitana* 9: 129-132.

Sánchez-Carrillo, S. & Alvarez-Cobelas, M. 2001. Nutrient dynamics and eutrophication patterns in a semiarid wetland: the effects of fluctuating hydrology. *Water, Air and Soil Pollution* 131: 97-118.

Sánchez-Carrillo, S., Alvarez-Cobelas, M. & Angeler, D.G. 2001. Settling seston in Las Tablas de Daimiel, a semiarid freshwater wetland. *Wetlands* 21: 67-79.

Sánchez-Carrillo, S., Álvarez-Cobelas, M., Cirujano, S., Riolobos, P., Moreno-Pérez, M. & Rojo, C. 2000. Rainfall-driven changes in the biomass of a semiarid wetland. *Verhandlungen der International Vereinigung für Limnologie* 27: 1690-1694.

Sargatal, J. & Fèlix, J. 1989. *Els Aiguamolls de l'Empordá*. Quaderns dels Indiketes 3. Figueres. 376 pp.

Vallès, C. (ed.). 1989. *Els Aiguamolls de l'Empordà. Aspectos ecològics, històric i socials del Parc Natural*. Quaderns dels Indiketes 3. Ed. ART. 376 pp.

Vv.Aa. 1987. *Bases Científicas para la protección de los Humedales Españoles*. Real Academia de Ciencias Exactas, Físicas y Naturales. Madrid.

Wagner, M. 1997. *Azolla*: A review of its biology and utilization. *The Botanical Review* 63: 2-26.

In: International Wetlands: Ecology, Conservation...   ISBN: 978-1-60456-999-5
Editor: Jose R. Herrara   © 2009 Nova Science Publishers, Inc.

Chapter 2

STRUCTURE AND MECHANICS
OF INTERMITTENT WETLAND COMMUNITIES:
BACTERIA TO ANACONDAS

D.D. Williams, C.M. Febria and T.A. Schriever
Department of Biological Sciences,
University of Toronto at Scarborough,
1265 Military Trail, Scarborough, Ontario, Canada M1C 1A4

ABSTRACT

Freshwater wetlands are highly productive and important features of terrestrial landscapes, yet knowledge of their biotas and understanding of their function has lagged behind that of other ecosystems. Superficially, their communities are known to include bacteria, protists, algae, fungi, higher plants, and invertebrate and vertebrate animals that come together to benefit from the rich resources and oftentimes relatively predator-free space that these habitats provide. The cost of these benefits requires adaptation to a highly variable, though often predictable, hydroperiod. This chapter will focus on wetland habitats that are truly intermittent in their water-balance, and will summarize what is known of the structure of the biota at different trophic levels and how they interact at a dynamic land/water interface. For example, fluctuations in flooding regime are known to cause fluxes in carbon, nitrogen and phosphorus in basin sediments as well as in the water column. Such nutrient pulses can lead to increased productivity followed by anoxia and fluxes of methane in these sediments; significantly, microbes that die during the drying period can represent an important source of carbon upon re-wetting at the same time that phosphorus and nitrogen enter the system. These fluxes produce a range of responses from both prokaryote and eukaryote components of the wetland community. Detritivore and herbivore consumers in wetlands are represented by a rich variety of protists and invertebrates, especially crustaceans and insects, many of which appear to be predictably represented, taxonomically, across wetland types. These groups provide the crucial trophic link between producers and the top predators – although the trophic status of some of the latter is variable. For example, many amphibians that breed in wetlands represent both prey and predator depending on the stage in their life cycle. While salamanders may input large amounts of energy via egg deposition, many larger vertebrate inhabitants of wetlands are seen as net removers of energy from these systems (grazing, predation, etc.), with very little return (chiefly wastes) – although intuitively, they must represent important links between aquatic and terrestrial compartments. However, it must be said that the role of vertebrates in nutrient cycling and energy flow in wetlands is not well understood, and that the consequences of, for example the global decline in amphibian populations, are largely unknown. Coverage of these topics will be global in extent. Future research needs to be focused on piecing together the components of intermittent wetland function through study of: 1) physiological and phenological response to natural (e.g., hydroperiod regime) and human-induced stress by individual taxa; 2) community response to these same factors; 3) trophic function, including bottom-up resources and top-down pressures; 4) interaction with, especially energy flow through, permanent sections of wetland, the riparian zone, adjacent woodlands and grasslands, and beyond.

INTRODUCTION

Wetlands are a dominant feature of the global environment, and may be regarded as the most important subset of our planet's temporary water environments (which range from water-filled leaf axils to river floodplains and arid-region lakes; see review by Williams, 2006). Temporary waters occur on all continents and in all climatic zones and, collectively, store vast quantities of that most precious of resources, fresh water. Quite apart from their role as natural reservoirs, freshwater wetlands support much of the world's aquatic and riparian biodiversity and are sites of important biological processes. For example, in a recent survey of 71 temporary ponds in England and Wales, Nicolet et al. (2004) found that almost 75% supported at least one nationally scarce macroinvertebrate and 8% supported at least one nationally scarce plant species. Similar findings in Spain, led Boix et al. (2001) to conclude that temporary waterbodies play a crucial role in the metapopulation dynamics of many species (especially of amphibians and corixid bugs). Further, Junk (1993) regards large floodplains, such as those of tropical South America, to be major sites of speciation – as plants and animals respond to the flood pulse via morphological, physiological, and other adaptation.

Despite such importance, the environs that support temporary fresh waters have been, and continue to be, under threat from human activities. Agriculture, urban sprawl, drainage, pollution, deforestation, and many other processes have taken their toll, worldwide. In northwestern Europe, for example, intensification of agriculture together with industrial development has resulted in a 60% reduction of wetland area (European Environment Agency 1999). In the United States, where the extent of alteration of freshwater habitats, particulary through channelization, is very great, loss of many invertebrate species has reached a critical level – as far back as 1993, for example, 213 (72%) of the 297 freshwater mussels native to the US and Canada were considered endangered (Williams et al., 1993). Fundamental to the protection and management of wetlands is a clear understanding of what species and communities they contain, how their biological processes (e.g., food webs) work and are maintained, and how all of these interact with local geomorphology, hydrology, and climate. This chapter reviews what is known of the microbial, plant, and animal components of wetlands, and attempts to relate these to community mechanics, chiefly trophic processes. It focuses on wetland habitats that are truly intermittent in their water-balance – that is, those in which the dry period, although it may be of varying duration, is cyclical and predictable.

MICROBIAL COMMUNITIES

Though small in size, heterotrophs and autotrophs play critical roles in the food web of intermittent wetlands. Between cycles of rewetting and desiccation, microbes are intricately involved in nutrient cycling within the water column and sediments. Protists are important predators of bacteria and may also be consumed within the foodweb. Algae, fungi, and mosses are known to collectively form mats together with bacteria (called cyanobacterial mats, or CBMs). Microbial and algal communities can be found in forms attached to the surfaces of macrophyte vegetation within a wetland and demonstrate key structural, reproductive and/or physiological mechanisms for surviving varying hydroperiods. Generally, the competition for resources may be high among primary producers (Williams et al., 2005; Janousek et al., 2007; Rolon et al., 2008) and benthic microbes (Sim et al., 2006a,b). Microbe diversity and community composition have been shown to vary among gradients of oxygen availability (Noll et al., 2005), nitrogen availability (Bastviken et al., 2003; Mentzer et al., 2006; Iribar et al., 2008), recovery from pollution (Boivin et al. 2007), gradients of salinity (Rejmankova and Komarkova, 2005; Sim et al., 2006a,b), between sediment-bound and free-living species (Iribar et al., 2008), as well as between species of bacteria and algae (Boivin et al., 2007). Here, we place emphasis on the dynamic linkages that occur between bacteria, algae, and macrophytes in wetlands with intermittent hydroperiods. Given the paucity of information available on the microbial and photosynthetic communities in these systems, we highlight the need for additional studies that address both the function and community structure of these understudied habitats.

Sediment microbial communities are structured in naturally-occuring intermittent wetlands in close relation to physicochemical factors and other competitors for resources including algae and macrophytes. Relatively little is known about the mechanisms that drive the lower trophic levels in these systems compared with larger organisms (e.g., invertebrates, amphibians). However, more studies have been published in recent years as molecular tools have advanced our ability to characterize spatial and temporal shifts in microbiota and nutrients within the water column and sediments. In a review by Gutknecht et al. (2006), the authors highlight the importance of linking soil and microbial ecology in freshwater wetlands and the environmental controls over processing rates and community composition. To date, research in this area is mostly experimental in scope and conducted at the levels of process or trophic level, nutrient cycle, specific enzyme activity, or molecular fingerprint. Functional descriptions of microbes are most common but little is known about the mechanisms driving communities in intermittent wetlands. Even less is known about the nature of viruses in these systems, however there is evidence that they are ecosystem-specific and are both diverse and abundant among freshwater sediments (Danovaro et al., 2008). Various techniques have been used to describe functional aspects of microbial community dynamics. These include Gram-staining of bacteria to differentiate major groups, enzyme activity analysis, or phospholipid fatty acid analysis (PLFA), which can provide quantitative insight to the community structure based on extracted structural materials of cells within a soil sample (Boon et al., 1996;

Mentzer et al., 2006; Balasooriya et al., 2008). In contrast, other quantitative tools such as DNA fingerprinting (e.g., terminal-restriction fragment length polymorphism, T-RFLP) describe community structure and composition of the microbial community in sediments (Baldwin et al., 2006; Thies, 2007). More recently, cloning of various RNA sequences has identified the main groups of prokaryotes present. The combination of approaches to characterizing the microbial communities has shown two major factors that influence taxon distribution and mechanisms controlling their abundance and diversity: flooding and nutrient cycling.

To address the question of whether the physical environment or biological processes define community structure, Magnusson and Williams (2006) examined a gradient of intermittent ponds with differing degrees of vegetation and invertebrate communities. They found physicochemical properties and native pond characteristics to be most crucial in defining community development. In another experimental manipulation, the variation and duration of inter-annual desiccation treatments were found to influence the rate of leaf litter decomposition and, consequently, processing rates by invertebrate shredders (Inkley et al., 2008). In this latter study, bags filled with leaf litter were placed in a wetland with a range of flooding regimes. While leaf litter dry mass did not change significantly across treatments, carbon and nitrogen content did, with the highest nitrogen levels being found in the intermittent treatment. In contrast, fungal and bacterial colonization rates were higher in the permanent and semi-permanent treatments where carbon content would be highest. Invertebrate shredders, especially caddisfly larvae, were higher in these treatments also. Such studies demonstrate how hydroperiod is one of the most critical physical factors affecting carbon budgets, community composition, and ecosystem functioning.

Using PLFA analysis, Boon et al. (1996) determined that prokaryotes dominate sediment in both permanently flooded and intermittently flooded wetlands. Despite changes in microbial types before and after flooding, the largest differences occurred between the permanent and intermittent sites. In a mesocosm study, Mentzer et al. (2006) assessed the effects of manipulated hydrological regime and increased nutrients on microbial communities. The treatments were used to simulate urban runoff as intermittent-, early season- or constant-flooding. Again using PLFAs, hydrological regime proved to be a dominant factor in controlling both the structure (as measured by lipids) and function (measured by enzyme activity) of the microbial communities. Interestingly, the communities responded differently to the isolated effects of flooding and nutrients. For example, whereas mycorrhizal fungi increased with intermittent flooding conditions, gram-negative anaerobic and gram-positive bacteria decreased. The study showed that, over time, intermittent conditions resulted in a decrease in specific enzyme activity of certain microbes although total activity did not change significantly. However, the study was not able to tease apart the individual responses of the microbial communities within the combination of nutrient addition and flooding regime. Such studies highlight the role that physical factors play in intermittent systems in shaping the structure and function of microbial communities. A key factor is flooding together with the resulting biogeochemical gradients that are created within the sediments and water column. Once wetland sediments are inundated with water, pulses of carbon, nitrogen, and phosphorus are immediately made available to microbes, algae and macrophytes, inducing competition for resources (Baldwin and Mitchell, 2000; Williams, 2006).

Another consequence of flooding is the creation of vertical gradients in oxygen. Noll et al. (2005) used rice-paddy soil microcosms to demonstrate how quickly oxygen can be depleted within a soil-water profile. From one hour to 168 days after flooding, they measured micro-level changes in community structure within the first 6 mm of sediment. Shifts in community structure were tracked using DNA fingerprinting (e.g., TRFLP) and cloning of the 16S RNA gene. Interestingly, species richness decreased dramatically within the first few hours of flooding, followed by transition to a different community and, later, to a more stable community structure. There was a succession from r-selected species immediately after flooding towards K-strategists over time – this same sequence has been demonstrated in wetland invertebrate communities (Williams, 2006). In a natural system, Boon et al. (1996) compared the microbial communities of one permanent and one intermittent wetland in Victoria, Australia. Based on measurements of the various microbially-derived lipids and extracellular materials, using PLFA, they noted that although high abundances of prokaryotic microbes were found in both types, there was an increased abundance of anaerobic methanogenic prokaryotes in the permanent wetland. It remains unclear as to whether this reflects a more stable vertical oxygen gradient in the sediment in the permanent system, or the inability of certain anaerobes to survive desiccation, or other factors. Microbial communities appear to be influenced not only by the availability of nutrients but also the presence of other taxa (e.g., macrophytes), season, and duration or frequency of flooding.

Also related to gradients in oxygen are the varying rates of nitrification or denitrification in wetland sediments. Sediments serve as a trap which can stimulate nitrification in oxic sediments and denitrification in anoxic sediments (Qiu and McComb, 1995, 1996; Baldwin and Mitchell, 2000; Baldwin et al., 2005). Phosphate cycling is closely tied to iron and can result in the release of phosphate in anoxic sediments, but this relationship is less clear in oxic sediments (Baldwin and Mitchell, 2000). Perhaps as critical as flooding of wetland sediments is the re-wetting of sediments after long periods of drying. Microbial cells lyse upon drying and, once flooded, quickly release nutrients back into the water column. Depending on the length of the dry period and how dry the sediments become, pulses of water that infiltrate into the sediment are key to diffusing and transferring nutrients from the overlying water. Microbes, algae, and macrophytes all compete for the nutrients that become available during flooding. Using stable isotope analysis in soil cores of varying compositions, Davidsson et al. (1997) determined that two types of wetland soils, sandy and peaty, showed contrasting forms of nitrogen within a vertical soil profile. In sandy soils, total nitrogen concentrations increased closer to the top of the soil profile whereas peaty soils showed higher total nitrogen concentrations at depth. Complete removal of nitrogen from the water was observed in both soil types, with oxygen and nitrogen being consumed completely within a shorter vertical distance into the peat-soil column compared with the sandy-soil column. This has implications for natural systems where flooded sediments act as a nitrogen sink. Understanding the role of nitrogen, in particular, is crucial for the management of wetland areas surrounded by urban or agricultural activities where fertilizers may be of use in high concentration, especially together with phosphorus. Casey and Klaine (2001) and Casey et al. (2001) investigated this for a riparian wetland that received pulses of nitrate and phosphate from an adjacent golf course. Artificial pulses of water loaded with nutrients were added to the wetland to simulate pulses associated with fertilizer use, and the fate of nitrate, phosphate and a conservative tracer, bromine, were tracked over time (Casey and Klaine, 2001). The authors determined that nitrate and phosphate were attenuated by the wetland relatively quickly, between 6 and 48 hours, with very little penetrating to the underlying groundwater aquifer. Water and phosphate likely stimulated, and were simultaneously taken up by, microbes, algae, and macrophytes. Wetland sediments were also effective in binding the phosphate. Taking a closer look at the processes influencing the attenuation of nitrogen in wetland sediments, Casey et al. (2001) tracked pulses of inorganic nutrient additions to the same intermittent wetland and found spatial differences in the rates of denitrification within the sediments. Rates were highest closest to the source of the nutrient pulse, and decreased into the wetland. There was also an interaction with carbon, showing that microbes were also carbon-limited, with denitrification rates at 75% of the rate observed among samples with added carbon (in the form of dextrose). Isolated wetlands are known to be a nutrient sink because they are hydrologically disconnected from other watercourses (Whigham and Jordan 2003), however the sediments, microbes, and vegetation have been shown to play a critical role in attenuating inorganic nutrients and fueling microbial communities, differing from systems with a more permanent hydroperiod (Toet et al., 2003; Mentzer et al., 2006; Orr et al., 2007).

CYANOBACTERIAL MATS, ALGAE AND MACROPHYTES

Cyanobacterial mats are a biological consortium of several cyanobacteria, diatom and algal species. Individually, each of these taxonomic groups plays an important role in wetland communities (Rejmankova et al., 2004; Janousek et al., 2007), but their combined roles are not well understood. The common forms present include *epiphyton* which grows on the submerged surfaces of macrophytes, *metaphyton* which floats at the water's surface, and *epipelon* which grows on wetland substrate (figure 1). Cosmopolitan cyanobacterial groups in wetlands include Chroococcales, Oscillatoriales, and Nostocales while common algae include Bacillariophyta (diatoms), Chlorophyta (green algae), and Cyanophyta.

Cyanobacteria and algae are intricately connected as biofilms (or CBMs) within wetlands. Very little has been published on the ecology of biofilms in freshwater intermittent wetlands, although a few studies have examined the succession of benthic microbes in restored saltmarshes (Underwood 1997; Janousek et al., 2007) and permanent freshwater wetlands (Domozych and Domozych, 2008). Biofilms represent a key microhabitat within wetland communities, as they are sites where nutrient cycling and primary production occur. Though there can be up to 50 different species in a particular grouping of cyanobacteria, only a handful of species may carry out key functions such as nitrification (Rejmankova et al., 2004). In a coastal system, Janousek et al. (2007) reported a succession of cyanobacteria and algae following flooding in a restored wetland habitat. Here, a diverse array of biofilms became re-established within seven months after flooding with diatoms being the most taxonomically-dominant group overall. The succession of 'microphytobenthos' did not appear to be related to physicochemical factors but perhaps microhabitats, such as mudflats, within the wetland.

a)

macrophytes

concentration of planktonic
communities,desiccation
& build-up of nutrients
(C,N,P) in sediments

dropping water level   metaphytes

epiphytes

b)

macrophytes

increased water level   metaphytes

increased light and
nutrient limitation

release of nutrients
(C, N, P) into the   epiphytes
water column

increasing anoxia,
denitrification &
phosphate release

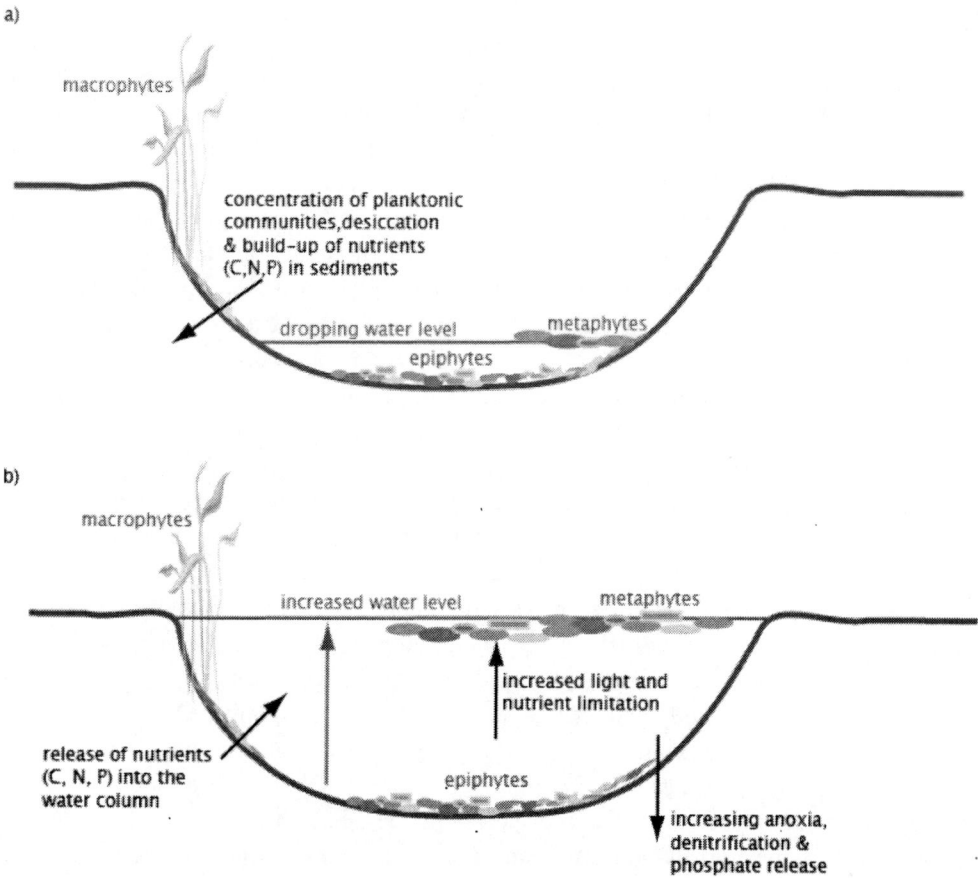

Figure 1. Summary of biogeochemical and microbial community gradients that are established in an intermittent wetland during: (a) dry phase and (b) rewetting phase. Arrows indicate direction of gradient across the water-sediment boundary or in the water column.

However, the tolerance thresholds of microbes to desiccation stress remain unclear. In one extreme case, the responses of Antarctic prokaryotic cyanobacteria and eukaryotic algae to freezing and desiccation were tested empirically (Sabacka and Elster, 2006). Antarctic wetlands are intermittent in their water balance, receiving pulses of melt water at the beginning of a short austral summer. Sabacka and Elster collected representative cyanobacterial and algal specimens from several types of continental and maritime wetlands and subjected each species to one of several freezing or desiccation treatments. Each microbial community was then isolated on agar plates where their growth was observed and quantified. Cyanobacteria were more resistant to both stressors than their algal counterparts. Cyanobacteria species showed differences in tolerance levels, which were attributed to habitat type. For example, filamentous cyanobacteria from continental Antarctic wetlands were more successful at tolerating lower temperatures than those from maritime wetlands or freshwater seepages. Cyanobacteria isolated from the seepages were least tolerant of desiccation and freezing, likely because this habitat had a more stable water level than the others, thus diminishing strong fluctuations in both temperature and drying.

Algae

Better studied in wetland basins is the phytoplankton community, suspended in the water column. The relative abundance and distribution of phytoplankton been shown to vary across time (Williams et al., 2005), water permanency and historical land-use (Casamatta et al., 2006). For autotrophs such as algae, seasonal succession is critical in systems with varying hydrological regimes. The role of algae is tightly coupled with our understanding of bacteria since they compete for similar resources. As water level and total volume decrease during the open water season, light and nutrients become increasingly limited, and populations become concentrated (figure 1). As a consequence, the abundance and distribution of filter-feeders in the water column may change. In southern Ontario, Andrushchyshyn et al. (2003) showed that seasonal phytoplankton abundance also reflected protist diversity in two vernal ponds located less than 500 m apart. The length of time during which the pond held water determined the seasonal levels of light and other physical parameters, ultimately resulting in very different protist populations. Resource limitation has also been shown to affect zooplankton communities, as Caceres et al. (2008) found in three vernal ponds in Michigan.

Using chlorophyll-a as the standard metric, primary production is commonly measured as a link to higher trophic levels such as protists and amphibians (Andrushchyshyn et al., 2003; Rubbo and Kiesecker, 2004; Andrushchyshyn et al., 2006; Sim et al., 2006b). Casamatta et al. (1999) surveyed 18 Ohio wetlands along a land-use gradient of ephemeral, non-impacted, to constructed. The survey found cyanophytes to be abundant among all wetland types (56% in temporary, 74% in non-impacted and 88.6% in constructed), but cryptophytes to be highest only among temporary wetlands (40.8% compared with 8.4% and 4.0% in non-impacted and constructed wetlands, respectively). While species richness indicators based on abundances of the six major algal divisions did not differ significantly across wetland type, a cluster analysis showed that four of the six temporary ponds, as well as four non-impacted sites, were clustered closely. Although this might suggest that key indicator algal groups are associated with wetland type, that study was a single survey and therefore did not incorporate any seasonal changes in the community. Temporal variation has been addressed in a study by Williams et al. (2005). These authors tracked algal dynamics at three sites in a pond in southern Ontario that represented a gradient of hydroperiod from permanently wet to intermittently flooded. Spatial and temporal shifts in algal community composition were identified for all three sites, with some strong similarities in species succession throughout the open water period. For example, the Chlorophyta was the first to dominate all three sites following the spring melt, with the highest biovolumes occurring in the early spring. All taxa were present at the permanent site whereas the intermittent sites were less rich. A combination of filamentous algae and diatoms increased at the intermittent sites into the summer, and euglenoids increased across all sites as seasonal temperatures increased. Overall, most algal species were present as attached forms and were gelatinous in structure – the latter thought to be an adaptation to dessication.

Physiological adaptations provide microbes with the ability to withstand the stress of desiccation. A key algal group, the green-algae desmids, secretes mucous-like substances called extracellular polymeric substances (EPS), which aid in adhesion to the substrate and gliding. In a summer study of a freshwater wetland in the Adirondack region of New York state, desmids constituted up to 23.7% of the total biofilm community and were first to colonize new habitats within the wetland (Domozych and Domozych, 2008). Bacterial cells were found attached to various desmids in the biofilm. The nature of the EPS ranged from homogeneous sheaths around the cell, to scattered patches or striated fibres that were composed mainly of polysaccharides (Kiemle et al., 2007). However, much has yet to be learned about spatial, seasonal and temporal changes associated with these assemblages, the individual species that make up these communities and the factors most influential in their establishment and distribution.

Macrophytes

Hydroperiod affects the ability of higher-plant species to germinate and also the distribution and cycling of nutrients within wetlands (Keddy, 2000; Deil, 2005; Balasooriya et al., 2008). Ecologically, macrophytes represent a rich wetland substrate, a food source for microorganisms, enhanced microhabitats, and a structural framework for invertebrates to access the water column (Williams, 2006). Decomposing macrophytes stimulate both bacterial production and photosynthetic activity (Neely and Wetzel, 1997). The distribution of vegetation is closely linked to the nature of wetland microbial communities present. For example, microbial community composition was linked to macrophyte composition at two contrasting sites within a wetland reserve in Belgium (Balasooriya et al., 2008). Notably, *Caricetum gracilis* dominated the flooded site whereas *Arrhenatheretum elatioris* was most common at the drier site, with sediments from the latter supporting an abundance of gram-negative bacteria. At both sites, fungi and gram-negative bacteria were more common than actinomycetes and other gram-positive bacteria. Such occurrences suggest that different hydrological regimes may drive both bacterial and macrophytes development along different paths according to local wetland conditions. This will likely result in modification to the invertebrate and vertebrate communities as well.

While 'keystone' wetland macrophyte taxa exist on a global scale (Deil, 2005), many regional intermittent wetlands have species that are locally cosmopolitan in their distribution and often found in permanent systems, for example, cattail (*Typha* sp.), reed (*Phragmites* sp.) and submerged forms (e.g., *Elodea* sp.). Further afield, related species fill similar wetland niches. In a survey of a gradient of tropical wetlands, Rolon et al. (2008) did not identify any key indicator species for intermittent wetlands but they did find several indicators for permanent wetlands. Overall, species richness and composition in intermittent systems seem to correlate closely with area and habitat diversity (Brose, 2001; Sraj-Krzic et al., 2007). Ray and Inouye (2006) tested the factors affecting colonization of arbuscular mycorrhizal fungi on *Typha latifolia*. They collected root samples from a constructed shallow wetland after a year of combined floodings from irrigation canals and precipiation. Roots were sampled several times in conjuction with the flooding schedules. Similar to patterns reflected in the microbial communities, the lowest levels of colonization occurred immediately after a change in water level. *T. latifolia* roots were colonized successfully by hyphae and mycorrhizae, with the best predictors of success being length of unflooded period and soil mosture: high soil moisture was associated with high hyphal colonization of roots. For mycorrhizae, oxygen availability in unflooded sediments and the biogeochemical consequences associated with water level appeared more critical.

Some macrophytes deposit seeds that may germinate under a range of flooding regimes (Deil, 2005). It is thought that macrophytes have evolved reproduction strategies that cope well with seasonal dynamics and inter-annual variability. As a result, intermittent wetlands may also accommodate rare species – however, these may come under threat from human activities. For example, Sim et al. (2006b) tested the tolerance of several macrophyte species to increasing levels of salinity, a common problem among Australian freshwater wetlands. Unsurprisingly, germination and reproductive success, plant biomass, and mean plant height decreased with increasing salinity. This indicates that native Australian wetland macrophyte communities may no longer be able to tolerate increased salinities. Where such species eliminations occur, their niches are likely to be filled with a lower diversity of tolerant species, with an accompanying change in micro- and macro-biota and wetland function.

PROTISTS TO INVERTEBRATES

The detritivores, herbivores, and predators of intermittent wetlands comprise a rich variety of protists and invertebrates. Unfortunately, relatively few studies exist for protists, although invertebrates, especially crustaceans and insects, are better known (e.g., Boix et al., 2001). Stout (1984) studied the protists of a wetland in New Zealand and found similarities with the biota of a shallow temporary pond: there was an autotrophic component of phytoflagellates and diatoms, and a bacterivorous component of zooflagellates, sarcodines (33 species), ciliates (57 species), and meiofauna. Of the 94 species of protist recorded, 45% were freshwater forms and 35% could be described as aquatic-terrestrial; the former dominated during winter when the soil was covered by several centimetres of water.

Typically, the limited data available for wetland protists have been focused on free-living ciliates. These are unicellular, eukaryotic, mainly phagotrophic organisms that play important metabolic roles in many freshwater ecosystems (Fenchel, 1987; Beaver and Crisman, 1989; Finlay and Fenchel, 2004). Ciliates and other protists are the most important grazers of microbes in aquatic environments, the only grazers of any importance in anoxic habitats, and usually predominate in sediments. Although ciliate biomass accounts for slightly less than 10% of total benthic invertebrate biomass, ciliate production may equal or even exceed invertebrate production (Finlay and Esteban, 1998a). Compared with meio- and macrofauna, ciliates consume more food, have a higher respiration rate per mass unit, have shorter generation times, and reproduce much faster, all of which contribute to their dominance in the turnover of pond organic matter. In an experimental ricefield in Italy, the ciliate community was found to be strongly influenced by environmental factors generated by the developing rice plants (Madoni, 1996). Early in the hydroperiod, when phytoplankton productivity was high both in the water column and at the sediment-water interface, grazing benthic species predominated. However, as the rice plants grew this productivity declined – likely due to shading and competition for nutrients – to be replaced by intense decomposition processes. The latter produced reducing conditions at the sediment-water interface, significantly reducing the number of ciliate species. In the water column, bacterivorous species were enhanced. In the final phase of rice growth, ciliate diversity and production were reduced across all groups. It is likely that some of the community dynamics seen in this ricefield may accompany the cycle of macrophytes growth in natural wetlands.

However, it appears that, despite their high dispersal abilities and often cosmopolitan distributions, protist species composition and richness can differ markedly between adjacent ponds (Finlay and Fenchel, 2004). For example, Andrushchyshyn et al. (2003) showed that in two intermittent ponds in southern Ontario one contained 50 species while the other contained 70 species (figure 2). Only 24 species were shared between the ponds despite them being only 500 m apart. Interestingly, the higher species count occurred in the pond with the shorter hydroperiod (34 days compared with 98 days for the second pond); the first pond also contained a greater number of rare species (27) than the second (13). Further, the 34-day pond was dominated by mid-sized ciliates (50-200 μm) whereas small species (< 50 μm) predominated in the 98-day pond.

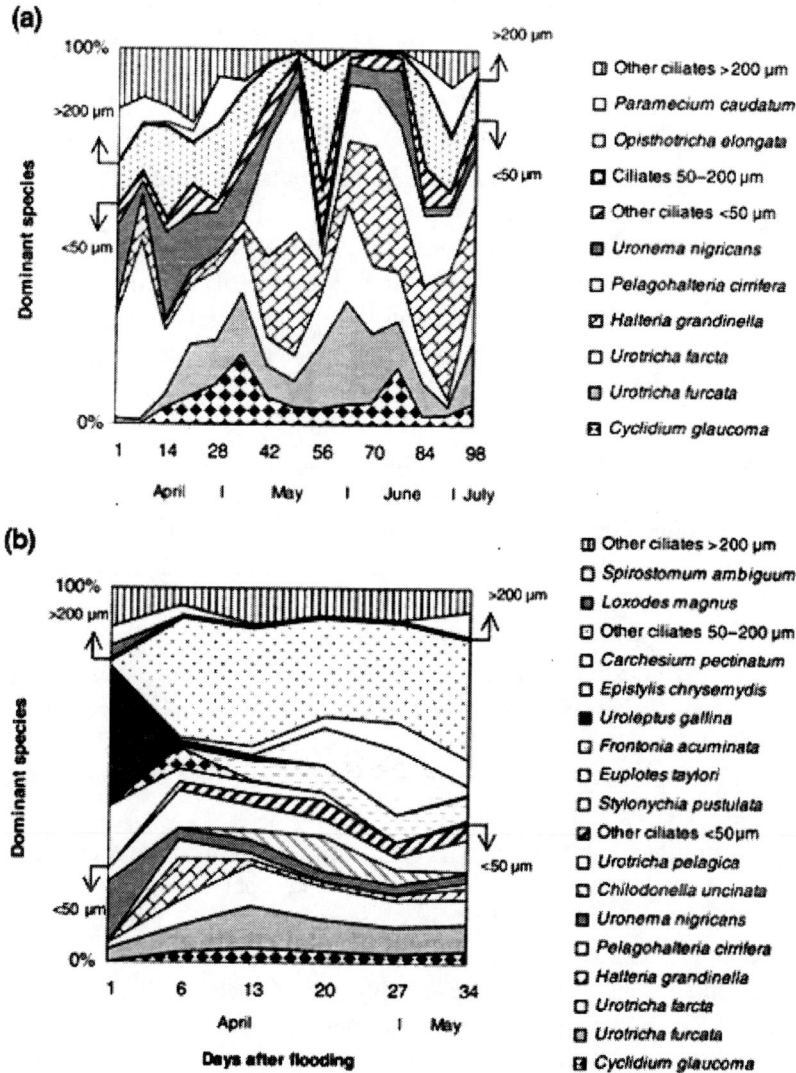

Figure 2. Relative seasonal development of the most abundant ciliate species in two intermittent ponds in southern Ontario, Canada: (a) 98-day hydroperiod pond; (b) 34-day hydroperiod pond (after Andrushchyshyn et al., 2003).

These same Ontario ponds were subject to experimental manipulation (in large, 4.2 m^2 circular enclosures) of their ciliate populations in an attempt to discover the relative importance of bottom-up versus top-down control. Pond physicochemical characteristics and bottom-up effects were found to be more important than top-down effects in governing ciliate community structure. The ciliates showed a bimodal seasonal pattern with abundances peaking early and late in the hydroperiods, and the communities showed a strong seasonal succession of species. The peak in early spring coincided with raised temperature and release of nutrients from the decomposition of dead aquatic vegetation and riparian leaf litter, whereas the later peak prior to drying was a concentration effect caused by shrinking water volume. Only 15% of the species were present throughout the hydroperiods and these included small bacterio- and algivorous forms (e.g., species of *Cyclidium, Halteria,* and *Urotricha*) that were numerically dominant at all times. Predatory taxa (e.g., species of *Pelagolacrymaria,* and *Lacrymaria*) appeared later in the hydroperiod. Large ciliates (e.g., belonging to the genera *Holophrya, Lacrymaria, Loxodes, Prorodon, Spirostomum,* and *Stentor*) appeared periodically in low abundances, but contributed significantly to total ciliate biomass.

Adding riparian leaf litter to the enclosures affected several physicochemical variables, increased bacterial abundance, and promoted the appearance of particular species – many of which are known to be associated with nutrient- or organic matter-enriched conditions. This treatment resulted in higher ciliate abundance (mainly small/medium-sized bacterivores, e.g. species of *Colpidium*) and lower ciliate diversity in mid-hydroperiod in one of the ponds. The removal of plant litter generally produced effects in the physicochemical variables that were opposite to those seen in the leaf litter addition, and resulted in a 15% decrease in the proportion of ciliate bacterivores in one pond. The effects of top-down manipulations (i.e., prevention of aerial colonization of insects, whose larvae may feed on ciliates) were minor. Several other studies have failed to observe cascading effects after invertebrate predator removal, including that of Pace and Funke (1991) who showed that top-down mechanisms were 'truncated' at the protist level, and Sipura et al. (2003) who attributed the absence of community-level cascades to consumer recycling of resources and trophic level heterogeneity. In contrast, observations and lab experiments on ciliate species from marine and permanent freshwater habitats have indicated tight correlations between the abundances of zooplankton predators (rotifers and microcrustaceans) and ciliates (Gilbert, 1989; Stoecker and Capuzzo, 1990; Gilbert and Jack, 1993).

Many treatment effects seen in the Ontario ponds were season-, and pond-specific. The measured environmental variables (including pond and treatments) explained half of the variation in ciliate abundance, one third of the species diversity, and one fifth of the species composition. Pond characteristics and the leaf litter additions were the most important factors for determining ciliate abundance (together with chlorophyll *a*), diversity (together with dissolved oxygen), and community composition (together with season) (Andrushchyshyn et al., 2006).

Variation in ciliate populations and communities among years has been shown to be high, with only 12.5% of species found in two Ontario ponds occurring in two consecutive years (A.K. Magnusson, pers. comm.). Interestingly, *Coleps hirtus* a global species of prorodontid, was predominant throughout the four year study of an Italian ricefield by Madoni (1996). This may be a consequence of its extremely flexible feeding behaviour, which allows it to ingest bacteria, algae, other protists, and even small metazoans.

The high within- and between-pond variation observed may be due to continuous, reciprocal interactions with microhabitats within the pond environment, alongside the ability of ciliates to rapidly and repeatedly exploit suitable, and novel, niches (see also Finlay and Esteban, 1998b). Because ciliates are an intermediate link between bacteria and zooplankton/benthos (McCormick and Cairns, 1991; Schmid-Araya and Schmid, 2000), high ciliate abundance and strong seasonal community development likely contribute significantly to the seasonal variation seen in other components of wetland food webs.

Taxonomically, many of the invertebrates living in intermittent waters appear to be predictably represented across wetland types, with crustaceans and insects predominating. The former are well represented by both micro- and macro-forms from the following groups: Branchiopoda, Ostracoda, Cladocera, Copepoda, Decapoda, and Peracarida. Among the insects, the Hemiptera, Coleoptera, Trichoptera, and Chironomidae (Diptera) are best represented. Aquatic mites (Hydracarina), molluscs, oligochaetes, and rotifers are also common.

Of the roughly 40,000 species of living crustaceans, about 10% live in inland waters including extreme habitats (e.g., polar and saline ponds) and wetlands. They may be conveniently subdivided into three 'functional' groups: the large, but primitive Branchiopoda; the large and more advanced Malacostraca; and the microcrustaceans. The latter comprise a variety of benthic, and especially planktonic forms, with many intermittent ponds supporting high species richness – e.g., 44 species of cladoceran and 7 species of calanoid copepod in coastal plain ponds in South Carolina (Mahoney et al., 1990); and 27 species of cladoceran and 17 species of copepod in temporary ponds in the Donana National Park in southwestern Spain (Serrano and Fahd, 2005). Microcrustaceans feed on a wide range of microorganisms and small particles, from diatoms and other algae to fine particulate organic matter (FPOM) on the bed and in the water column, and some are micropredators.

Malacostracans include peracarids (amphipods) and the larger decapods (crayfishes and shrimp) which are primarily benthic in habit, feeding omnivorously and opportunistically. Wetlands commonly support species that have diverse drought-coping adaptations – such as burrowing in the bed to access moist sediments (e.g., species of the amphipod genus *Crangonyx*; Williams and Hynes, 1977) or to reach retreating watertables (e.g., the crayfish *Fallicambarus fodiens*; Williams et al., 1974).

Branchiopods are sometimes regarded as 'flagship' species, their occurrence reliably signalling that a waterbody is not permanent as their eggs require a period of drying before they will hatch. Their relatively large size and high visibility in the early, clear water stage of the hydroperiod reinforce this signal. Typically, only one or two species, representing each of the three large-bodied orders (Anostraca, Notostraca, Conchostraca [some authorities place the latter, together with the Cladocera, in the new order Diplostraca]), occur in any one pond. However, in a single temporary pond in western Morocco, Thiery (1991) found six species of anostracan, two notostracans, and two conchostracans. Similarly, Petrov and Cvetkovic (1997) found seven species of large branchiopod in a single pond in Yugoslavia. Accomodation of these higher numbers of species appears to be based on temporal segregation related to hydrology, water temperature, and water chemistry requirements (Eder et al., 1997). The diets of these large branchiopods indicate that they are chiefly non-selective algal and detrital filter feeders, taking suspended particles and smaller plankton from the water column together with FPOM and other items stirred up from the sediment surface. In India, the conchostracan *Leptestheriella maduraiensis* has been observed to ingest food in similar proportions as in the overlying pond water, with phytoplankton (Bacillariophyceae, Chlorophyceae, and Cyanophyceae) forming the major part of the gut contents (48.9%), followed by detritus (38.6%), zooplankton (11.1%), and miscellaneous items (1 .4%) (Royan, 1976). The notostracan *Lepidurus arcticus* has been observed feeding on the anostracan *Branchinecta paludosa* in tundra ponds. From an evolutionary perspective, it is believed that persistence of these large and defenseless crustaceans in temporary freshwaters has been in response to avoidance of predatory fishes. They do fall prey to large insect predators such as dytiscid beetle adults and larvae, but only during the time in the hydroperiod when both groups overlap; typically, intermittent pond communities are dominated by crustaceans early in the hydroperiod and by insects later on (Williams, 2006).

In keeping with other taxa belonging to intermittent wetland communities, the crustacean component is heavily influenced by the length of the hydroperiod. As one might expect, species richness tends to decline in ponds that are dry for much of the year, however it also declines in ponds that are nearly permanent. Data from ponds in North America (table 1) indicate that crustacean richness is greatest in waterbodies with a hydroperiod length of between 150 and 250 days – which lends support to the intermediate disturbance hypothesis (Grime, 1973; Morin, 1999).

A low-resolution view of the major groups of aquatic insects (including the Collembola) known to occur in intermittent wetland ponds is presented in figure 3. The horizontal axis in this figure represents a relative scale of habitat permanence. Habitats such as permanent lakes are placed at the high permanence end, while rain-filled rockpools and the water contained in phytotelmata (e.g., pitcherplants) are placed at the low permanence end. Between these extremes are intermittent ponds. Particularly well suited to the latter are the Odonata, Hemiptera, Coleoptera, Diptera, and Trichoptera. Also present are the Ephemeroptera and Collembola, and occasionally Plecoptera, Neuroptera, and Lepidoptera (typically Pyralidae). Higher resolution of the Diptera shows that both nematocerans and brachycerans are present, with tipulids, culicids, ceratopogonids, chironomids, and ephydrids well represented and, to a lesser degree, ptychopterids, psychodids, chaoborids, stratiomyids, tabanids, empidids, syrphids, sciomyzids, sepsids, sphaerocerids, and anthomyids. Dolochopodids, dryomyzids, and scathophagids are also sometime found. Some interesting questions can be generated from the abundance/presence/absence data of these and other insect taxa in intermittent waters. For example, why is it that stoneflies (Plecoptera) are very rare in temporary ponds yet certain species survive very well in temporary streams? – it seems they are less able to adapt to standing water than to drought. Further, neuropterans seem not to have developed the capability of living in temporary habitats even though they are found in a range of permanent waters. Again, why are mayflies (Ephemeroptera) not widespread in intermittent ponds despite occasionally being locally abundant? Clearly, though, there needs to be more extensive surveying of habitats in order to confirm these tentative global characteristics of temporary water communities and to extend/confirm the known ranges of taxa along the permanent-temporary water habitat gradient. In addition, diet analyses need to be performed on many wetland species. Existing information has been gleaned from very diverse sources, and while it appears that most of the recognized functional feeding groups (*sensu* Merritt and Cummins, 1996) are represented within wetland insect communities the trophic statuses of individual species needs to be more precisely known so that food webs can be more accurately constructed.

Table 1. Occurrence of crustaceans in North American temporary waters in relation to length of the hydroperiod (based on data in Batzer *et al.* 1999)

		Mean length of hydroperiod (days)					
		<10	10-40	40-70	70-150	150-250	250-330
Number of studies		2	2	4	10	11	17
Taxa:							
Branchiopoda							
Anostraca	*Artemina*					*	*
	Chirocephalopsis				A		
	Branchinecta					*	
	Eubranchipus		*	*	*	*	
	Streptocephalus					C	
	Thamnocephalus					*	
Notostraca	*Triops*					*	
Conchostraca	*Caenestheriella*					*	
	Eocyzicus					*	
	Eulimnida					*	
	Leptestheria					*	
	Limnadia					C	
	Lynceus			*	A	*	*
Cladocera	*Alona*					C	
	Alonella				*	*	*
	Bosmina						R
	Chydorus				C	*	*
	Chydoridae				C		
	Ceriodaphnia				C		
	Daphnia		*	*	*	C	*
	Diaphanosoma					C	
	Macrothrix					C	C
	Moina					C	
	Pleuroxus						*
	Polyphemus					C	
	Pseudosida					C	
	Scapholeberis	*	*		*	C	*
	Sididae						R
	Simocephalus			R	C	C	*
Ostracoda	*Candona*				A	*	
	Cypria				A		
	Cypricercus				C		
	Cypridopsis						*
	Cyprinotus					*	
	Cyclocypris					*	
	Eucypris				C		
	Limnocythere					*	
	Megalocypris					C	
	Pleocypris					*	
	Potamocypris					*	
Copepoda							

Table 1. (Continued)

		Mean length of hydroperiod (days)					
		<10	10-40	40-70	70-150	150-250	250-330
Number of studies		2	2	4	10	11	17
Calanoida					*	*	*
	Diaptomus				C		
	Leptodiaptomus			C	C	C	
Harpacticoida						*	*
	Canthocamptus				A		
Cyclopoida		*	*	*	*	*	*
	Acanthocyclops				C		
	Cyclops				A		
Siphonostomatoida	*Argulus*					*	*
Malacostraca							
Isopoda	*Caecidotea*				C	*	*
	Oniscus				R		
Amphipoda	*Crangonyx*					*	C
	Gammarus						C
	Hyallela					*	C
Decapoda	*Palaemonetes*					C	*
	Cambarus				C	*	R
	Procambarus					*	C
Key: A = Abundant, C = Common, R = Rare, * = Present (not quantified)							

A low-resolution view of the major groups of aquatic insects (including the Collembola) known to occur in intermittent wetland ponds is presented in figure 3. The horizontal axis in this figure represents a relative scale of habitat permanence. Habitats such as permanent lakes are placed at the high permanence end, while rain-filled rockpools and the water contained in phytotelmata (e.g., pitcherplants) are placed at the low permanence end. Between these extremes are intermittent ponds. Particularly well suited to the latter are the Odonata, Hemiptera, Coleoptera, Diptera, and Trichoptera. Also present are the Ephemeroptera and Collembola, and occasionally Plecoptera, Neuroptera, and Lepidoptera (typically Pyralidae). Higher resolution of the Diptera shows that both nematocerans and brachycerans are present, with tipulids, culicids, ceratopogonids, chironomids, and ephydrids well represented and, to a lesser degree, ptychopterids, psychodids, chaoborids, stratiomyids, tabanids, empidids, syrphids, sciomyzids, sepsids, sphaerocerids, and anthomyids. Dolochopodids, dryomyzids, and scathophagids are also sometime found. Some interesting questions can be generated from the abundance/presence/absence data of these and other insect taxa in intermittent waters. For example, why is it that stoneflies (Plecoptera) are very rare in temporary ponds yet certain species survive very well in temporary streams? – it seems they are less able to adapt to standing water than to drought. Further, neuropterans seem not to have developed the capability of living in temporary habitats even though they are found in a range of permanent waters. Again, why are mayflies (Ephemeroptera) not widespread in intermittent ponds despite occasionally being locally abundant? Clearly, though, there needs to be more extensive surveying of habitats in order to confirm these tentative global characteristics of temporary water communities and to extend/confirm the known ranges of taxa along the permanent-temporary water habitat gradient. In addition, diet analyses need to be performed on many wetland species. Existing information has been gleaned from very diverse sources, and while it appears that most of the recognized functional feeding groups (*sensu* Merritt and Cummins, 1996) are represented within wetland insect communities the trophic statuses of individual species needs to be more precisely known so that food webs can be more accurately constructed.

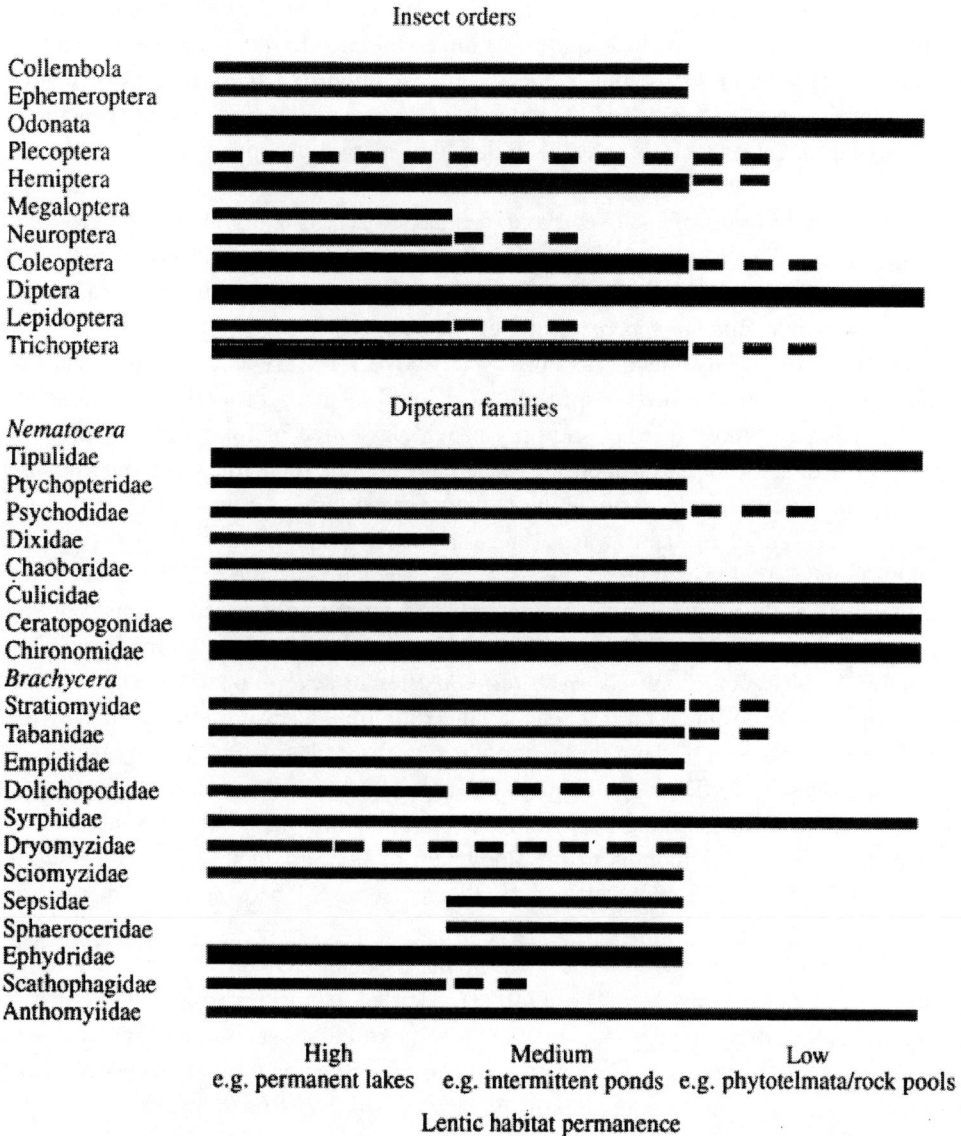

Insect orders

Collembola
Ephemeroptera
Odonata
Plecoptera
Hemiptera
Megaloptera
Neuroptera
Coleoptera
Diptera
Lepidoptera
Trichoptera

Dipteran families

Nematocera
Tipulidae
Ptychopteridae
Psychodidae
Dixidae
Chaoboridae
Culicidae
Ceratopogonidae
Chironomidae
Brachycera
Stratiomyidae
Tabanidae
Empididae
Dolichopodidae
Syrphidae
Dryomyzidae
Sciomyzidae
Sepsidae
Sphaeroceridae
Ephydridae
Scathophagidae
Anthomyiidae

High	Medium	Low
e.g. permanent lakes	e.g. intermittent ponds	e.g. phytotelmata/rock pools

Lentic habitat permanence

Figure 3. Overview of the relative importance of different insect orders and dipteran families across a gradient of standing water habitat permanence (thick line indicates 'well represented', thin line indicates 'moderately well represented', broken line indicates 'present but not significantly so'; information taken mainly from Merritt and Cummins, 1996; and Williams and Feltmate, 1992).

The relationship between a habitat permanence gradient and community structure can best be examined when a dataset contains a high level of taxonomic resolution. Such an intruiging insect dataset was provided by Wissinger et al. (1999), who compared the insect species found in 41 subalpine wetland basins in Colorado. These basins were divided into three groups, based on their number of open-water days: 40-64; 74-116; and greater than 130. The taxon counts were 10, 26, and 50, respectively. Clearly, numerically, these data support a trend of decreasing species richness with decreasing length of the hydroperiod. However, examination of the taxa in the >130 day-hydroperiod group, which was designated 'permanent' by the authors, reveals quite a number of taxa frequently seen in temporary waters, such as the hemipterans; the beetle genera *Helophorus, Agabus, Acilius, Dytiscus, Rhantus*; the dipterans *Bezzia* and *Stratiomyia*, and several chironomid genera. The interesting point that emerges about the >130 day-hydroperiod group is that although none of these basins has dried out completely in the past 50 years, they are only free of ice and snow cover for about four months of each year, and most are frozen to within just a few centimetres of their beds. It could be, therefore, that they do not truly represent permanent water habitats, and that their short window of opportunity for colonisation attracts pioneering taxa, resulting in a community with distinct *temporary* water elements.

In addition to the largely aquatic taxa described above, there are a number of other insects associated with intermittent waters, especially at the end of the hydroperiod, when decaying aquatic vegetation and moribund aquatic prey provide a source of food. Little attention has been paid to these, and other arthropod taxa (see below), thus their roles in temporary water ecosystems are poorly understood, but perhaps important. Groups of note are the grylloblattids (rock crawlers), ants, wasps, and various terrestrial beetles. For example, Lude et al. (1999) found that at least nine species of ant were able to survive frequent inundation on Alpine floodplains. In particular, *Formica selysi* regularly colonized relatively young, unvegetated gravel islands and bars, and survived flooding by forming swimming rafts when its nest entrances were compromised. Each raft consisted of several dozen workers, a queen, and brood, and remained intact until it reached the shoreline. Clearly, this ant species found suitable foraging opportunities on these periodically flooded areas. Other hymenopterans associated with seasonally flooded sites include wasps and bees. For example, Visscher et al. (1994) recorded the ground-nesting bee *Calliopsis pugionis* emerging from sites that had been underwater for more than 3 months in the floodplain of the San Jacinto River. The authors suggested that the flooding regime influenced the sex ratios of bees as they emerged from diapause.

Carabid beetles are known to be commonly associated with damp environments, including wetlands, floodplains, etc. Basta (1998) recorded 94 species from a single marsh in the Czech Republic, where population densities of some of the most abundant species seemed tied to fluctuations in water level. Carabids were also identified as a major component of the riparian fauna on four alpine floodplains in Bavaria, where most of their prey were aquatic species. In particular, on the Isar floodplain, river-derived invertebrates represented 89% of carabid prey, primarily emerging chironomids (fed upon by small species of *Bembidion*) and stoneflies (favoured by *Nebria picicornis*) (Hering and Plachter, 1997). Juliano (1985) found a variety of carabids of the genus *Brachinus* associated with different pond types in Arizona. *Brachinus lateralis* dominated the margins of more permanent ponds, whereas high-elevation temporary ponds were dominated by *B. mexicanus*. *Brachinus javalinopsis* and *B. lateralis* co-dominated the margins of a low-elevation temporary pond, but only *B. mexicanus* was found in dry pond basins. All three species were believed to share at least two potentially limiting resources: food for the adults (carrion and other arthropods), and water beetle pupae (required as hosts for their ectoparasitoid larvae).

The arachnids most commonly found associated with freshwater wetlands can be divided into three informal groups: water mites, which are primarily associated with the hydroperiod; soil mites, which are more likely to be associated with the moist sediments of the basin after the water has evaporated; and spiders, which, apart from some aquatic and semi-aquatic forms, live in the riparian zone, but also may move onto the drying bed to scavenge.

Water mites are considered those belonging to five somewhat unrelated subgroups of small arachnids. The Hydrachnida is the most familiar, but some forms belonging to the Oribatida, Halacaridae, Mesostigmata, and Acaridida (the latter three being of minor importance) have invaded fresh waters, giving rise to species that are now fully adapted to an aquatic existence. Water mites may be extremely abundant in ponds and the littoral zones of shallow lakes, for example, 2,000 individuals m^{-2}, representing as many as 75 species. Many mites represent important micropredators in temporary waters, and some have coevolved with major insect groups (especially the Diptera), both parasitising their bodies and using their adult stages as vessels of dispersal to new habitats (Smith and Cook, 1991). In North America, families/subfamilies of hydrachnids commonly found in temporary waters include the: Hydrachnidae, Eylaidae, Piersigiinae, Hydryphantinae, Thyadinae, Tiphyinae, Pioninae, and Arrenurinae. In general, hydrachnids are able to survive in temporary waters by one of two means: physiological endurance, or avoidance - for example, larvae of the Eylaidae and Hydrachnidae can remain attached to their adult insect hosts for the entire duration of the dry phase (Wiggins et al., 1980; Smith and Cook, 1991).

The ecology of oribatid mites is not well known, partially due to problematic, taxonomy. However, they are extremely abundant in most forested ecosystems and often comprise around 50% of the total microarthropod fauna. As such, oribatids are believed to play important roles in the decomposition of organic materials, modification of the physical and chemical textures of soil, the cycling of nutrients, and the conservation of healthy soil environments (Wallwork, 1983). Habitat features thought to influence their population biology include habitat complexity and soil micropore size, soil humidity and organic content, soil temperature, surface vegetation, precipitation, and the activity of other soil microfauna (Wauthy et al., 1989). Because forest soils are often contiguous with wetland basins, it is likely that oribatid mites contribute to biological processes (largely unknown) in these sediments. That there are characteristic assemblages of these mites in temporary waters is illustrated by the fact that oribatids have been found useful in distinguishing between different stages of degradation of freshwater mires in the area of Berlin (Kehl, 1997). In coastal freshwater habitats in Antarctica, two species of terrestrial oribatids, *Edwardzetes elongatus* and *Trimaloconothrus flagelliformis*, have become adapted to survive prolonged submergence on aquatic mosses – apparently in response to niches unoccupied by truly aquatic forms (Pugh, 1996).

There is only one species of truly aquatic spider. *Argyroneta aquatica* (Agelenidae), found in ponds of central and northern Europe, builds an underwater air-store from silk fashioned in the form of an inverted vase. It supplies this store with air brought from the water surface, and uses it as a medium in which to externally digest prey captured under water. Many other spiders, however, frequent the margins of both permanent and temporary ponds and streams. All members of the Pisauridae, for example, depend on being close to water, and some, such as *Dolomedes* dive while hunting for tadpoles, aquatic insects, and small fishes. There is a rich diversity of pisaurids in Australia, for example, including: *Dendrolycosa* (which also extends from New Guinea to India), *Dolomedes* (cosmopolitan), *Hygropoda* (also in southeast Asia, Venezuela, New Guinea, Madagascar, and central Africa), *Perenthis* (also in India, Burma, Papua New Guinea, and Japan), and the endemics *Inola* and *Megalodolomedes*. (Main et al., 1985). Another family of common riparian spiders is the Tetragnathidae – the long-jawed orb weavers. The genus *Tetragnatha* has a wide global distribution, with *T. versicolor* and *T. elongata* being circumboreal. Williams et al. (1995) calculated that, as a minimum estimate, individuals of these two species captured 0.2% of the total number of insects emerging from a small river in Canada, particularly chironomids and mayflies. Tetragnathid webs are common on any vegetation overhanging temporary ponds and streams, where their prey also includes emerging mosquitoes and other long-legged dipterans. The diversity of spiders associated with wetlands can be very high, for example, van Helsdingen (1996) recorded a total of 63 species from two Irish floodplains.

A number of other, non-insect arthropods are associated with the end of the hydroperiod. Whereas zoogeography will dictate local composition, this 'clean-up crew' may include millipedes, centipedes, symphylans, pauropods, pseudoscorpions, harvestmen (Opiliones), diplurans, bristletails (Archaeognatha), and silverfishes (Thysanura) (Williams, 2006).

VERTEBRATES

Numerous vertebrate species exploit productive temporary waters for food, reproduction, thermoregulation, and for fresh drinking water. For example, Paton (2005) estimated 86% of mammals, 73% of amphibians, 62.5% of reptiles, and 34.5% of birds in the northeastern US either forage or breed in seasonal forest pool habitats. Those numbers reflect the importance of wetland habitats for both the transitory animal and the obligate vertebrate user. Vertebrates function as the top predators as well as occupying the lower trophic levels in temporary ponds. Some animals may stop at a water body to rest, feed, or drink not spending more than a few days time (i.e., migrating birds), while others rely on the pond for part of their life cycle (e.g., the majority of amphibians), thus establishing vertebrates as a strong link between aquatic and terrestrial habitat boundaries. Multiple abiotic and biotic factors determine community structure for vertebrate groups using wetlands. Despite the significant and diverse uses of intermittent wetlands by vertebrates, the bulk of research studies have concentrated on amphibians. The following account briefly reviews what is known of other vertebrate taxa, and is partitioned into fishes, amphibian, reptilian, birds, and mammals.

Fishes

Most intermittent wetland ponds are fishless, an important feature for many organisms that are vulnerable to predation, but which has resulted in a lack of attention by ichthyologists. Fairly recently, Snodgrass et al. (1996) investigated the extent of fish occupancy in 63 isolated Carolina bay wetlands in South Carolina, USA. In comparison to previous estimates, their study found 21% fish occupancy in the ponds surveyed, which is higher than the less than 10% estimated by Sharitz and Gibbons (1982) and Richardson and Gibbons (1993) for the same area. The Snodgrass et al. study recognized some obvious, but not well reported patterns of community structure (for fish pond usage): fishes were more closely associated with ponds that dry less frequently and were located closer to other aquatic habitats. However, some fishes have adaptations to withstand living without water for some time. For example, *Lepidogalaxias salamandroides* (salamanderfish) lives in intermittent ponds that dry completely in south-west Western Australia (Berra and Allen, 1991). This unique species lies burrowed 2-60 cm into the sand during the dry months (Allen and Berra, 1989) only to reappear at the onset of pond filling. Pusey and Bradshaw (1996) examined the gut contents from six fish species, including *L. salamandroides*, occupying seasonally inundated waters in the southern acid peat flats of this same region of Australia and found a diversity of prey, comprising chironomid larvae, planktonic Entomostraca, midge adults, Collembola, ants, spiders, Thysanoptera, calanoid copepods, and Cladocera. They noted decreasing diet overlap among species over the course of the hydroperiod. A unique group of fishes, the Snake-heads (Channidae, genus *Channa*) are found in temporary ponds, beels and haors (natural depressions), and swamps in Bangladesh. These fishes can live for months buried down to a meter deep in the mud, where they subsist on stored fat (Rahman, 1989). Similarly, African lungfishes are well known for their ability to live in habitats that dry temporarily. These fishes (*Protopterus* spp.) live through the dry season by either creating a subterranean cocoon, making burrows through the substrate down to the water table, or burrowing into the moist mud after their lake, river, or swamp has dried up (Greenwood 1986).

It is well known that, worldwide, fishes use the inundated floodplains along rivers and streams for spawning and nursery habitat (Copp, 1989; Rahman, 1989; Riemer, 1991; Winemiller and Kelso-Winemiller, 1996; Homes et al., 1999; Henning et al., 2007). However, the temporary nature of these habitats has limited their study (Henning et al., 2007). The floodplain is of transitory use because juveniles and adults need to recede to the main channel once water levels decline. However, when in residence, fry benefit from rapid growth and development made possible by the warmer temperatures and high nutrients that result from the mixing and accumulation of fertile silt, decaying vegetation, and other resources washed over the river banks. Cucherousset et al. (2007) examined the integration of biotic and abiotic factors influencing habitat use, colonization timing, and fish assemblages of a temporarily flooded grassland along a canal in northwest France. Fish assemblage and density changed in response to the extent of flooding (water level), temperature, and zooplankton abundance in flooded grassland habitat. Similarly, Poizat and Crivelli (1997) monitored fish movement into and out of flooded marshes alongside the Fumemorte Canal in southern France for 3 years. They found that species-dependent movement patterns into seasonally flooded marshes and hydrology highly influenced community structure. Unfortunately, many river drainages are now disconnected from their floodplains as a result of flood control structures, such as dams and levees (Junk et al., 1989; Sparks 1995) preventing fishes from using these productive waters.

Fishes also readily occupy permanent pools remaining in intermittent streams, often colonizing different pools following flood disturbance (Fausch and Bramblett, 1991). Ephemeral karst lakes and wetlands in south-central Kentucky, having a similar hydrological cycle to intermittent streams, can also harbor fishes in small isolated pools when the main lake is dry. Vessels and Jack (2001) measured the effect of a Golden Shiner (*Notemigonus crysoleucas*) fish population on the zooplankton community of an isolated pool in such a lake. Growth rates and densities of larger zooplankton species (e.g., *Bosmina* and *Anthocyclops*) were negatively affected by the presence of these fish, but there was no effect on smaller invertebrates.

In summary, fish assemblage structure seems to be controlled by length of hydroperiod, degree of isolation from a permanent water source, and colonization rates, rather than by biotic factors, such as competition (Pusey and Edward, 1990; Snodgrass et al., 1996, Baber et al., 2002). Fishes in intermittent waters may exploit the abundant prey base of tadpoles and invertebrates, while benefiting from low predation risk (Snodgrass et al. 1996). Whereas, fish assemblages are relatively well studied in inundated floodplain areas they have been less studied in other intermittent water bodies. Such habitats are highly capable of supporting breeding fish populations when the latter either have an adaptation to withstand pond drying, or when an escape outlet to a permanent water source is available.

Amphibians

Many diverse and abundant assemblages of amphibians depend on temporary freshwater bodies for reproduction, growth, and development. Aquatic-breeding amphibians spatially and temporally partition their use of both permanent and temporary ponds by initiating breeding at different times during the hydroperiod. In central and eastern northern North American, predictable signs of spring are the calls of early spring breeders such as *Pseudacris crucifer* (Spring Peeper) and *Rana sylvatica* (Wood Frog) which enter ponds just after snow melt. Other species breed later in the season, after water temperatures have risen (e.g., *Rana clamitans*, *Hyla versicolor*). Species composition thus varies throughout a season and across years because of breeding phenology, temperature, hydrology, population dynamics, and natural and anthropogenic habitat changes.

Amphibian species richness is generally not related to wetland size or pond size (Snodgrass et al., 2000; Oertili et al., 2002), but a study conducted by Beja and Alcazar (2003) contradicts this generalization. The latter authors surveyed 57 ponds in southwest Portugal which varied in surrounding agricultural land use and hydroperiod, and found that species richness increased with increases in pond area and hydroperiod. Findlay and Houlahan (1997) combined amphibian and reptile species richness in their study of 30 southeastern Ontario, Canada wetlands and found, similarly, that species richness increased with increased wetland area. Unfortunately, Findlay and Houlahan did not provide separate data for these two taxa to test if the pattern was driven by reptile species richness. Amphibian species richness is negatively related to fish presence (Werner et al., 2007). Along with a change in the number of species present in a pond of a particular size, the species assemblage also changes. Ponds of different size host unique assemblages, a consequence of the fact that while some species are cosmopolitan in occurrence across pond size, others occur only in ponds of a specific size.

Pond size is often related to pond hydroperiod. Pond hydroperiod is viewed as the most important factor determining amphibian assemblage structure (Snodgrass et al., 2000) because it is a critical determinant of reproductive success and species survival. Interestingly, species richness tends to be highest in ponds with hydroperiods of intermediate length. Both historical and current hydrological regimes are important in determining amphibian assemblage structure (Church 2008). Frogs and salamanders are able to adjust timing of metamorphosis in response to pond drying (Semlitsch and Gibbons, 1985; Semlitsch 1987; Wilbur, 1987), for example *Scaphiopus couchii* (Couch's Spadefoot) has evolved the remarkable ability to metamorphose in as little as 8 days in rapidly drying desert ponds in southwestern Texas (Newman 1987, 1989). Particular life stages of frogs, toads, and salamanders may use separate ponds for reproduction, feeding, and over-wintering depending on the hydrology, surrounding habitat, and the predator component in the pond. Presence of fishes is a major factor controlling amphibian assemblage structure in pond systems as well. Other factors controlling assemblage composition and structure are: pond age (Ward and Blaustein 1994, Homes et al. 1999); competition for resources (Finlay and Vredenburg, 2007); canopy cover; surrounding landscape; and within pond vegetation structure (Paton, 2005).

Assemblage structure and species richness are determined by the interaction of interspecific competition, predation, and pond drying (Wellborn et al., 1996; Eason and Fauth, 2001). Predation, as traditionally defined, is a strong determinant of amphibian species composition in temporary habitats and in its most common form involves amphibians being eaten by larger organisms (especially large insects [e.g., dytiscid beetles], salamanders, and fishes) at early embryo stages. Because of the interaction of predation and hydroperiod, many amphibian species have to trade-off between predation pressure and hydrological constraints, often choosing temporary waters over more permanent bodies with potential predators. Many accounts of feeding in tadpoles show that they exhibit macrophagy (Petranka and Kennedy, 1999, and sources within), but they also exhibit strong predation pressure on other conspecific amphibian larvae (Woodward, 1982; Petranka et al., 1994, Petranka et al., 1998). Furthermore, in Southern Appalachian vernal pools, intraspecific tadpole predation has been shown to be a strong force in structuring larval assemblages (Petranka et al., 1998).

Amphibians thus function as both predators and prey within wetland systems, and therefore play an important role in the food web. Species living in intermittent water bodies connect aquatic and terrestrial systems because energy flow is bidirectional. Energy gained in the larval stage is carried out of the aquatic system during metamorphosis to the terrestrial system, and adults will return terrestrially gained energy in the form of egg deposition to the aquatic system. For example, salamander egg masses were shown to provide an easily assimilated source of energy for pond communities in southern Illinois (Regester et al., 2006). Egg mass energy input has been found to be greatest in large ponds and varies with species composition. Given that networks of egg masses can exceed 10,000 eggs, this form of energy is an extremely important source of nutrients for caddisfly larvae, leeches, and turtles (Colburn, 2004), as well as other invertebrate and vertebrate predators (Stebbins and Cohen, 1995). Although, amphibian larvae occupy lower trophic levels, are small and often overlooked, they can control systems both from the bottom-up and from the top-down. For example, *Rana sylvatica* tadpoles were shown to induce high mortality, through predation, on a higher trophic level consumer, Spotted Salamander larvae (*Ambystoma maculatum*) in a natural pond experiment in North Carolina (Petranka et al., 1998). Also, microhabitat segregation in the placement of egg clutches has been observed in ponds containing both *Rana* tadpoles and salamander larvae (T.A. Schriever, personal observations). Bottom-up control is demonstrated through high resource use of phytoplankton, protozoans, and green algae (Seale, 1980). Seale's study was conducted in a permanent pond, but the effects on nitrogen levels, algal biomass, and rates of primary production are likely similar for temporary waters. In addition to anuran tadpoles, salamander larvae have been found to influence zooplankton communities (Wilbur, 1997). Yet, larval diets of amphibians are still poorly known (Altig and Johnston, 1989; Altig et al., 2007).

The importance of amphibian larvae in wetland ecosystem function, however, lacks quantification and remains largely unexplained – although, for example, high amounts of energy can be assumed to remain in ponds as a result of larval mortality. Despite the lack of research on larval diets and energy flow, it is intuitive that amphibian larvae must play diverse and important functional roles in wetlands. Other vertebrates, such as reptiles, use the abundance of metamorphosing and adult amphibians as a major food source. Significant declines in amphibian specialist predators, such as the Gartersnake *Thamnophis elegans*, have been noted after a drastic decline in their main prey item, *Rana muscosa* (the Mountain Yellow-legged Frog) in the California Sierra Nevada (Jennings et al., 1992; Finlay and Vredenburg, 2007).

Created wetlands and ponds constructed for agricultural purposes provide important habitat for amphibians. Although created wetlands may be recolonized by source populations, they often exhibit changes in vertebrate community structure and pond hydrology that are different from natural ponds. Vasconcelos and Calhoun (2006) studied populations of Wood Frogs and Spotted Salamanders in a series of created wetlands and natural seasonal potholes on a forested island in mid-coast Maine from 1999 to 2004 to assess the success of created ponds. The latter had open canopies with intermediate-length or permanent hydroperiods in contrast to the closed canopy and intermittent hydroperiods of the previous, natural ponds. These changes allowed Green Frogs (*Rana clamitans*) to colonize, since their larvae over-winter in permanent ponds, which resulted in nearly 100% mortality of Wood Frog egg masses from direct predation by Green Frogs.

Rice fields provide temporary habitat for reptiles, birds, amphibians, mammals, and aquatic invertebrates as they contain water for the growing season and then are drained to harvest the crop (Lawler, 2001). Species diversity and composition vary according to farming (organic vs. pesticide use) and irrigation (intermittently irrigated vs. continuously flooded) practices, and whether fishes are present (Lawler, 2001). In Japan, some organisms (such as the nymphs of the giant water bug, *Kirkaldyia deyrolli*) are specialized predators within hydrological unstable rice fields. In three separate rice plots, Ohba et al. (2008) showed that tadpoles were the main prey item for first to third instar (younger) nymphs. Significantly, nymphs that specialized on tadpoles developed faster than nymphs that fed on a combination diet of tadpoles and frogs, or on just odonata nymphs.

Reptiles

In general, reptiles are less dependent on intermittent systems than amphibians. However, some reptiles such as turtles, snakes, and alligators use intermittent waters for summer habitat, breeding, and for food. Some may capitalize on the abundance of emerging amphibians during particular times of the year and rarely visit these habitats for the rest of the year. The mostly aquatic Black Swampsnake (*Seminatrix pygaea*) reaches high densities in ponds in the southeastern US (Gibbons and Dorcas, 2004; Winne et al., 2005), where it feeds on leeches, amphibians, and small fishes (Mills et al., 2000; Gibbons and Dorcas, 2004). Paton (2005) estimated that 75% of native snake species in the northeast US use temporary pools for foraging or basking. Roe et al. (2003) noted the frequent movement between, and extensive time spent in, wetlands of varying hydroperiods by the Copperbelly Watersnake (*Nerodia erythrogaster*) and the Northern Watersnake (*Nerodia sipedon*) in northwestern Ohio and southern Michigan. *Thamnophis sauritus* (Eastern Ribbonsnake) has been observed eating frogs and fishes along the margins of temporary ponds in Nova Scotia (Bell et al., 2007). These same authors suggested that Eastern Ribbonsnakes in the Nova Scotia population depend on temporary ponds in order to acquire an adequate number of prey in an area with a short growing season. A well known occupant of temporary shallow waters in the vast lowlands of tropical South America is the Green Anaconda (*Eunectes murinus*). Well known yes, but few detailed studies on its ecology have been published. Through a four year radio-telemetry and mark and recapture study, Rivas (2000) found Green Anacondas to spend the majority (86%) of their time in shallow, densely vegetated water or near water (14%) in the Venezuelan llanos. Here, they fed on birds (e.g., Jacana, whistling duck), large mammals (e.g., white-tailed deer, capybaras), and large reptiles (e.g., Side-necked turtles, Spectacled caiman). During the wet season, these snakes distribute themselves widely across the inundated floodplain but during the dry season large groups congregate in the few remaining water bodies (Rivas 2000).

Turtles also frequent temporary ponds. Milam and Melvin (2001) recorded high rates of pond usage by *Clemmys guttata* (Spotted Turtle) in two populations in Central Massachusetts from 1993 to 1995. Their radio-telemetry data confirmed Spotted Turtles dispersed from hibernacula directly to temporary ponds, where they stayed for approximately 2 to 4 months for basking, swimming, mating, and foraging on amphibians, macroinvertebrates, and algae. Distance between hibernacula and seasonal ponds ranged from 20-550 m over upland forest habitat, leading one to assume that intermittent water bodies are much sought after habitats. Graham (1995) and Joval (1996) noted similar movements in *C. guttata*, but added that some individual turtles used temporary ponds as hibernation sites as well. Movement out of ponds coincided with a decrease in water level. The American Alligator (*Alligator mississippiensis*) is common and can be found at high densities in large permanent lakes, rivers, swamps, marshes, and small temporary ponds throughout the southeastern US. Bondavalli and Ulanowicz (1999) found *A. mississippiensis* to have a significant positive effect upon its prey (frogs) in Big Cypress National Preserve, Florida because it removes the frogs' main predators, snakes. 'Gator holes', holes dug by alligators, during the dry season provide habitat for fishes, frogs, and snakes enhancing wetland biodiversity. American Alligators have a diverse diet consisting of crayfishes, tadpoles, salamander larvae, frogs, fishes, turtles, small mammals, and birds which varies seasonally in accordance with prey abundance (Platt et al. 1990). Distance from the nearest permanent aquatic habitat does not reliably predict density of adult alligators, however density has been found to be positively correlated with the log function of pond area (Altrichter and Sherman 1999). Similarly, the Nile crocodile (*Crocodylus niloticus*) uses temporary streams in south-eastern Mauritania (Shine et al., 2001) and seasonal rivers and ponds in Zimbabwe (Kofron, 1993) for basking and foraging. Nile crocodiles may benefit with increased habitat conservation because of the tourist industry in Mauritania after a study recognized the range extent and abundance of sight-seeing opportunities in Tagant Plateau (Tellería et al., 2008). Temporary wetlands in the southeastern Coastal Plain of the US are possibly overlooked as usable habitat for reptiles (Dodd, 1992).

Studies are needed to determine diversity, richness, and trophic separation across the hydroperiod gradient to better understand the functional and structural roles of reptiles in temporary waters.

Birds

Surprisingly little information is available for intermittent pond/wetland use by birds (Silveira 1998, Hanowski et al., 2006; Scheffers et al., 2006), however, Silveira gives a rather extensive summary of bird usage studies or inventories conducted in California. Most information on bird usage focuses on waterbirds and has a management slant in particular wetland types (e.g., 'managing wetlands for waterbirds'; Reid 1993). Research on waterfowl tends to investigate population dynamics, feeding ecology, and duck production capabilities. The Prairie Pothole Region of the northern Great Plains of North America is known as a 'duck factory' because it harbors a vast array of water bodies with varying hydroperiods (Baldassarre and Bolen 1994) that are important for growth and breeding. The potholes that fill for only a few weeks after snowmelt are important for breeding pairs; ponds that last until late summer are used as feeding grounds because of high invertebrate abundance and adequate nesting cover, whereas ponds that hold water into the winter are particularly valuable for breeding activity. Floodplains also offer valuable breeding habitat for waterbirds. For example, the Paroo River floodplain of northwestern New South Wales and southwestern Queensland maintains a wide diversity of temporary freshwater and saline water bodies hosting a productive invertebrate fauna that supports waterbirds during the breeding season (Timms, 1997). Recently, Scheffers et al. (2006) found higher avian species richness, diversity, and abundance within a 50 m radius of 25 ephemeral ponds than in the surrounding hardwood forests on the Cumberland Plateau in Tennessee. They postulated that higher bird abundance and richness resulted from high invertebrate and insect abundance over ponds, as well as diverse terrestrial invertebrates in moist soils around pond edges. However, community composition around temporary ponds did not differ from surrounding forest habitat in Tennessee or in northern Minnesota (Scheffers et al., 2006; Hanowski et al., 2006). Temporary ponds may be more often used by some non-breeding birds such as the Northern Waterthrushes (*Seiurus noveboracensis*). Hunt et al. (2005) surveyed *S. noveboracensis* overwintering habitat in southwestern Puerto Rico reporting the highest abundance in mangrove swamps, but dry coastal shrub and savanna areas were also used.

Birds forage in terrestrial habitats surrounding ponds and canopy areas above the ponds, gaining nutrients from both aquatic and terrestrial sources. Some of the nutrients may be returned to the ponds via guano, which adds significant amounts of nitrogen and phosphorus to the system, especially when a large roosting colony of birds is present (McColl and Burger, 1976). Likewise, birds remove aquatic-derived nutrients from the system and deposit them on land. Migrating birds likely transfer nutrients great distances across diverse landscapes, contributing to distant food webs. Nutrient transfers can be tracked using stable isotopes, measuring chemical/biological concentrations (e.g., P, N, Ca, Na, Mg, fecal coliform), and measuring organic matter content before and after migration.

Mammals

Relatively little information exists on the use by, and importance of temporary waters to, small mammals. However recently, a few studies have recognized intermittent ponds as important habitat for bats and small mammals in the northeast US. Paton (2005) summarized vertebrate use in seasonal forest ponds in the northeastern US, noting that all 9 species of bat in the study range foraged heavily over ponds. Francl (2008) studied summer bat use of 17 woodland seasonal ponds in northern Wisconsin and the upper peninsula of Michigan that differed in hydroperiod and size. She concluded that the bats preferentially use ponds based on canopy cover and amount of water, with more bat activity noted over large and small ponds than over medium-sized ponds. Owen et al. (2003) captured more northern myotis bats over ephemeral ponds than riparian habitats in the central Appalachians of West Virginia. However, other small mammals do not share similar temporary pond use or association. For instance, in central Massachusetts, Brooks and Doyle (2001) found no difference in the abundance of shrews near temporary ponds compared with those in adjacent upland habitat. However, in western Kentucky, Brown and Fuller (2006) noted higher abundances of female white-footed mice in dry, non-flooding areas, but higher male abundance in floodplain habitat.

Beavers are recognized for their ability to alter stream and wetland ecosystems by modifying the hydrology, rates of decomposition, and plant and animal community composition and structure (Weller, 1981; McDowell and Naiman, 1986; McCall et al., 1996; Snodgrass, 1997; Snodgrass and Meffe, 1998; Russell et al., 1999). They have been referred to as 'allogenic' (i.e., alteration of the environment by creating structures) ecosystem engineers (Alper, 1998). However, they may damage one habitat and simultaneously create new valuable breeding areas for amphibian species (Cunningham et al., 2007), feeding grounds for waterbirds, and habitat for colonization by crustaceans (Alper, 1998). An *a priori* model developed by Cunningham et al. (2007) predicted high amphibian species richness in ponds modified by beavers in Acadia National Park in Maine, US. Empirically, they reported species-specific occupancy trends correlated with beaver activity (e.g., bullfrogs occurred only in active beaver wetlands). The lentic ecosystems created by beavers are temporary in nature because they are used for a time and then abandoned. A neglected beaver dam may break, releasing water and allowing the impounded land to drain, sometimes creating small pools.

Primate species richness is high in Amazonian forests, however primate community structure, composition, and abundance in relation to seasonally inundated floodplain forests and unflooded forest has received little attention (Haugaasen and Peres, 2005). Peres (1997) and Haugaasen and Peres (2005) conducted line-transect surveys in western Amazonia reporting higher primate species richness in never flooded (terra firma) forest habitat, but higher density and biomass in seasonally flooded (várzea) forests. Seasonally plentiful fruit availability can entice primates to the floodplains, resulting in an increase of species richness at certain times in the year (Haugaasen and Peres, 2005).

Large mammals are highly mobile, resulting in less frequent or only very brief occurrence (e.g., for drinking) in intermittent wetlands. However, in Africa, Cape buffalo, hippopotamus and other large semi-aquatic mammals often use wetlands for grazing and thermoregulation. Moose (*Alces alces*) are well-known feeders on wetland plants. Attracted to fresh grazing and prey, larger mammals are sometimes more likely to be found in dry basins than when they are filled. Dry floodplains are used by deer, turkeys (Sparks, 1995), and mice (Brown and Fuller, 2006) then abandoned for higher ground or move into the trees when inundated with flood waters. In Africa, antelope such as *Tragelaphus scriptus* (bushbuck) and *Kobus leche* (lechwe) have life cycles that are timed so that they give birth to their young as floodwaters recede and new pasture is exposed (Welcomme, 1979). The contributions of large mammals to wetland communities and food webs lack quantification, although nutrient addition via feces and stirring up of basin sediments, together with aiding dispersal and colonization of elements of the flora and fauna are obvious occurrences.

CONCLUDING REMARKS AND FUTURE NEEDS

Microbial and macrophyte community shifts have been observed in response to changes in hydroperiod. Evidence suggests that microbial communities may change from *r*- to k-strategist dominated following flooding events and also be influenced by the available oxygen within sediments. As with most ecological studies at the microbiological scale, a key question is how to scale findings and inferences up to the ecosystem level? For example, how can microbial processing rates be directly related to hydrological processes, and watershed, regional, and global scales? Several models have been proposed to explain general patterns in wetland development and community structure (e.g., Euliss et al., 2004), however, these summaries rarely incorporate intermittent hydroperiods, or the contribution of microbiota.

A further aspect lacking in the current literature is that of temporal variability. Few studies have tracked changes within a single ecosystem (Noll et al., 2005) or over many years, thus long-term changes in wetlands are difficult to observe and interpret. The majority of predictions are based on short-term studies, often conducted in constructed wetland systems or in experimental or mesocosm setting. Our understanding of the microbial, fungal, and algal components of freshwater wetlands is quite limited, and that of viruses non existent. In order to truly understand the response of these systems to stress, baseline studies on intermittent wetlands, particularly in different biogeographical regions, are necessary. Several studies over the years have attempted to define thresholds of tolerance (to, for example, salinization) for components of the biota, including bacteria (Baldwin et al., 2006), however with little understanding of the pre-existing communities, interpretation and management are precarious.

While the invertebrate components of the wetland biota are better known, this knowledge is largely descriptive and often lacking in taxonomic resolution. Properties such as community succession and colonization dynamics are reasonably well understood, as are aspects of predation by large invertebrates – the same cannot be said for amphibian and reptile predators, however. Clearly, the study of wetland protists is in its infancy, yet what is known points to high species richness and an important trophic function, channeling resources from the bacteria and algae level to the level of invertebrates and above.

Conservation of ponds of various sizes and hydroperiods is vitally important to sustain vertebrate diversity (Semlitsch and Bodie, 1998; Oertli et al., 2002; Beja and Alcazar, 2003, Zedler 2003; Vignoli et al., 2007; Francl, 2008). Although, we have a good grasp on the amphibian response to hydroperiod in terms of species richness, diversity, competition, and are gaining more information on predator-prey interaction, we lack understanding of food web dynamics in relation to hydroperiod and vertebrate communities in general. We have extensive knowledge of food web dynamics in lakes, streams, and rivers, yet we lack basic knowledge of energy flow through vertebrates in wetlands and their connections to terrestrial systems. Existing studies have tended to concentrate on either the aquatic or the terrestrial component in isolation. As we have noted, energy flows in both directions between these adjacent systems, and there is an urgent need to study them together. Suitable technologies exist for accomplishing this: for example, stable isotope analysis (SIA) has routinely been used on invertebrates, fishes, birds, and mammals yet it has rarely been used to investigate the trophic ecology of amphibians and reptiles (Vanderklift and Ponsard, 2003; Dalerum and Angerbjörn, 2005). Further, most SIA studies have been carried out in streams, lakes, and rivers (Post, 2002), pointing to a real need for their application in wetlands and other lentic environments. SIA can be used to reveal the major energy sources used by vertebrates in intermittent wetlands and whether these sources are allochthonous or autochthonous. Estimating the number of trophic levels in such wetlands would be a necessary start to understanding species diversity, habitat complexity, and productivity. Kupfer et al. (2006), for example, estimated that a tropical permanent pond and its adjacent forest each had five trophic levels. Quantification of energy flow is extremely important given the rate of habitat loss and destruction, and species introductions. Significantly, when fishes invade historically fishless habitats, dynamics, such as nutrient cycling, food web structure, and invertebrate composition and diversity are changed (Finlay and Vredenburg, 2007). Characterization of food web dynamics and energy flow across systems exhibiting a range of hydroperiods would be most informative. Knowing the amount of energy transferred and its pathways between aquatic and terrestrial systems will enable the real costs of wetland destruction to be determined, but also will allow new constructed wetlands to be configured to achieve optimal operation.

REFERENCES

Allen, G. R., and Berra, T. M. (1989). Life history aspects of the West Australian salamanderfish, *Lepidogalaxias salamandroides* Mees. *Records of the Western Australia Museum* 14: 253-267.

Alper, J. (1998). Ecosystem ‚Engineers' shape habitats for other species. *Science* 280: 1195-1196.

Altig, R. and G. F. Johnston. 1989. Guilds of anuran larvae: relationships among developmental modes, morphologies, and habitats. *Herpetological Monographs* 3: 81–109.

Altig, R., M. R. Whiles, and C. L. Taylor. 2007. What do tadpoles really eat? Assessing the trophic status of an understudied and imperiled group of consumers in freshwater habitats. *Freshwater Biology* 52: 386–395.

Altrichter, M., and Sherman, P. M. (1999). Distribution and abundance of the American Alligator (*Alligator mississippiensis*) in the Welder Wildlife Refuge, Texas. *Texas Journal of Science* 51:139-146.

Andrushchyshyn, O., Magnusson, A.K., and Williams, D.D. (2003). Ciliate populations in temporary freshwater ponds: seasonal dynamics and influential factors. *Freshwater Biology* 48: 548-564.

Andrushchyshyn, O., Magnusson, A.K., and Williams, D.D. (2006). Responses of intermittent pond ciliate populations and communities to *in situ* bottom-up and top-down manipulations. *Aquatic Microbial Ecology* 42: 293-310.

Baber, M. J., Childers, D. L., Babbitt, K. J., and Anderson, D. H. (2002). Controls on fish distribution and abundance in temporary wetlands. *Canadian Journal of Fisheries and Aquatic Sciences* 59: 1441-1450.

Balasooriya W.K., Denef K., Peters J., Verhoest N.E.C. and Boeckx P. (2008). Vegetation composition and soil microbial community structural changes along a wetland hydrological gradient. *Hydrology and Earth System Sciences* 12: 277-291.

Baldassarre, G.A., and Bolen, E.G. (1994). *Waterfowl Ecology and Management.* New York, New York, USA: John Wiley and Sons.

Baldwin D.S. and Mitchell A.M. (2000). The effects of drying and re-flooding on the sediment and soil nutrient dynamics of lowland river-floodplain systems: A synthesis. *Regulated Rivers - Research Management* 16: 457-467.

Baldwin D.S., Rees G.N., Mitchell A.M. and Watson G. (2005). Spatial and temporal variability of nitrogen dynamics in an upland stream before and after a drought. *Marine and Freshwater Research* 56: 457-464.

Baldwin D.S., Rees G.N., Mitchell A.M., Watson G. and Williams J. (2006). The short-term effects of salinization on anaerobic nutrient cycling and microbial community structure in sediment from a freshwater wetland. *Wetlands* 26: 455-464.

Basta, J. (1997). Contribution to knowledge of the ground-beetle fauna (Coleoptera: Carabidae) in the environs of Brno. *Acta Museum Moraviae Scientiarum Biologica* 82: 207-213.

Bastviken S.K., Eriksson P.G., Martins I., Neto J.M., Leonardson L. and Tonderski K. (2003). Potential nitrification and denitrification on different surfaces in a constructed treatment wetland. *Journal of Environmental Quality* 32: 2414-2420.

Batzer, D.P., Rader, R.B., and Wissinger, S.A. (Eds). (1999). *Invertebrates in Freshwater Wetlands of North America.* New York, N.Y.: John Wiley and Sons.

Beaver, J.R., and Crisman, T.L. (1989). The role of ciliated protozoa in pelagic freshwater ecosystems. *Microbial Ecology* 17: 111-136.

Beja, P., and Alcazar, R. (2003). Conservation of Mediterranean temporary ponds under agricultural intensification: an evaluation using amphibians. *Biological Conservation* 114: 317-326.

Bell, S. L.M., Herman, T.B., and Wassersug, R.J. (2007). Ecology of *Thamnophis sauritus* (Eastern Ribbon Snake) at the Northern Limit of its Range. *Northeastern Naturalist* 14: 279–292.

Berra, T.M., and Allen, G.R. (1991). Population Structure and Development of *Lepidogalaxias salamandroides* (Pisces: Salmoniformes) from Western Australia. *Copeia* 1991: 845-850.

Bondavalli, C. and Ulanowicz, R.E. (1999). Unexpected effects of predators upon their prey: the case of the American Alligator. *Ecosystems* 2: 49-63.

Boivin M.E.Y., Greve G.D., Garcia-Meza J.V., Massieux B., Sprenger W., Kraak M.H.S., Breure A.M., Rutgers M. and Admiraal W. (2007). Algal-bacterial interactions in metal contaminated floodplain sediments. *Environmental Pollution* 145: 884-894.

Boix, D., Sala, J., and Moreno-Amich, R. (2001). The faunal composition of Espolla Pond (NE Iberian Peninsula): the neglected biodiversity of temporary waters. *Wetlands* 21: 577-592.

Boon P.I., Virtue P. and Nichols P.D. (1996). Microbial consortia in wetland sediments: A biomarker analysis of the effects of hydrological regime, vegetation and season on benthic microbes. *Marine and Freshwater Research* 47: 27-41.

Brooks, R.T., and Doyle, K.L. (2001). Small mammal diversity and abundance in vernal pool habitat. *Northeastern Naturalist* 8: 137-148.

Brose, U. (2001). Relative importance of isolation, area and habitat heterogeneity for vascular plant species richness of temporary wetlands in east-German farmland. *Ecography* 24: 722-730.

Brown, T.T., and Fuller, C.A. (2006). Stress and parasitism of white-footed mice (Peromyseus leucopus) in dry and floodplain environments. *Canadian Journal of Zoology* 84:1833-1839.

Caceres C.E., Tessier A.J., Andreou A. and Duffy M.A. (2008). Stoichiometric relationships in vernal pond plankton communities. *Freshwater Biology* 53: 1291-1302.

Casey R.E., and Klaine S.J. (2001). Nutrient attenuation by a riparian wetland during natural and artificial runoff events. *Journal of Environmental Quality* 30: 1720-1731.

Casey R.E., Taylor M.D. and Klaine S.J. (2001). Mechanisms of nutrient attenuation in a subsurface flow riparian wetland. *Journal of Environmental Quality* 30: 1732-1737.

Casamatta D.A., Beaver J.R. and Fleischman D.J. (1999). A survey of phytoplankton taxa from three types of wetlands in Ohio. *Ohio Journal of Science* 99: 53-56.

Church, D.R. (2008). Role of current versus historical hydrology in amphibian species turnover within local pond communities. *Copeia* 1: 115-125.

Colburn, E.A. (2004). *Vernal pools: Natural history and conservation.* Blacksburg, VA., USA: McDonald and Woodward Pub. Co.

Copp, G.H. (1989). The habitat diversity and fish reproductive function of floodplain ecosystems. *Environmental Biology of Fishes* 26: 1-27.

Cucherousset, J., Carpentier, A., and Paillisson, J. –M. (2007). How do fish exploit temporary waters throughout a flooding episode? *Fisheries Management and Ecology* 14:269-276.

Cunningham, J.M., Calhoun, A.J.K., and Glanz, W.E. (2007). Pond-Breeding Amphibian Species Richness and Habitat Selection in a Beaver-Modified Landscape. *The Journal of Wildlife Management* 71: 2517–2526.

Dalerum, F., and Angerbjörn, A. (2005). Resolving temporal variation in vertebrate diets using naturally occurring stable isotopes. *Oecologia* 144: 647-658.

Danovaro R., Corinaldesi C., Filippini M., Fischer U.R., Gessner M.O., Jacquet S., Magagnini M. and Velimirov B. (2008). Viriobenthos in freshwater and marine sediments: a review. *Freshwater Biology* 53: 1186-1213.

Davidsson T.E., Stepanauskas R. and Leonardson L. (1997). Vertical patterns of nitrogen transformations during infiltration in two wetland soils. *Applied and Environmental Microbiology* 63: 3648-3656.

Deil U. (2005). A review on habitats, plant traits and vegetation of ephemeral wetlands - a global perspective. *Phytocoenologia* 35: 533-705.

Dodd, C.K. (1992). Biological diversity of a temporary pond herpetofauna in north Florida sandhills. *Biodiversity and Conservation* 1: 125-142.

Domozych D.S. and Domozych C.R. (2008). Desmids and biofilms of freshwater wetlands: Development and microarchitecture. *Microbial Ecology* 55: 81-93.

Eason, G. W. Jr., and Fauth, J. E. (2001). Ecological correlates of anuran species richness in temporary pools: a field study in South Carolina, USA. *Israel Journal of Zoology* 47: 347-365.

Eder, E., Hödl, W., and Gottwald, R. (1997). Distribution and phenology of large branchiopods in Austria. *Hydrobiologia* 359: 13-22.

Euliss N.H., Labaugh J.W., Fredrickson L.H., Mushet D.M., Laubhan M.R.K., Swanson G.A., Winter T.C., Rosenberry D.O. and Nelson R.D. (2004). The wetland continuum: A conceptual framework for interpreting biological studies. *Wetlands* 24: 448-458.

European Environment Agency. (1999). *Environment in the European Union at the Turn of the Century*. Environmental Assessment Report, No. 2. Copenhagen Denmark: European Environment Agency.

Fenchel, T. (1987). *Ecology of Protozoa – the biology of free-living phagotrophic protists*. New York, N.Y.: Springer Verlag.

Finlay, B.J., and Esteban, G.F. (1998a). Freshwater protozoa: biodiversity and ecological function. *Biodiversity and Conservation* 7: 1163-1186.

Finlay, B.J., and Esteban, G.F. (1998b). Planktonic ciliate species diversity as an integral component of ecosystem function in a freshwater pond. *Protist* 149: 155-165.

Finlay, B.J., and Fenchel, T. (2004). Cosmopolitan metapopulations of free-living eukaryotes. *Protist* 155: 237-244.

Findlay, C. S., and Houlahan, J. (1997). Anthropogenic correlates of species richness in Southeastern Ontario wetlands. *Conservation Biology* 11: 1000-1009.

Finlay, J.C., and Vredenburg, V.T. (2007). Introduced trout sever trophic connections in watersheds: consequences for a declining amphibian. *Ecology* 88: 2187-2198.

Francl, K. E. (2008). Summer bat activity at woodland seasonal pools in the Northern Great Lakes Region. *Wetlands* 28: 117-124.

Fausch, K.D., and Bramblett, R.G. (1991) Disturbance and Fish Communities in Intermittent Tributaries of a Western Great Plains River. *Copeia* 1991: 659-674.

Gibbons, J.W., and Dorcas, M.D. (2004). *North American Watersnakes: A Natural History*. The University of Oklahoma Press, Norman, OK. 438pp.

Greenwood, P.H. (1986). The natural history of African lungfishes. *Journal of Morphology* supplement 1: 163-179.

Hanowski, J., Danz, N., and Lind, J. (2006). Response of breeding bird communities to forest harvest around seasonal ponds in northern forests, USA. *Forest Ecology and Management* 229: 63–7.

Haugaasen, T., and Peres, C.A. (2005). Primate assemblage structure in Amazonian flooded and unflooded forests. *American Journal of Primatology* 67: 243-258.

Henning, J. A., Gresswell, R. E., and Fleming, I. A. (2007). Use of seasonal freshwater wetlands by fishes in a temperate river floodplain. *Journal of Fish Biology* 71: 476–492.

Hering, D., and Plachter, H. (1997). Riparian ground beetles (Coleoptera: Carabidae) preying on aquatic invertebrates: a feeding strategy in alpine floodplains. *Oecologia* 111: 261-270.

Homes, V. Hering, D., and Reich, M. (1999). The distribution and macrofauna of ponds in stretches of an alpine floodplain differently impacted by hydrological engineering. *Regulated Rivers - Research and Management* 15: 405-417.

Hunt, P.D, Reitsma, L., Burson, S.L. III, and Steeles, B.B. (2005). Spatial and Temporal Distribution of Northern Waterthrushes Among Nonbreeding Habitats in Southwestern Puerto Rico. *Biotropica* 37: 697-701.

Gilbert, J.J. (1989). The effect of *Daphnia* interference on a natural rotifer and ciliate community: short-term bottle experiments. *Limnology and Oceanography* 34: 606-617.

Gilbert, J.J., and Jack, J.D. (1993). Rotifers as predators on small ciliates. *Hydrobiologia* 255: 247-253.

Grime, J.P. (1973). Competitive exclusion in herbaceous vegetation. *Nature* 242: 344-347.

Gutknecht J.L.M., Goodman R.M. and Balser T.C. (2006). Linking soil process and microbial ecology in freshwater wetland ecosystems. *Plant and Soil* 289: 17-34.

Inkley M.D., Wissinger S.A. and Baros B.L. (2008). Effects of drying regime on microbial colonization and shredder preference in seasonal woodland wetlands. *Freshwater Biology* 53: 435-445.

Iribar A., Sanchez-Perez J.M., Lyautey E. and Garabetian F. (2008). Differentiated free-living and sediment-attached bacterial community structure inside and outside denitrification hotspots in the river-groundwater interface. *Hydrobiologia* 598: 109-121.

Janousek C.N., Currin C.A. and Levin L.A. (2007). Succession of microphytobenthos in a restored coastal wetland. *Estuaries Coasts* 30: 265-276.

Jennings, W.B., Bradford, D.F., and Johnson, D.F. (1992). Dependence of the Garter Snake *Thamnophis elegans* on Amphibians in the Sierra Nevada of California. *Journal of Herpetology* 26: 503-505.

Joyal, L.A. (1996). Ecology of Blanding's (*Emydoidea blaniiingi*) and spotted (*Clemmys guttata*) turtles in southern Maine: population structure, habitat use, movements, and reproduction biology. Unpubl. master's thesis, Univ. of Maine, Orono.

Juliano, S.A. (1985). Habitat associations, resources, and predators of an assemblage of *Brachinus* (Coleoptera: Carabidae) from southeastern Arizona. *Canadian Journal of Zoology* 63: 1683-1691.

Junk, W.J. (1993). Wetlands of tropical South America. In: D. Whigham, D. Dykyjova and S. Hejny (Eds), pp. 679-739. *Wetlands of the World.* Dordrecht, The Netherlands: Kluwer Academic Publishers.

Junk, W. J., Bayley, P. B., and Sparks, R. E. (1989). The flood pulse concept in river-floodplain systems. *Canadian Journal of Fisheries and Aquatic Sciences* 106: 110-127.

Keddy P.A. (2000). *Wetland ecology: Principles and conservation.* Cambridge University Press.

Kehl, C. (1997). Die Hornmilbenzönosen (Acari, Oribatida) unterschiedlich stark degradierter Moor-standorte in Berlin und Brandenburg. Ph.D. thesis, Free University of Berlin.

Kiemle S.N., Domozych D.S. and Gretz M.R. (2007). The extracellular polymeric substances of desmids (Conjugatophyceae, Streptophyta): chemistry, structural analyses and implications in wetland biofilms. *Phycologia* 46: 617-627.

Kofron, C. P. (1993). Behavior of Nile Crocodiles in a Seasonal River in Zimbabwe. *Copeia* 1993: 463-469.

Kupfer, A., Langel, R., Scheu, S., Himstedt, W., and Maraun, M. (2006). Trophic ecology of a tropical aquatic and terrestrial food web: insights from stable isotopes (^{15}N). *Journal of Tropical Ecology* 22: 469-476.

Lawler, S. P. (2001). Rice fields as temporary wetlands: a review. *Israel Journal of Zoology* 47: 513-528.

Lude, A., Reich, M. and Plachter, H. (1999). Life strategies of ants in unpredictable floodplain habitats of Alpine rivers (Hymenoptera: Formicidae). *Entomologia Generalis* 24: 75-91.

Madoni, P. (1996). The contribution of ciliated Protozoa to plankton and benthos biomass in a European rice-field. *Journal of Eukaryotic Microbiology* 43: 193-198.

Main, B.Y., Davies, V.T. and Harvey, M.S. (1985). *Zoological Catalogue of Australia, Volume 3: Arachnida: Mygalomorphae, Araneomorphae in part; Pseudoscorpionida; Amblypygi and Palpigradi.* Bureau of Flora and Fauna, Canberra.

Mahoney, D.L., Mort, M.A., and Taylor, B.E. (1990). Species richness of calanoid copepods, cladocerans and other branchiopods in Carolina Bay temporary ponds (South Carolina, USA). *American Midland Naturalist* 123: 244-258.

McCall, T.C., Hodgman, T.P., Diefenbach, D.R., and Owen, R.B.Jr. (1996). Beaver populations and their relation to wetland habitat and breeding waterfowl in Maine. *Wetlands* 16: 163-172.

McColl, J. G., and Burger, J. (1976). Chemical inputs by a colony of Franklin's Gulls nesting in cattails. *American Midland Naturalist* 96: 270-280.

McCormick, P.V., and Cairns, J. (1991). Effects of micrometazoa on the protistan assemblages of a littoral food web. *Freshwater Biology* 26: 111-119.

McDowell, D.M., and Naiman, R.J. (1986). Structure and function of a benthic invertebrate stream community as influenced by beaver (*Castor Canadensis*). *Oecologia* 68: 481-489.

Mentzer J.L., Goodman R.M. and Balser T.C. (2006). Microbial response over time to hydrologic and fertilization treatments in a simulated wet prairie. *Plant and Soil* 284: 85-100.

Merritt, R.W., and Cummins, K.W. (1996). *An Introduction to the Aquatic Insects of North America.* Dubuque, Iowa: Kendall/Hunt Publishing Company.

Milam, J.C., and Melvin, S.M. (2001). Density, Habitat Use, Movements, and Conservation of Spotted Turtles (Clemmys guttata) in Massachusetts. *Journal of Herpetology* 35: 418-427.

Mills, M.S., Poppy, S.M, Mills, A.M, Ryan, T.J, and Dorcas, M.E. (2000). Seminatrix pygaea diet. *Herpetological Review* 31: 47.

Morin, P.J. (1999). *Community Ecology.* Oxford: Blackwell Science Inc.

Neely R.K. and Wetzel R.G. (1997). Autumnal production by bacteria and autotrophs attached to *Typha latifolia* L detritus. *Journal of Freshwater Ecology* 12: 253-267.

Newman, R. A. (1989). Developmental plasticity of *Scaphiopus couchii* tadpoles in an unpredictable environment. *Ecology* 70: 1775-1787.

Newman, R. A. (1987). Effects of density and predation on *Scaphiopus couchii* tadpoles in desert ponds. *Oecologia* 71: 301-307.

Nicolet, P., Biggs, J., Fox, G., Hodson, M.J., Reynolds, C., Whitfield, M., and Williams, P. (2004). The wetland plant and macroinvertebrate assemblages of temporary ponds in England and Wales. *Biological Conservation* 120: 261-278.

Noll M., Matthies D., Frenzel P., Derakshani M. and Liesack W. (2005). Succession of bacterial community structure and diversity in a paddy soil oxygen gradient. *Environmental Microbiology* 7: 382-395.

Oertli, B., Joye, D.A., Castella, E., and Juge, R., Cambin, D., and Lachavanne, J-B. (2002). Does size matter? The relationship between pond area and biodiversity. *Biological Conservation* 104: 59-70.

Ohba, S., Miyasaka, H., and Nakasuji, F. (2008). The role of amphibian prey in the diet and growth of giant water bug nymphs in Japanese rice fields. *Population Ecology* 50: 9–16.

Orr C.H., Stanley E.H., Wilson K.A. and Finlay J.C. (2007). Effects of restoration and reflooding on soil denitrification in a leveed midwestern floodplain. *Ecological Applications* 17: 2365-2376.

Owen, S.F., Menzel, M.A., Ford, W.M., Chapman, B.R., Miller, K.V., Edwards, J.W., and Wood, P.B. (2003). Home-range Size and Habitat Used by the Northern Myotis (Myotis septentrionalis). *American Midland Naturalist* 150: 352–359.

Pace, M.L., and Funke, E. (1991). Regulation of planktonic microbial communities by nutrients and herbivores. *Ecology* 72: 904-914.

Paton, P.W.C. (2005). A review of vertebrate community composition in seasonal forest pools of the northeastern United States. *Wetlands Ecology and Management* 13: 235-246.

Peres, C.A. (1997). Primate community structure at twenty western Amazonian flooded and unflooded forests. *Journal of Tropical Ecology* 13: 381-405.

Petranka, J.W., and Kennedy, C.A. (1999). Pond tadpoles with generalized morphology: is it time to reconsider their functional roles in aquatic communities? *Oecologia* 120: 621-631.

Petranka, J.W., Hopey, M.E., Jennings, B.T., Baird, S.D., and Boone J. (1994). Breeding habitat segration of wood frogs and American toads: the role of interspecific tadpole predation and adult choice. *Copeia* 1994: 691-697.

Petranka, J.W., Rushlow, A.W., Hopey, M.E. (1998). Predation by tadpoles of *Rana sylvatica* on embryos of *Ambystoma maculatum*: implication of ecological role reversals by *Rana* (predator) and *Ambystoma* (prey). *Herpetologica* 54: 1-13.

Petrov, B., and Cvetkovic, D.M. (1997). Community structure of branchiopods (Anostraca, Notostraca and Conchostraca) in the Banat province of Yugoslavia. *Hydrobiologia* 359: 23-28.

Platt, S.G., Brantley, C.G., and Hastings, R.W. (1990). Food habits of juvenile American alligators in the upper Lake Pontchartrain estuary. *Northeast Gulf Science* 11: 123-130.

Poizat, G., and Crivelli, A.J. (1997). Use of seasonally flooded marshes by fish in a Mediterranean wetland: timing and demographic consequences. *Journal of Fish Biology* 51: 106-119.

Post, D. M. (2002). Using stable isotopes to estimate trophic position: models, methods, and assumptions. *Ecology* 83: 703–718.

Pugh, P.J.A. (1996). Edaphic oribatid mites (Cryptostigmata, Acarina) associated with an aquatic moss on sub-antarctic South Georgia. *Pedobiologia* 40: 113-117.

Pusey, B.J., and Bradshaw, S.D. (1996). Diet and dietary overlap in fishes of temporary waters of southwestern Australia. *Ecology of Freshwater Fish* 5: 183-194.

Pusey, B.J. and Edward, D.H. (1990). Structure of fish assemblages in waters of the southern acid peat flats, south-western Australia. *Australian Journal of Marine and Freshwater Research* 41: 721-734.

Qiu S. and McComb A.J. (1995). Planktonic and microbial contributions to phosphorus release from fresh and air-dried sediments. *Marine and Freshwater Research* 46: 1039-1045.

Qiu S. and McComb A.J. (1996). Drying-induced stimulation of ammonium release and nitrification in reflooded lake sediment. *Marine and Freshwater Research* 47: 531-536.

Rahman, A. K. A. (1989). *Freshwater fishes of Bangladesh.* Nayapaltan, Dhaka: Zoological Society of Bangladesh.

Ray A.M. and Inouye R.S. (2006). Effects of water-level fluctuations on the arbuscular mycorrhizal colonization of Typha latifolia L. *Aquatic Botany* 84: 210-216.

Reid, F.A. (1993) Managing wetlands for waterbirds. *Transactions of the North American Wildlife and Natural Resources Conference* 58: 345-350.

Rejmankova E., Komarek J. and Komarkova J. (2004). Cyanobacteria - a neglected component of biodiversity: patterns of species diversity in inland marshes of northern Belize (Central America). *Diversity and Distributions* 10: 189-199.

Rejmankova E. and Komarkova J. (2005). Response of cyanobacterial mats to nutrient and salinity changes. *Aquatic Botany* 83: 87-107.

Rivas, J.A. (2000). Life history of the green anaconda (*Eunectes murinus*) with emphasis on its reproductive biology. Unpublished Ph. D. dissertation at the University of Tennessee. 287p.

Roe, J.H., Kingsbury, B.A., and Herbert, N.R. (2003). Wetland and upland use patterns in semi-aquatic snakes: Implications for wetland conservation. *Wetlands* 23: 1003–1014.

Rolon A.S., Lacerda T., Maltchik L. and Guadagnin D.L. (2008). Influence of area, habitat and water chemistry on richness and composition of macrophyte assemblages in southern Brazilian wetlands. *Journal of Vegetation Science* 19: 221-228.

Rubbo M.J. and Kiesecker J.M. (2004). Leaf litter composition and community structure: Translating regional species changes into local dynamics. *Ecology* 85: 2519-2525.

Royan, J.P. (1976). Studies on the gut contents of *Leptestheriella maduraiensis* (Conchostraca: Branchiopoda) Nayar and Nair. *Hydrobiologia* 51: 209-212.

Russell, K.R., Moorman, C.E., Edwards, J.K., Metts, B.S., and Guynn, D.C. Jr. (1999). Amphibian and reptile communities associated with Beaver (Castor Canadensis) ponds and unimpounded streams in the piedmont of South Carolina. *Journal of Freshwater Ecology* 14: 149-158.

Sabacka M. and Elster J. (2006). Response of cyanobacteria and algae from Antarctic wetland habitats to freezing and desiccation stress. *Polar Biology* 30: 31-37.

Scheffers, B.R., Harris, J.B.C., and Haskell, D.G. (2006). Avifauna associated with ephemeral ponds on the Cumberland Plateau, Tennessee. *Journal of Field Ornithology* 77: 178-183.

Schmid-Araya, J.M., and Schmid, P.E.. (2000). Trophic relationships: integrating meiofauna into a realistic benthic food web. *Freshwater Biology* 44: 149-163.

Seale, D.B. (1980). Influence of amphibian larvae on primary production, nutrient flux, and competition in a pond ecosystem. *Ecology* 61: 1531-1550.

Semlitsch, R.D. (1987). Relationship of pond drying to the reproductive success of the salamander *Ambystoma talpoideum. Copeia* 1987: 61-69.

Semlitsch, R.D. and Bodie, J.R. (1998). Are small, isolated wetlands expendable? *Conservation Biology* 12: 1129-1133.

Serrano, L., and Fahd, K. (2005). Zooplankton communities across a hydroperiod gradient of temporary ponds in the Donana National Park (SW Spain). *Wetlands* 25: 101-111.

Sharitz, R.R., and Gibbons, J.W. (1982). *The ecology of southeastern shrub bogs (pocosins) and Carolina bays: a community profile.* U.S. Fish and Wildlife Service, Division of Biological Services, Washington, D.C. FWS/OBS-82/04. 93 pp.

Shine, T., Böhme, W., Nickel, H., Thies, D.F., and Wilms, T. (2001) Rediscovery of relict populations of the Nile crocodile *Crocodylus niloticus* in south-eastern Mauritania, with observations on their natural history *Oryx* 35: 260–262.

Silveira, J.G. (1998). Avian uses of vernal pools and implications for conservation practice. In C. W. Witham, E. T. Bauder, D. Belk, W. R. Ferren Jr., and R. Ornduff (Eds.), *Ecology, Conservation, and Management of Vernal Pool Ecosystems: Proceedings from a 1996 conference* (pp. 92-106). Sacramento, CA: California Native Plant Society.

Sim L.L., Chambers J.M. and Davis J.A. (2006a). Ecological regime shifts in salinised wetland systems. I. Salinity thresholds for the loss of submerged macrophytes. *Hydrobiologia* 573: 89-107.

Sim L.L., Davis J.A. and Chambers J.M. (2006b). Ecological regime shifts in salinised wetland systems. II. Factors affecting the dominance of benthic microbial communities. *Hydrobiologia* 573: 109-131.

Sipura, J., Lores, E., and Snyder, R.A. (2003). Effects of copepods on estuarine microbial plankton in short-term microcosms. *Aquatic Microbial Ecology* 33: 181-190.

Smith, I.M., and Cook, D.R. (1991). Water mites. In: J.H. Thorpe and A.P. Covich (Eds), pp. 523-592 *Ecology and Classification of North American Freshwater Invertebrates.* San Diego, CA: Academic Press.

Snodgrass, J.W. (1997). Temporal and spatial dynamics of beaver-created patches as influenced by management practices in a southeastern North American landscape. *Journal of Applied Ecology* 34: 1043-1056.

Snodgrass, J. W., Komoroski, M. J., Bryan Jr., A. L., and Burger, J. (2000). Relationships among isolated wetland size, hydroperiod, and amphibian species richness: implications for wetland regulations. *Conservation Biology* 14: 414-419.

Snodgrass, J. W., Bryan Jr., A. L., Lide, R. F., and Smith, G. M. (1996). Factors affecting the occurrence and structure of fish assemblages in isolated wetlands of the upper coastal plain, U.S.A. *Canadian Journal of Fisheries and Aquatic Sciences* 53: 443-454.

Snodgrass, J. W., and Meffe, G.K. (1998). Influence of beavers on stream fish assemblages: Effects of pond age and watershed position. *Ecology* 79:928-942.

Sparks, R. E. (1995). Need for Ecosystem Management of Large Rivers and Their Floodplains. *BioScience* 45: 168-182.

Sraj-Krzic N., Germ M., Urbanc-Bercic O., Kuhar U., Janauer G.A. and Gaberscik A. (2007). The quality of the aquatic environment and macrophytes of karstic watercourses. *Plant Ecology* 192: 107-118.

Stebbins, R.C., and Cohen, N.W. (1995). *A natural history of amphibians.* Princeton, NY:Princeton University Press.

Stoecker, D.K., and Capuzzo, J.M. (1990). Predation on protozoa: its importance to zooplankton. *Journal of Plankton Research* 12: 891-908.

Stout, J.D. (1984). The protozoan fauna of a seasonally inundated soil under grassland. *Soil Biology and Biochemistry* 16: 121-125.

Tellería, J. L., Mamy Ghaillani, H. E., Fernández-Palacios, J. M., Bartolomé, J. and Montiano, E. (2008). Crocodiles *Crocodylus niloticus* as a focal species for conserving water resources in Mauritanian Sahara. *Oryx* 42: 292-295.

Thiery, A. (1991). Multispecies coexistence of branchiopods (Anostraca, Notostraca and Spinicaudata) in temporary ponds of Chaouia Plain (western Morocco): sympatry or syntopy between usually allopatric species. *Hydrobiologia* 212: 117-136.

Thies J.E. (2007). Soil microbial community analysis using terminal restriction fragment length polymorphisms. *Soil Science Society of America Journal* 71: 579-591.

Timms, B. V. (1997). A comparison between saline and freshwater wetlands on Bloodwood Station, the Paroo, Australia, with special reference to their use by waterbirds. *International Journal of Salt Lake Research* 5: 287-313.

Toet S., Huibers L., Van Logtestijn R.S.P. and Verhoeven J.T.A. (2003). Denitrification in the periphyton associated with plant shoots and in the sediment of a wetland system supplied with sewage treatment plant effluent. *Hydrobiologia* 501: 29-44.

Underwood G.J.C. (1997). Microalgal colonization in a saltmarsh restoration scheme. *Estuarine Coastal and Shelf Science* 44: 471-481.

Vanderklift, M. A. and Ponsard, S. (2003). Sources of variation in consumer-diet δ^{15}N enrichment: a meta-analysis. *Oecologia* 136: 169-182.

van Helsdingen, P.J. (1996). The spider fauna of some Irish floodplains. *Irish Naturalists' Journal* 25: 285-293.

Vasconcelos, D., and Calhoun, A.J.K. (2006). Monitoring created seasonal pools for functional success: a six-year case study of amphibian response, Sears Island, Maine. *Wetlands* 26: 992–1003.

Vessels, N., and Jack, J. D. (2001). Effects of fish on zooplankton community structure in Chaney Lake, a temporary Karst wetland in Warren county, Kentucky. *Journal of the Kentucky Academy of Science* 62: 52-59.

Vignoli, L., Bologna, M.A., and Luiselli, L. (2007). Seasonal patterns of activity and community structure in an amphibian assemblage at a pond network with variable hydrology. *Acta Oecologia* 31: 185-192.

Visscher, P.K., Vetter, R.S., and Orth, R. (1994). Benthic bees? Emergence phenology of *Calliopsis pugionis* (Hymenoptera: Andrenidae) at a seasonally flooded site. *Annals of the Entomological Society of America* 87: 941-945.

Wallwork, J.A. (1983). *The Distribution and Diversity of Soil Fauna*. London: Academic Press.

Ward, D., and Blaustein, L. (1994). The overriding influence of flash floods on species: area curves in ephemeral Negev desert pools: a consideration of the value of island biogeography theory. *Journal of Biogeography* 21: 595-603.

Wauthy, G.M., Noti, M., and Dufrene, M. (1989). Geographic ecology of soil oribatid mites in deciduous forests. *Pedobiologia* 33: 399-416.

Welcomme, R.L. (1979). *Fisheries Ecology of Floodplain Rivers*. London: Longman.

Wellborn, G.A., Skelly, D.K., and Werner, E.E. (1996). Mechanisms creating community structure across a freshwater habitat gradient. *Annual Review of Ecology and Systematics* 27: 337-363.

Weller, M.W. (1981). *Freshwater marshes: Ecology and wildlife management*. Minneapolis, MN, USA: University of Minnesota Press.

Werner, E.E., Skelly, D.K., Relyea, R.A., and Yurewicz, K.L. (2007). Amphibian species richness across environmental gradients. *Oikos* 116:1697-1712.

Whigham, D.F., and Jordan, T.E. (2003). Isolated wetlands and water quality. *Wetlands* 23: 541-549.

Wiggins, G.B., Mackay, R.J. and Smith, I.M. (1980). Evolutionary and ecological strategies of animals in annual temporary pools. *Archiv für Hydrobiologie (Supplement)* 58: 97-206.

Wilbur, H.M. (1987). Regulation of Structure in Complex Systems: Experimental Temporary Pond Communities. *Ecology* 68: 1437-1452.

Wilbur, H.M. (1997). Experimental ecology of food webs: complex systems in temporary ponds. *Ecology* 78: 2279-2302.

Williams, D.D. (2006). *The biology of temporary waters*. Oxford University Press, Oxford.

Williams, D.D., and Feltmate, B.W. (1992). *Aquatic Insects*. Wallingford, Oxford: CAB International.

Williams, D.D., and Hynes, H.B.N. (1977). The ecology of temporary streams II. General remarks on temporary streams. *Internationale Revue gesamten Hydrobiologie* 62: 53-61.

Williams D.D., Nalewajko C. and Magnusson A.K. (2005). Temporal variation in algal communities in an intermittent pond. *Journal of Freshwater Ecology* 20: 165-170.

Williams, D.D., Williams, N.E., and Hynes, H.B.N. (1974). Observations on the life history and burrow construction of the crayfish *Cambarus fodiens* (Cottle) in a temporary stream in southern Ontario. *Canadian Journal of Zoology* 52: 365-370.

Williams, D.D., Ambrose, L.G. and Browning, L.N. (1995). Trophic dynamics of two sympatric species of riparian spider (Araneae: Tetragnathidae). *Canadian Journal of Zoology* 73: 1545-53.

Williams, J.D., Warren, M.L., Cummings, K.S., Harris, J.L., and Neves, R.J. (1993). Conservation status of freshwater molluscs of the United States and Canada. *Fisheries* 18: 6-22.

Winemiller, K.O., and Kelso-Winemiller, L.C. (1996). Comparative ecology of catfishes of the Upper Zambezi River floodplain. *Journal of Fish Biology* 49: 1043–1061.

Winne, C.T., Dorcas, M.E., and Poppy, S.M. (2005). Population Structure, Body Size, and Seasonal Activity of Black Swamp Snakes (*Seminatrix pygaea*). *Southeastern Naturalist* 4: 1-14.

Wissinger, S.A., Bohonak, A.J., Whiteman, H.H., and Brown, W.S. (1999). Subalpine wetlands in Colorado. In: D.P. Batzer, R.B. Rader and S.A. Wissinger (Eds), *Invertebrates in Freshwater Wetlands of North America* pp. 757-790. New York, N.Y.: John Wiley and Sons.

Zedler, P.H. (2003). Vernal pools and the concept of "isolated wetlands". *Wetlands* 23: 597-607.

In: International Wetlands: Ecology, Conservation...   ISBN: 978-1-60456-999-5
Editor: Jose R. Herrara   © 2009 Nova Science Publishers, Inc.

Chapter 3

BIOMASS AND PRODUCTIVITY OF PLANT COMMUNITY IN A RAINFED MONSOONAL WETLAND ECOSYSTEM WITH SPECIFIC EMPHASIS ON ITS TEMPORAL VARIABILITY

Rachna Chandra[1,], B. Anjan Kumar Prusty[1,2,†] and P. A. Azeez[1,‡]*

[1]Environmental Impact Assessment Division, Sálim Ali Center for Ornithology and Natural History, Anaikatty, Coimbatore – 641108, India
[2] Gujarat Institute of Desert Ecology, P.O. Box # 83,
Opp. Changleswar Temple, Mundra Road, Bhuj - 370 001, Gujarat, India

ABSTRACT

The biomass and productivity of macrophytes on a spatio-temporal scale was studied in the monsoonal wetland system of Keoladeo National Park (KNP), Bharatpur, India during July 2005 through April 2006. Macrophyte samples were collected from 7 blocks of the wetland, which remain inundated for most of the months in the year. Species richness, assemblage of all the macrophytes and aboveground net primary productivity (NPP, 2.71 g m^{-2} day^{-1} on dry weight basis) were examined. 37 species of macrophytes were recorded in the quadrats sampled. Of the macrophytes studied, the order of the major species in terms of their biomass per unit area was *Pseudoraphis spinescens* > *Cynodon dactylon* > *Sporobolus* sp. > *Paspalum distichum* > *Vallisneria natans* > *Alternanthera* sp. > *Utricularia* sp. > *Hemiadelphis polyspermus* > *Najas minor*. The maximum average biomass of macrophytes was recorded during November 2005 (92.21 g m^{-2}) and the minimum during January 2006 (27.52 g m^{-2}). Schluter's test for the macrophyte species showed positive association with each other (W = 147.79, *P < 0.05*). Univariate analysis of variance showed significant variation in biomass among the months, the blocks and the categories of macrophytes. The water depth during the study period varied from 0.5 – 1.15 meter. Water depth seems to be the significant explanatory

---

[*] Corresponding author. Email: rachnaeia@gmail.com
[†] anjaneia@gmail.com
[‡] azeezpa@gmail.com

variable for aquatic vegetation assemblage in our study, conjointly with other factors including location specific history of the plant communities. The biological diversity often observed in the park is strongly related to hydro period.

Keywords: Keoladeo National Park, biomass, productivity, macrophytes, wetlands

INTRODUCTION

Fresh water wetlands, covering almost 10% of the world's surface, as natural or artificial reservoirs are integral parts of river basins and are one of the most productive ecosystems of the earth. These "waterlogged wealth" (Maltby 1986) are also among the most threatened ecosystems. Conservation of wetlands is of utmost significance as they hold rich floral and faunal biodiversity and harbour large number of endangered and threatened species. In view of the accelerating and potentially catastrophic irreversible loss of biodiversity setting right the conservation priorities are crucial and are a matter of concern in the case of wetlands. Wetlands as crucial and important ecosystems of the world provide diverse and very valuable ecological services to human kind and other dependent species. They help in improving water quality, controlling floods, regulating global carbon levels by sequestering them, and providing habitat for plants and animals uniquely adapted to wet conditions (Mitsch & Gosselink 2000). Nutrient dynamics is known to play major role in the growth and distribution of wetland vegetation, which respond to nutrient additions by increased storage of nitrogen (N) and phosphorous (P) in plant tissues and by increased Net Primary Productivity (NPP). The increase in production in turn affects ecosystem processes including decomposition (Davis 1991), accumulation of soil organic matter, and organic carbon export. When natural wetlands receive high nutrient loadings, ecosystem processes, such as plant productivity and nutrient cycling, change in measurable ways. The structure of the plant community may change as slower growing native species are replaced by faster growing ones that take advantage of high nutrient levels to boost growth (Davis 1991).

The importance of aquatic macrophytes to the functioning of freshwater ecosystems has been underscored during the last few decades (Carignam & Kalff 1982, Denny 1985, Downing 1986, Cyr & Downing, 1988, Randall et al. 1996) because they harbour high biodiversity and influence biomass production, nutrient cycling and community dynamics (Kateyo 2007). In situations of pollution macrophytes, occupying vital position in structure and functioning of aquatic ecosystems (Boston & Perkins 1982), are among the most affected group serving as a "polishing system" (Matagi et al. 1998) of water quality. The diversity and existence of wetland plants are very important in reducing nutrient enrichment of wetland systems (Kao et al. 2003). Study of aquatic vegetation provides information on the state of the environment in the water bodies they thrive in (Grasmuck et al. 1995). Being an important component of aquatic ecosystems changes in the macrophytes community composition or abundance of individual species provide crucial and valuable information about the ecosystem functioning and dynamics (Scott et al. 2002).

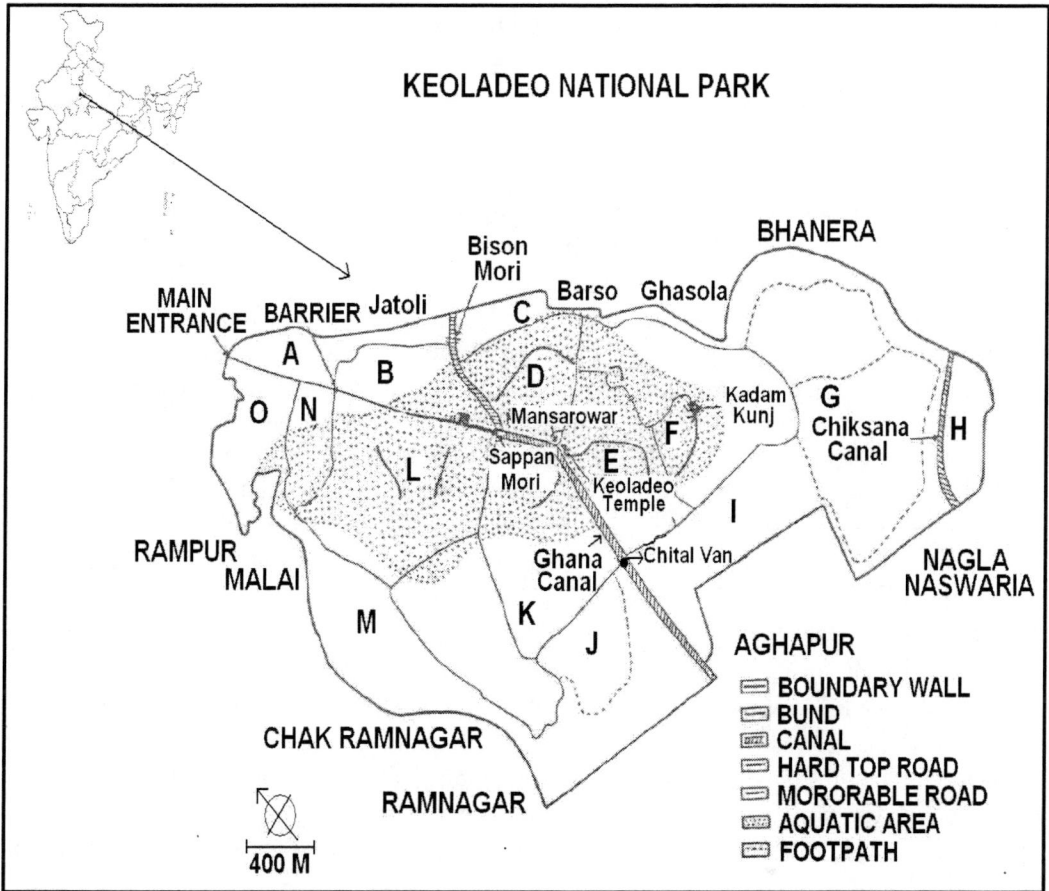

Figure 1. The study area.

The species diversity of aquatic vegetation changes with respect to factors such as water depth, pollution level, nutrient concentration (Pokorny 1994). The death and decay of various species and dynamics of nutrients also influence macrophyte growth to a large extent. As far as the growth and productivity is concerned, open systems (with inflow and outflow of water) maintain a regulated growth. However, in certain cases such as Keoladeo National Park (KNP), the system is of accumulation type concerning the nutrient dynamics. This wetland system has only inflow of water resulting in the gradual enrichment of nutrients and no notable outflows. Hence, the growth and productivity pattern is ought to be different from other open wetland systems, where there are both inflows and outflows. Biomass and productivity estimation being one of the criteria for checking the ecological balance helps in understanding the status of the wetlands. The present study was devised to assess the above ground biomass, productivity and species richness of aquatic macrophytes, and their seasonality in the KNP wetland system.

STUDY AREA

Keoladeo National Park (KNP, Figure 1), a world heritage site and a wetland well-known for its diverse and multiple ecosystems, is one of the finest bird sanctuaries in the world (Naturetrek, 2005). The KNP (27° 7.6' to 27° 12.2' N and 77° 29.5' to 77° 33.9' E), one of the unique waterfowl habitats in India existing for more than 250 years (Azeez et al. 1992, Azeez et al. 2000, Prusty et al. 2007b), is one of the early Ramsar sites (Mathur et al. 2005). It is a wintering and breeding ground for several avian species (both migratory and resident) that directly depend on this wetland system for food. The KNP gets water from the catchments of two non-perennial rivers (Figure 2). It is situated at the southeastern corner of the Bharatpur city located almost equidistant from New Delhi and Jaipur on either side. The climate of the park is sub-humid to semi-arid (Pal et al. 2000). During the period of study, the average temperature varied from 8.4°C (in December 2005) to 39.5°C (in April 2006, Figure 3.) and rainfall ranged between 33.8 mm (in March 2006) and 659.9 mm (in July 2005, Figure 4). The Park received a total of 13.62 million m^3 water during the whole study duration in two lots (in July 2005 and in September 2005). The water depth in the wetlands of the KNP was 0.5 – 1.15 meter (Figure 5) during the study period.

The park is about 29 sq km, of which 8.5 sq km central depression is the wetland (the central dotted area in the map, Figure 1), and the rest grassland and woodland (Prusty & Azeez 2007). KNP as a whole is divided into 15 blocks / compartments named alphabetically from A to O for the ease of management and eco-tourism. Being a monsoonal wetland, as mentioned earlier, the park depends mainly on the monsoonal water flow from the catchments of two non-perennial rivers viz. Banganga and Gambhir (Figure 2). These rivers supply water to the Ajan dam, an earthen dam located half a kilometer away from the KNP boundary, from which water is released to the park every year. The rivers and the Ajan dam thus play a critical role in the biodiversity of KNP. The river Banganga being a mountain torrent with a coarse textured bed of sand mixed gravel carries huge amount of suspended and dissolved solids / sediments from its headwaters to the plains (Kumar et al. 1995). These solids / sediments partly are subsequently carried to the park by the supply channel (Kumar & Vijayan 1988). The catchment of KNP is largely agricultural which during the post monsoon period is used for raising crops. By and large, water is released twice a year from the Ajan bund. The water level in the compartments of the park is regulated through sluice gates placed at strategic locations. October onwards, the water level in the wetland compartments starts declining (Azeez et al. 2000) and many of the blocks largely go dry by the end of April leaving small puddles, pools and ditches. During the dry months, the park management pumps out groundwater to just meet the water requirements of the denizens of the KNP. Being mostly seasonal and dynamic the wetland system of the KNP supports three major/dominant types of vegetation communities, namely i) suspended and submerged macrophytes such as *Hydrilla verticillata* and *Najas minor*, ii) floating vegetation such as *Nymphaea* sp. and *Nymphoides* sp., and iii) emergent vegetation such as *Paspalum distichum* and *Cyperus alopecuroides* (Prusty 2007, Prusty et al. 2007a, Prusty et al. 2008).

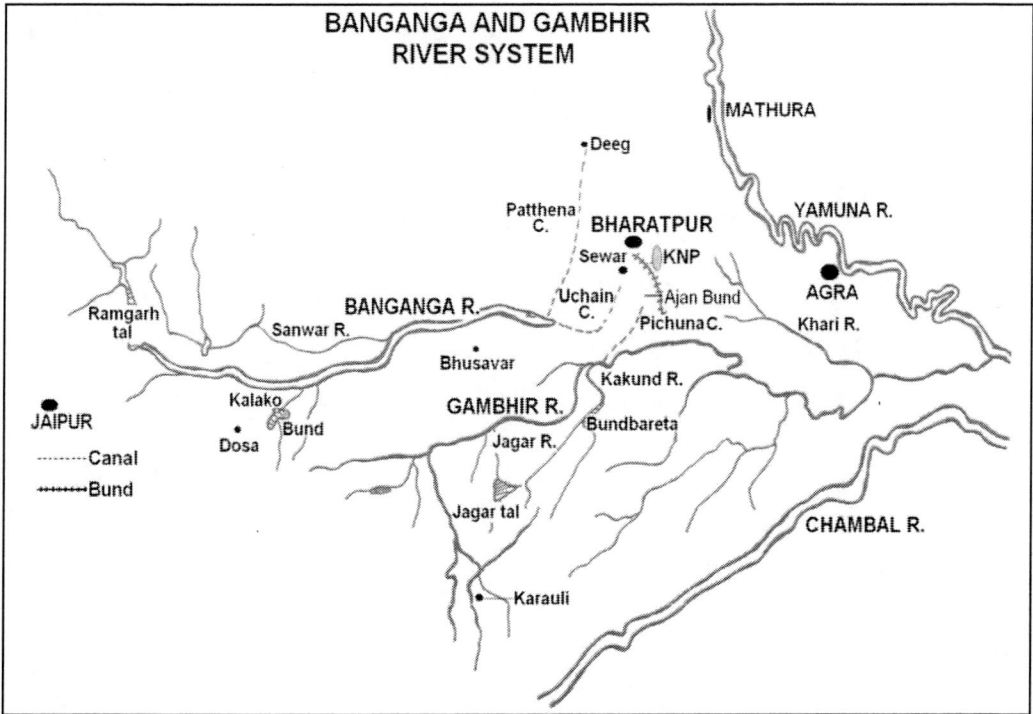

Figure 2. Banganga and Gambhir river system, the water source for the KNP.

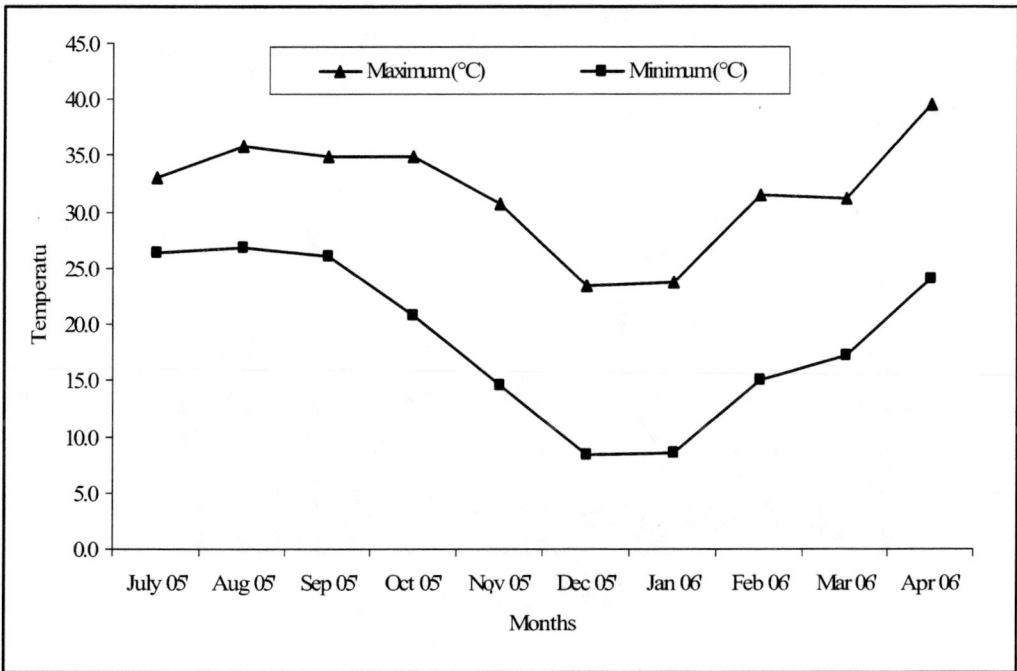

Figure 3. Average maximum and minimum temperature during the study.

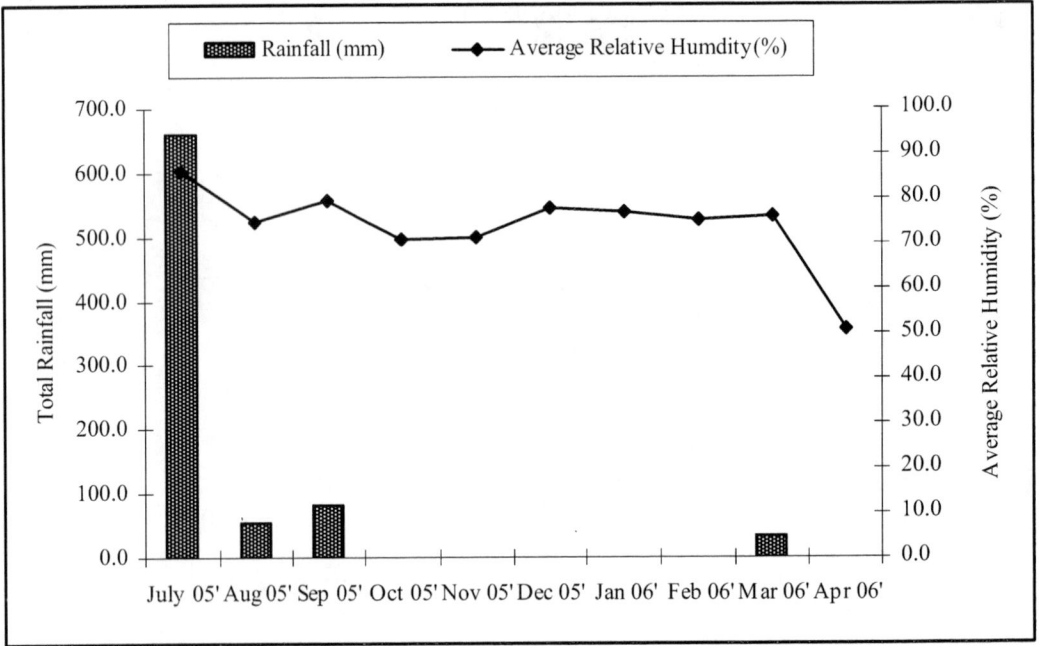

Figure 4. Total rainfall and average relative humidity during the study.

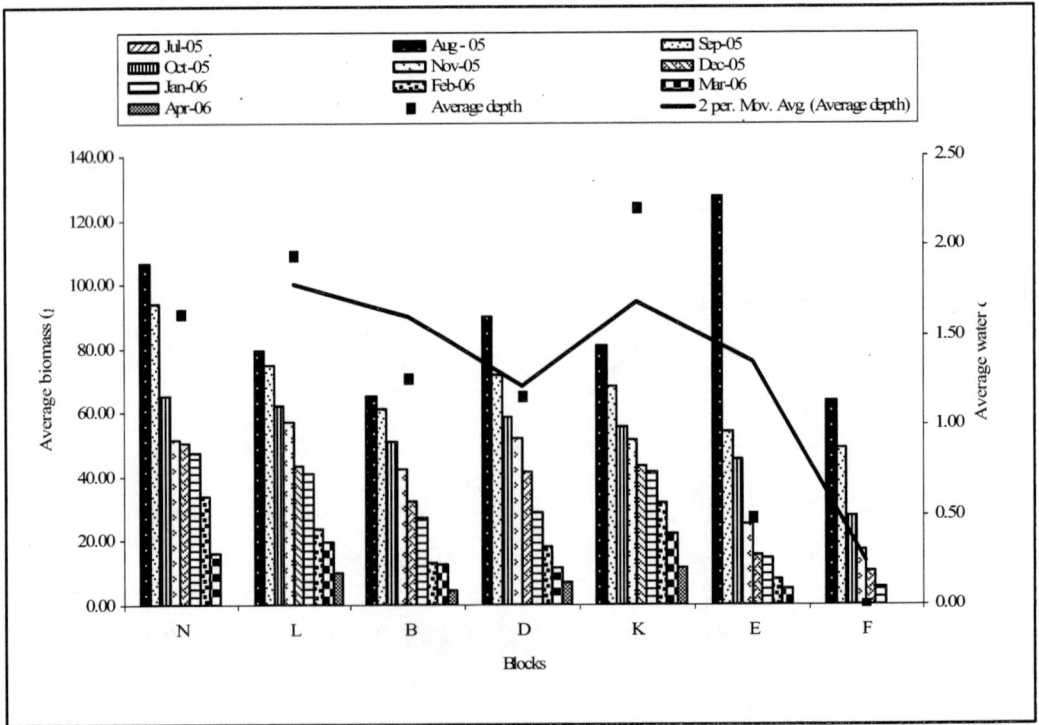

Figure 5. Block wise average water depth recorded from monthly surveyed quadrats.

METHODS

Macrophyte Sampling

The macrophyte sampling was carried out during July 2005 through April 2006 in the seven wetland blocks (B, D, E, F, K, L and N) following standard protocols. To estimate the productivity, macrophytes were sampled regularly on a fixed date of each month following harvest method (Westlake 1963). For estimating the above ground biomass, samples were collected from pre-fixed monthly sampling points in the wetland blocks, which remain inundated from July till April. A total of 21 sampling points were selected, i.e. 4 each in D and L blocks, 3 each in B, F and K blocks, and 2 each in E and N blocks. Macrophytes were sampled using quadrats (0.5 x 0.5 m) laid at each point, the coordinates of which were noted using a handheld Global Positioning System (GPS, Magellan). Additional 139 quadrats were randomly selected for biomass each during November 2005 and March 2006, and were also considered to document the species assemblage and richness.

After collecting the plant samples (above-ground parts), species identification was done referring the published reports and illustrations, and each species was subsequently segregated. On the basis of their habit, the macrophytes were categorized into algae, rooted submerged, free submerged, free floating, rooted floating, emergent and terrestrial. Specimens of unidentified species were preserved as per Jain and Rao (1976) and were coded. Help from the Botanical Survey of India, Coimbatore was also sought for identification of certain species. The samples segregated in the field were brought immediately to the KNP field laboratory and washed under tap water to remove any detritus, soil and other such matters. Since the study was intended to estimate the above ground biomass, all the underground parts of the plants, if any, were cut and discarded. The samples were then weighed to record the fresh weight of each species per quadrat. Subsequently the samples were labeled, dried in a hot air oven (Make-Genuine) at 105°C till constant weight. Weight of the oven dried samples was recorded from which average monthly biomass per unit area (g m^{-2}) was estimated following Sharma (2004). Macrophytic Net Primary Productivity (NPP) was calculated from the biomass by summation of the monthly increments (Vijayan 1991). The monthly total average biomass of each species (recorded in 0.5 m x 0.5 m quadrats) was calculated by dividing the total biomass of the species in all the quadrats by the total number of quadrats sampled

Environmental Parameters

Water level / depth were recorded at each location during the macrophyte sampling. In addition, electrical conductivity (EC, µS/cm), the level of dissolved oxygen (DO), total hardness (as $CaCO_3$), calcium (Ca), chlorides (Cl), salinity, total alkalinity, and free carbon dioxide in the water samples from the blocks were estimated (Table 1). The chemical characterization of the water samples was performed in the laboratory following standard protocol (Allen 1989, APHA 1989, Tandon 2005). The values were expressed in mg/lit unless otherwise stated. However, limiting the scope of the chapter the details on water quality are not reported.

Table 1. Physical and Chemical characteristics of the wetland water

pH	6.8-9.1
Total Dissolved Solids (mg/lit)	138-376
Dissolved Oxygen (mg/lit)	5.7-6.9
Total Hardness (as $CaCO_3$, mg/lit)	81-158
Calcium (mg/lit)	34-57
Chlorides (mg/lit)	53-99
Salinity (mg/lit)	95-179
Total Alkalinity (mg/lit)	102-190
Free CO_2 (mg/lit)	31-64
Electrical Conductivity (µs/cm)	200-362

Data Analyses

The Test of Association among species was performed following Schluter (1984), and Ludwig and Reynolds (1988). Tests were performed to evaluate species-wise as well as category-wise associations. Univariate test to assess variations in the biomass among the categories, the blocks and the months was performed adopting General Linear Model (GLM). When significant differences were found, identification of the differing means was achieved by a test of Least Significant Difference (LSD) as a post-hoc analysis at an alpha = 0.05 level. Hierarchical cluster analysis was performed on the data and dendrograms were prepared to demonstrate grouping of macrophyte species as well as categories. All statistical analyses were performed using the SPSS 11.0 (Norusis 1990).

RESULTS

In total, 37 species of macrophytes belonging to 34 genera within 22 families were reported. Of these, 22 species were actual / true hydrophytes. Several algal species (algae) were seen in most of the months. Among the blocks, the maximum average biomass was recorded from K block (223.2 g m^{-2}) during November 2005 and minimum (1.8 g m^{-2}) from E block during September 2005 (Table 2). Among the months, November 2005 had maximum average biomass (92.21 g m^{-2}) and January 2006 had minimum average biomass (27.52 g m^{-2}, Figure 6). The NPP was found to be 2.71 g m^{-2} per day (dry weight). Species richness was highest in L block (27) followed by D (23), E (18), F (12), N (10) and K (09) blocks. Among months March and April were the richest in species having 20 species each.

The order of major species in terms of their average biomass per unit area during the study was *Pseudoraphis spinescens* (133.30 g m^{-2}) > *Cynodon dactylon* (97.90 g m^{-2}) > *Sporobolus sp.* (35.37 g m^{-2}) > *P. distichum* (22.70 g m^{-2}) > *Vallisneria natans* (21.28 g m^{-2}) > *Alternanthera* sp. (15.35 g m^{-2}) > *Utricularia sp.* (14.09 g m^{-2}) > *Hemiadelphis polyspermus* (13.64 g m^{-2}) > *N. minor* (13.04 g m^{-2}). *Gnaphalium purpureum* and *Grangea maderaspatana* were seen having the lowest average biomass (0.02 g m^{-2}) among all the reported species. The emergent species constituted highest amount of biomass compared to submerged and floating species. With increase in biomass of the emergent species, decline in the biomass of submerged and floating plants was discernible. *P. spinescens* and *P. distichum* were the dominant emergent species while *V. natans* and *N. minor* were the dominant rooted-submerged species. *Azolla pinnata*, *Lemna perpusilla* and *Spirodela polyrhiza* were the common free-floating plants. Among the rooted-floating plants, *Nymphoides* sp. and flourished during certain seasons. The monthly increment in biomass was highest (857.3 g m^{-2}) during March 2006 and lowest (26.8 g m^{-2}) during February 2006.

Table 2. Monthly variation in average biomass in KNP

Month	Average Biomass and range (g m^{-2})	N	L	B	D	K	E	F
July 05'	Average Biomass	48.00	26.00	28.00	32.45	19.87	20.00	65.33
	Range of average biomass	5.6 – 40	0.4 - 43.2	0.8 - 40.4	0.4 - 48	2.4 - 14.4	0.4 - 20.4	1.2 - 75.2
Aug 05'	Average Biomass	0.00	13.50	18.40	26.10	124.67	0.00	3.33
	Range of average biomass	0.00	0.8 - 28	2 - 31.6	0.8 - 57.6	9.2 - 312.8	0.00	0.8 - 9.2
Sep 05'	Average Biomass	0.00	53.00	13.00	76.80	81.73	1.80	9.47
	Range of average biomass	0.00	0.4 - 88.8	0.4 - 32.2	0.4 - 93.6	245.20	3.60	2.4 - 20
Oct 05'	Average Biomass	26.00	37.60	53.73	66.60	204.00	12.60	83.47
	Range of average biomass	1.6 - 33.2	0.4 - 35.2	3.2 - 63.6	1.6 - 147.6	470.80	2.8 - 9.2	1.2 - 122
Nov 05'	Average Biomass	28.40	67.80	70.93	129.43	223.20	69.37	56.27
	Range of average biomass	1 - 21.6	2 - 39.2	10 - 142	7.6 - 119.7	2.4 - 440	2.4 - 52.3	5.6 - 102.8
Dec 05'	Average Biomass	19.00	69.30	26.80	42.40	68.00	12.00	16.60
	Range of aver biomass	5.6 - 26.8	0.4 - 136	5.6 - 36	22 - 116.4	204.00	4.8 - 19.2	20.8 - 29
Jan 06'	Average Biomass	11.40	19.40	12.00	41.50	56.47	35.00	12.00
	Range of average biomass	22.80	0.4 - 48	36.00	8 - 92.8	169.40	70.00	7.2 - 28.8
Feb 06'	Average Biomass	64.20	37.50	18.27	22.90	4.53	36.80	31.07
	Range of average biomass	32 – 60	2.4 - 61.6	19.6 - 35.2	0.4 - 55.6	13.60	1.2 - 36	2 - 36.8
Mar 06'	Average Biomass	64.53	94.17	37.20	113.30	13.17	22.13	142.80
	Range of average biomass	2.4 - 147.2	0.8 - 46.4	0.4 - 35.6	0.4 - 86.8	12.8 - 26.7	0.4 - 12.4	0.8 - 124.8
Apr 06'	Average Biomass	5.80	74.60	75.07	81.60	0.00	104.20	153.87
	Range of average biomass	11.60	0.4 - 51.2	1.2 - 202.8	0.4 - 105.6	0.00	0.4 - 159.6	2 - 165.6

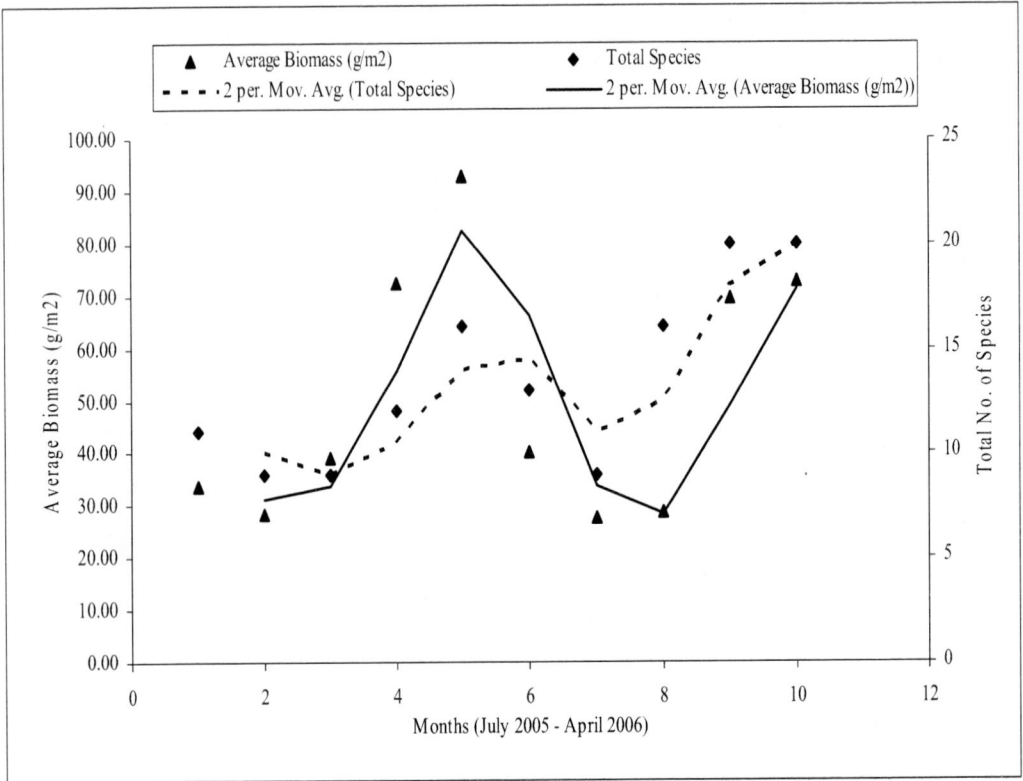

Figure 6. Monthly variation in average biomass and total number of macrophytes.

Macrophytes species had positive association with each other (W = 147.79, P < 0.05). However, as far as the categories are concerned, no significant association among them could be seen. The Pearson's correlation matrix (Two-tail), in the case of category, showed positive correlation between free submerged and rooted floating categories (P < 0.05). The biomass varied significantly among months, among blocks as well as among categories (GLM-ANOVA, P < 0.05, Table 3). However, the post-hoc analysis showed that the biomass during the month of August 2005 was significantly distinct from the biomass observed during October 2005 (LSD, P < 0.05, Table 4). Likewise, the biomass of macrophyte species during the months of July 2005, August 2005 and September 2005 was significantly distinct from November 2005 (P < 0.05, Table 4). In the case of blocks, the macrophyte biomass in N, L and B blocks was significantly distinct from that of K block (LSD, P < 0.05, Table 5). In the case of macrophyte categories algal biomass was found significantly distinct from the biomass of free submerged, free floating, rooted floating, emergent and terrestrial species (LSD, P < 0.05, Table 6). Rooted floating, free submerged, free floating and rooted floating species were apparently distinct from emergent and terrestrial ones.

Table 3. Univariate analysis of variance of macrophytic biomass

Tests of Between-Subjects Effects

Dependent Variable: BIOMASS

Source	Type III Sum of Squares	df	Mean Square	F	Sig.
Corrected Model	664578.163[a]	489	1359.056	2.475	.000
Intercept	69461.907	1	69461.907	126.507	.000
MONTH	14419.536	9	1602.171	2.918	.002
BLOCK	8792.965	6	1465.494	2.669	.014
CATEGORY	60697.476	6	10116.246	18.424	.000
MONTH * BLOCK	37245.393	54	689.730	1.256	.105
MONTH * CATEGORY	110028.570	54	2037.566	3.711	.000
BLOCK * CATEGORY	152379.167	36	4232.755	7.709	.000
MONTH * BLOCK * CATEGORY	281015.055	324	867.330	1.580	.000
Error	538094.980	980	549.077		
Total	1272135.050	1470			
Corrected Total	1202673.143	1469			

a. R Squared =.553 (Adjusted R Squared = .329).

Table 4. Matrix for LSD (Post-Hoc) tests on macrophytes biomass in each month

Months	July 05'	Aug 05'	Sep 05'	Oct 05'	Nov 05'	Dec 05'	Jan 06'	Feb 06'	Mar 06'	Apr 06'
July 05'	NS									
Aug 05'	NS	NS								
Sep 05'	NS	NS	NS							
Oct 05'	NS	-6.1	NS	NS						
Nov 05'	-7.3	-8.2	-7.1	NS	NS					
Dec 05'	NS	NS	NS	5.9	7.9	NS				
Jan 06'	NS	NS	NS	6.5	8.6	NS	NS			
Feb 06'	NS	NS	NS	5.8	7.9	NS	NS	NS		
Mar 06'	NS	NS	NS	NS	NS	NS	-5.6	NS	NS	
Apr 06'	-5.8	-6.6	-5.5	NS	NS	-6.4	-6.9	-6.4	NS	NS

LSD based on observed means, NS: Not Significant ($P < 0.05$).

Rescaled Distance Cluster Combine

```
            0         5        10        15        20        25
            +---------+---------+---------+---------+---------+

Macrophyte Species

Gnaphalium purpureum
Grangea maderaspatana
Vetiveria zizanioides
Cassia tora
Echinochloa colonum
Cochlearia cochlearioides
Coldenia procumbens
Scirpus littoralis
Phyla nodiflora
Glinus lotoides
Scirpus tuberosus
Ipomoea aquatica
Potentilla supina
Glinus oppositifolia
Panicum paludosum
Gnaphalium pulvinatum
Nymphoides sp.
Lemna perpusilla
Cyperus alopecuroides
Eclipta prostata
Ceratophyllum demersum
Limnophyton obtusifolium
Neptunia oleracea
Paspalidium punctatum
Spirodela polyrhiza
Azolla sp.
```

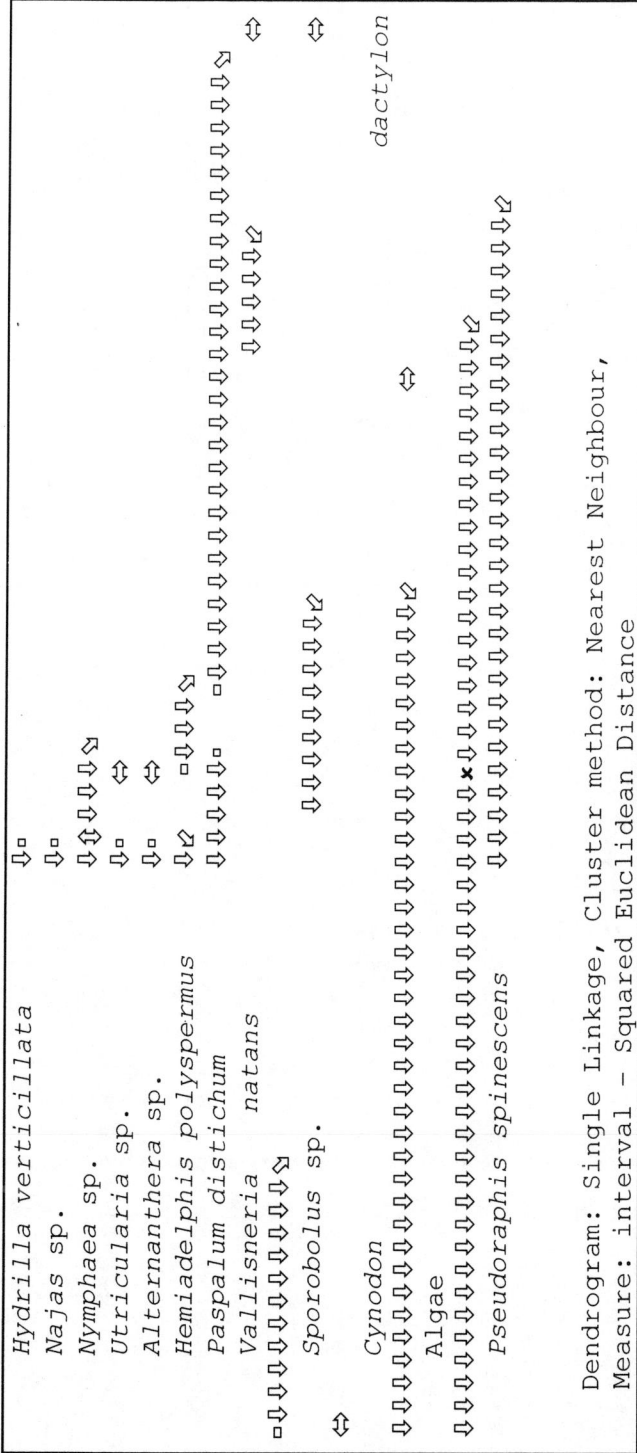

Figure 7. Hierarchical cluster analysis on biomass of macrophyte species.

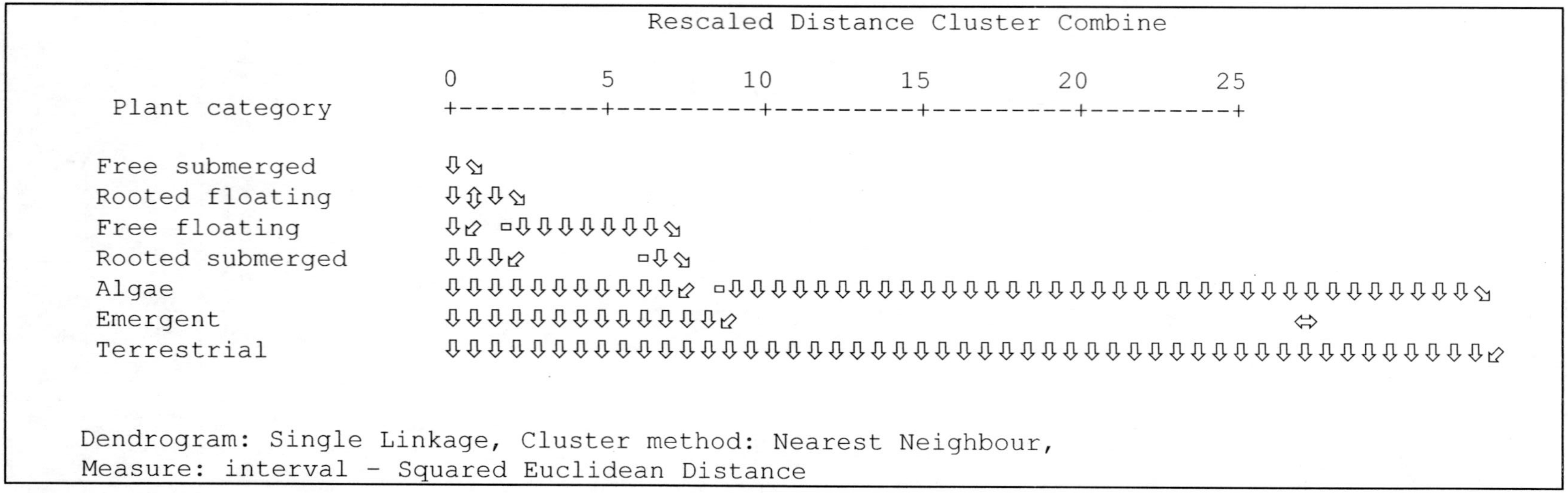

```
                    Rescaled Distance Cluster Combine

                    0         5        10        15        20        25
                    +---------+---------+---------+---------+---------+
Plant category

Free submerged
Rooted floating
Free floating
Rooted submerged
Algae
Emergent
Terrestrial

Dendrogram: Single Linkage, Cluster method: Nearest Neighbour,
Measure: interval - Squared Euclidean Distance
```

Figure 8. Hierarchical cluster analysis with respect to macrophyte category biomass.

Table 5. Matrix for LSD (Post-Hoc) tests on macrophyte biomass in each block

Blocks	N	L	B	D	K	E	F
N	NS						
L	NS	NS					
B	NS	NS	NS				
D	NS	NS	NS	NS			
K	-7.4	-4.5	-6.3	NS	NS		
E	NS	NS	NS	NS	6.9	NS	
F	NS	NS	NS	NS	NS	NS	NS

LSD based on observed means, NS: Not Significant ($P < 0.05$).

Table 6. Matrix for LSD (Post-Hoc) tests on biomass in each macrophyte categories

Category	A	RS	FS	FF	RF	E	T
A	NS						
RS	NS	NS					
FS	6.9	NS	NS				
FF	7.1	NS	NS	NS			
RF	7.9	NS	NS	NS	NS		
E	-6.3	-10.3	-13.3	-13.5	-14.3	NS	
T	-9.3	-13.3	-16.2	-16.4	-17.2	NS	NS

LSD based on observed means, NS: Not Significant ($P < 0.05$). A: Algae, RS: Rooted submerged, FS: Free submerged, FF: Free floating, RF: Rooted floating, E: Emergent, T: Terrestrial.

The hierarchical cluster analysis of plant species with respect to biomass showed 6 clusters in total (Figure 7). Algal species distinctly formed one cluster. *C. dactylon, P. spinescens* and *Sporobolous* sp. formed three distinct and separate clusters. The 5th cluster was formed by *P. distichum* and *V. natans,* and rest of the species formed another cluster. The cluster analysis with respect to categories (Figure 8) resulted, in total, 5 clusters. Terrestrial, emergent species, algal species and rooted submerged categories formed one distinct cluster each, and the fifth one was formed by the rest of the categories.

DISCUSSION

The present study represents nearly 41% of the total aquatic macrophytes reported during the late 1980s from KNP (Prasad 1988, 1989). Several species such as *Eichhornia crassipes, Potamogeton* sp., *Oryza rufipogon* though could not be recorded in the quadrats were seen elsewhere in the wetland blocks of the KNP. As reported in the present study and also earlier studies (Vijayan 1991) the vegetation of KNP wetland was largely dominated by emergent macrophytes, mainly *P. spinescens*, together with *P. distichum, P. punctatum, Limnophyton obtusifolium, C. alopecuroides, Panicum paludosum, Scirpus tuberosus, Scirpus littoralis, Echinochloa colonum.* The next category was the rooted submerged (especially *V. natans* and *Najas* sp.). However, the present study revealed that the dominating species (*P. distichum*) of 1980's (Middleton 1994), currently ranks second in biomass preceded by *P. spinescens*

(133.30 g m^{-2}). During the present investigation *P. distichum* seemed to have an average biomass of 22.7 g m^{-2} only. Such changes may be due to changes in the duration of flooding that might have directly affected the individual emergent species due to species tolerance of anoxia (van der Brink et al. 1995). It is known that only few emergent species are able to germinate under standing water conditions (Kadlec 1962, van der Valk & Davis 1978, Seabloom et al. 1998). Barber and Behrens (1985) ascribe the increment in biomass of macrophyte species to the rising water temperature from winter through summer months. Keddy and Elis (1985), and Galinato and van der Valk (1986) report the regeneration niches to be dependent on water depth as well as water temperature. Further, the quick release of high quantity of water causes mortality in the dominating species and openings in the emergent macrophyte cover (Harris & Marshall 1963). Once the water recedes and temperature increases, during the subsequent period of low-water level annual species that are totally reliant on the mud-pools emerge and dominate (Salisbury 1971). The ability of emergent macrophytes to reach the water surface depends much on the light quality and quantity, which in turn is regulated by the water depth and water quality. Different plant species may require different time for germination. Hence, it is understood that the different species can be expected to have different depth tolerances under stable water regime (Seabloom et al. 1998). The present study reports lesser biomass for submerged (fully aquatic) species. Water depth plays a crucial role in the photosynthesis of macrophytes as molecular diffusion of free carbon dioxide is much slower in water than in air. Similarly availability of light also is likely to change with depth as well as water quality aspects such as turbidity.

Several anthropogenic changes in the catchment also play important role in plant community development in the wetland. Irrigation canals, construction of barrages on the rivers, alteration along the course of the rivers and channels, and agricultural activities along the river system cause siltation and increase in suspended and dissolved solids in water, which may affect growth of macrophytes. Such anthropogenic activities may also lead to eutrophication and spread of aquatic exotics such as *E. crassipes* (water hyacinth). It is reported that under suitable temperature and water levels *E. crassipes* spreads double within 2.5 days which may lead to reduced photosynthetic activity (light intensity) and reduced dissolved oxygen thus, hampering the growth of other species especially the submerged ones. Shrestha and Janauer (2000) have also reported the diversity of aquatic macrophytes being affected by siltation and promoted / cultural eutrophication.

The study in KNP conducted almost two decades back reported higher water levels (1.0-1.5 meters average depth, Vijayan 1991). However, the water depth in the wetland blocks during the present survey was significantly lower (0.5 – 1.15 meter). The KNP wetland system apparently getting shallower, perhaps, is due to siltation, accumulation of decomposing biomass and colonization by terrestrial species such as *Prosopis juliflora*. Roberts (1994) and Tabacchi (1995) opine that plant communities are more likely to respond to the history of water levels than the water level at the time of survey. On the other hand Casanova and Brock (2000) are of the view that depth plays less important / significant role in temporary or seasonal wetlands. Nevertheless, much of the variations in water depth can occur in a short time frame due to high rates of evaporation, or when a wetland is purposely drained or filled. It can happen in a little longer period by being colonized by water tolerant terrestrial or littoral species.

The macrophyte biomass and species richness in the KNP varied among the blocks. The invasion of *P. juliflora* an exotic terrestrial species, in almost all the wetland blocks, was quite discernible. Variation in species richness in each block as well as the *P. juliflora* invasion may be due to the increasing speed of colonization that encourages homogenization of the plant community and also the water regime in each specific location that either allows the survival of macrophytes or for the colonization by less water requiring / terrestrial species. During the present investigation the K block supported 27 macrophytes species where as the N block was found to support only 10 species indicating perhaps the variation in water quality, water depth, quality of the bottom substratum, or the past history and the management activities happening in each blocks. The temporary arrangement of ground water pumping by the park authorities keeps the K block inundated through out the year, while the N block faces a total drain out at the end of April.

The present study recorded the maximum biomass during November, which starts declining from December until February (perhaps due to colder weather). This trend is reversed by a gradual increase in biomass since March as the drier months approach. The increase in biomass during September to November is because of the growth of pure aquatic forms either from seeds, roots, tubers or culms. This could also probably be due to the direct contact of the roots of emergent species with the bottom substratum, which is a rich repository of minerals and nutrients (Mathew et al. 2002, 2003, Prusty & Azeez 2007). The direct availability and subsequent uptake of minerals and nutrients stimulate the macrophytic growth. During December that witnesses gradually receding water level in the wetland system, the average biomass also decreased. The species richness increased reaching a maximum (20) during March and April 2006. This may be attributed to the germination of more number of semi-aquatic / hydric soil tolerant species happening when the water level comes down. The gradual increase in ambient temperature during the months of March and April changes the wetland into open mudflats triggering the growth of other non-aquatic annual species. The frequency of water inputs / flood disturbances affects species richness (Casanova & Brock 2000). It is reported that an intermediate (2-3 times) frequency of flooding events creates opportunities for species to get established thus, preventing competitive exclusion (Bornette & Amoros 1996). The present study recorded fewer numbers of species during sudden water input (released from the dam after the rainfall).

The present study notes the association of rooted submerged species (primarily those such as *Najas* sp.) with water depth and its certain qualities. For example *Najas* sp. is found associated with deeper and clearer water. Though less studied, the depth of the bed sediment plays a role in the occurrence and association of the species especially rooted submerged. In the present study during March and April, when the KNP wetland system undergoes drying, the association between emergent and free floating species (Bini et al. 1999) was quite prominent. However, perhaps it may be coincidental as the floating species passively drift to deeper locations, where emergents grow, as the periphery of the wetland gets dry. Murphy (2002) opines that water depth is a significant factor for aquatic species association in freshwater systems worldwide. Water depth seemed to be significant explanatory variable for aquatic vegetation assemblage in our study as well. Water depth, the duration of inundation, and the frequency rate of filling and drying contribute to the plant community development (Bunn et al. 1997). In the case of temporary or monsoonal wetland as that of KNP, the plant community development is largely the result of germination and establishment from a long-lived, dormant seed and vegetative propagules (from the seed bank) that survive the drought conditions. In such cases, as reported by Casanova and Brock (2000), the community zonation may differ from year to year and season to season as water levels fluctuate in space and time thus making it difficult to predict the composition of plant communities in relation to depth. It also suggests that depth is not always the criterion / indicator of community composition for zonation. Issues such as cattle grazing in the park, harvesting grass and other such biomass during dry seasons might also affect the plant community.

The present study showed absence of some of the species as reported earlier from KNP (Vijayan 1991) and notable variation in biomass for species that existed then and now. Many species that were recorded earlier in this wetland did not appear during the present study, which may be due to the anthropogenic changes / alternations in the KNP microenvironment. The study for a longer duration would have brought up such species as some species may, if buried in the bed deeper in the sediment, take longer time to emerge upon getting appropriate microenvironment. The biological diversity often observed in the park is strongly related to hydro period. Development of plant communities under natural, fluctuating water regime is a consequence of several factors, including livestock grazing and individual plant tolerance or response to the stresses of flood and drought, characteristics of the catchment and the nature of the substratum of the wetland.

ACKNOWLEDGEMENTS

We are grateful to the Director, SACON for the facilities for carrying out this research work. The authors also thank Mr. KCA Arun Prasad, IFS, then Director of KNP, Bharatpur, Rajasthan for permitting us to undertake the study in KNP and for providing logistic support for the purpose. The help, excellent assistance, and conventional insights of Mr. Brijendra Singh during the fieldwork are highly acknowledged. Dr. PN Joshi of Gujarat Institute of Desert Ecology commented on the earlier version of this manuscript.

REFERENCES

Allen, S. E. (1989). *Chemical analysis of Ecological materials* (368 pp.). Oxford, London, Blackwell Scientific Publications.

APHA. (1989). *Standard methods for the examination of water and wastewater*, 17th Edition. Washington DC: American Public Health Association.

Azeez, P. A., Nadarajan, N. R. & Mittal, D. D. (2000). The impact of a monsoonal wetland on ground water chemistry. *Pollution Research*, 19(2): 249-255.

Azeez, P. A., Ramachandran, N. K. & Vijayan, V. S. (1992). The socioeconomics of the villagers around Keoladeo National Park, Bharatpur, Rajasthan. International *Journal of Ecology and Environmental Sciences*, 18: 1-15.

Barber, B. J. & Behrens, P. J. (1985). Effects of elevated temperature on seasonal in situ productivity of *Thalassia testudinum*. Bankx ex Konig and Syringodium filiforme Kutzing. *Aquatic Botany*, 22: 61-69.

Bini, L. M., Thomas, S. M., Murphy, K. J. & Camargo, A. F. M. (1999). Aquatic macrophytes distribution in relation to water and sediment conditions in the eltaipu Reservoir, Brazil. *Hydrobiologia*, 415:147-154.

Bornette, G. & Amoros, C. (1996). Disturbance regimes and vegetation dynamics: role of floods in riverine wetlands. *Journal of Vegetation Science,* 7: 615-622.

Boston, H. L. & Perkins, M. A. (1982). Water column impacts of macrophyte decomposition beneath fiberglass screens. *Aquatic Botany*, 14: 15-27.

Bunn, S. E., Boon, P. I., Brock, M. A. & Schofield, N. J. (1997). *National Wetlands R & D Program Scoping Review*. Land and Water Resources Research and Development Corporation Occasional Paper 01/97 Canberra.

Carignam, R. & Kalff, J. (1982). Phosphorus release by submerged macrophytes: significance to epiphyte and photoplankton. *Limnology and Oceanography*, 27: 419-427.

Casanova, M. T. & Brock, M. A. (2000). How do depth, duration and frequency of flooding influence the establishment of wetland plant communities? *Plant Ecology*, 147: 237-250.

Cyr, H. & Downing, J. A. (1988). Empirical relationships of photofaunal abundance to plant biomass and macrophyte bed characteristics. *Canadian Journal of Fisheries and Aquatic Sciences*, 45: 976–984.

Davis, S. M. (1991). Growth, decomposition, and nutrient retention of *Cladium jamaicense* Crantz and *Typha domingensis* Pers. In the Florida Everglades. *Aquatic Botany,* 62: 115-133.

Denny, P. (1985). *The Ecology and Management of African Wetland Vegetation*. Dr W. Junk Publishers, The Hague.

Downing, J. A. (1986). A regression technique for the estimation of epiphytic invertebrate populations. *Freshwater Biology*, 16: 161–173.

Galinato, M. I. & van der Valk, A. G. (1986). Seed germination traits of annuals and emergents recruited during drawdowns in the Delta Marsh, Manitoba, Canada. *Aquatic Botany*, 26: 89-102.

Grasmuck, N., Haury, J., Leglize, L. & Muller, S. (1995). Assessment of the bio-indicator capacity of aquatic macrophytes using multivariate analysis. *Hydrobiologia*, 300/3001: 115-122.

Harris, S. W. & Marshall, W. H. (1963). Ecology of water-level manipulations on a northern marsh. *Ecology*, 44: 331-343.

Jain, S. K. & Rao, R. R. (1976). *A handbook of field and herbarium methods*. Today and Tommorow printers and publishers. 89 pp.

Kadlec, J. A. (1962). Effects of a drawdown on a waterfowl impoundment. *Ecology*, 43: 267-281.

Kao, J. T., Titus, J. E. & Zhu, W.X. (2003). Differential nitrogen and phosphorous retention by five wetland plant species. *Wetlands*, 23(4): 979-987.

Kateyo, E. M. (2007). Biodiversity of an interface zone of a nutrient-deficient lake (Nabugabo) in Uganda: macrophytes. *African Journal of Ecology*, 45(2): 130-134

Keddy, P. A. & Ellis, T. H. (1985). Seedling recruitment of 11 wetland plant species along a water level gradient: shared or distinct responses? *Canadian Journal of Botany*, 63: 1876-1879.

Kumar, C. R. A. & Vijayan, V. S. (1988). On the fish fauna of Keoladeo national Park, Bharatpur (Rajasthan). *Journal of Bombay Natural History Society*, 85 (1): 44-49.

Kumar, C. R. A., Ramachandran, N. K. & Asthana, A. (1995). Composition, abundance and distribution of fish in Banaganga-Gambhir river system and source of fish to the Keoladeo National Park, Bharatpur. *Journal of Bombay Natural History Society*, 92 (1): 30-39.

Ludwig, J. A. & Reynolds, J. F. (1988). *Statistical Ecology: A primer on methods and computing* (337 pp.). New York, John Wiley and Sons.

Maltby, E. (1986). *Waterlogged Wealth* (200 pp.). London, Earthscan.

Matagi, S. V., Swai, D. & Muganbe, R. (1998). A review of heavy metal removal mechanisms in wetlands. *African Journal of Tropical Hydrobiology and Fisheries*, 8: 23-35.

Mathew, M., Mohanraj, R., Azeez, P. A. & Pattabhi, S. (2003). Speciation of heavy metals in bed sediments of wetlands in urban Coimbatore, India. *Bulletin of Environmental Contamination and Toxicology*, 70: 800-808.

Mathew, M., Sathishkumar, M., Azeez, P. A., Sivakumar, R. & Pattabi, S. (2002). Sediment quality of wetlands in Coimbatore, Tamilnadu, India. *Bulletin of Environmental Contamination and Toxicology*, 68: 389-393.

Mathur, V. B., Sinha, P. R. & Mishra, M. (2005). *Keoladeo National Park World Heritage Site*, Technical Report No.5, UNESCO-IUCN-Wild Life Institute of India, Dehradun, India.

Middleton, B. A. (1994). Management of monsoonal wetlands for Greylag (*Anser anser* L.) and Barheaded Geese (*Anser indicus* L.) in the Keoladeo National Park, India. *International Journal of Ecology and Environmental Sciences*, 20: 163-171.

Mitsch, W. J. & Gosselink, J. G. (2000). *Wetlands*, Third edition (pp. 156-164). New York, John Wiley and Sons. Inc.

Murphy, K. J. (2002). Plant communities and plant diversity in software lakes of Northern Europe. *Aquatic Botany*, 73: 287-324.

Naturetrek (2005). *India – Bharatpur & Chambal* (pp. 06). Naturetrek Tour Dossier 2006. August 05.

Norusis, M. J. (1990). *SPSS/PC+4.0 Base Manual – Statistical Data Analysis*. SPSS Inc.

Pal, D.K., Bhattacharyya, T., Deshpande, S.B., Sarma, V.A.K. & Velayutham, M. (2000). *Significance of minerals in soil environment of India* (68 pp.), NBSS Review Series 1. Nagpur, India, National Bureau of Soil Science & Land Use Planning.

Pokorny, J. (1994). Development of aquatic macrophytes in shallow lakes and ponds. IWRB Publication, 32: 36-43.

Prasad, V. P. (1988). Wetland angiosperms of Keoladeo National Park, Bharatpur. *Journal of Economic and Taxonomic Botany*, 12: 457-466.

Prasad, V. P. (1989). Flora of Keoladeo National park, Bharatpur. *Journal of Economic and Taxonomic Botany*, 13: 729-750.

Prusty, B. A. K. (2007). *Distribution of select nutrients and metals in the soils of s wetland-terrestrial ecosystem complex: A case study of Keoladeo National Park, Bharatpur, India* (188 pp.), Ph D Thesis, Bharthiar University, Coimbatore, India.

Prusty, B. A. K. & Azeez, P. A. (2007). Alkali and alkaline earth metals in the soil profile of a wetland-terrestrial ecosystem complex in India. *Australian Journal of Soil Research*, 45(7): 533-542.

Prusty, B. A. K., Azeez, P. A. & Jagadeesh, E. P. (2007a). Alkali and transition metals in macrophytes of a wetland system. *Bulletin of Environmental Contamination and Toxicology*, 78(5): 405-410.

Prusty, B. A. K., Chandra, R., Azeez, P. A. & Sharma, L. L. (2007b). New additions to the ichthyofauna of Keoladeo National Park, a world heritage site in India. *Zoos' Print Journal*, 22(10): 2848-2852.

Prusty, B. A. K., Chandra, R. & Azeez, P. A. (2008). Vertical and temporal variation of Zn in the soils of a multiple habitat system. *Journal of Food, Agriculture and Environment*, 6(2): 381-387.

Randall, R. G., Minns, C. K., Cairns, V. W. & Moore, J. E. (1996). The relationship between an index of fish production and submerged macrophytes and other habitat features at three littoral areas in the Great Lakes. *Canadian Journal of Fisheries and Aquatic Sciences*, 53: 35-44.

Roberts, J. (1994). *Riverbanks, plants and water management* (pp. 1-7). I: Roberts J and Oliver R (Eds.), The Murrumbidgee, Past and Present. CSIRO Water Resources, Griffith.

Salisbury, E. (1971). The pioneer vegetation of exposed muds and its biological features. Phil. Trans. Roy. Soc. London, 259: 207-255.

Schluter, D. (1984). A variance test for detecting species associations, with some example applications. *Ecology*, 65: 998-1005.

Scott, W. A., Adamson, J. K., Rollinson, J. & Parr, T. W. (2002). Monitoring of aquatic macrophytes for detection of long-term changes in river systems. *Environmental Monitoring and Assessment*, 73: 131-153.

Seabloom, E. W., van der Valk, A. G. & Moloney, K. A. (1998). The role of water depth and soil temperature in determining initial composition of prairie wetland coecoclines. *Plant Ecology*, 138: 203-216.

Sharma, P. D. (2004). *Ecology and management* (640 pp). Meerut, India, Rastogi Publications.

Shrestha, P. & Janauer (2000). Species diversity of aquatic macrophytes in lake Phewa and lake Rupa of Pokhara valley, Nepal. *International Journal of Ecology and Environmental Sciences*, 26: 269-280.

Tabacchi, E. (1995). Structural variability and invasions of pioneer plant communities in riparian habitats of the middle Ardour River (SW France). *Canadian Journal of Botany*, 73: 33-44.

Tandon, H. L. S. (2005). *Methods of Analysis of soils, plants, waters, fertilizers & organic manures* (204 pp.), 5th Edition. New Delhi, Fertilizer Development and Consultation Organization.

van der Brink, F. W. B., van der Velde, G., Bossman, W. W. & Coops, H. (1995). Effects of substrate parameters on growth responses of eight helophyte species in relation to flooding. *Aquatic Botany*, 50: 79-97.

van der Valk, A. G. & Davis. C. B. (1978). The role of seed banks in the vegetation dynamics of prairie glacial marshes. *Ecology*, 59: 322-335.

Vijayan, V. S. (1991). *Keoladeo National Park Ecology Study (1980 – 1990), final report* (337 pp.). Mumbai, Bombay Natural History Society.

Westlake, D. F. (1963). Comparisons of plant productivity. *Biological Review*, 38: 385-425.

Chapter 4

CATCHMENT CHANGES INFLUENCE THE ARRIVAL PATTERN OF FISHES INTO A MONSOONAL WETLAND SYSTEM

B. Anjan Kumar Prusty[*,1,2], *Rachna Chandra*[†1], *M. Shah Hussain*[‡3] *and P. A. Azeez*[§1]

[1]Environmental Impact Assessment Division, Sálim Ali Center for Ornithology and Natural History, Anaikatty, Coimbatore – 641 108, India
[2]Gujarat Institute of Desert Ecology, P.O. Box # 83,
Opp. Changleswar Temple, Mundra Road, Bhuj-370 001, Gujarat, India
[3]Center for Environmental Management of Degraded Ecosystems,
University of Delhi, New Delhi – 110 007, India

ABSTRACT

We examined the arrival pattern of fishes into the wetland system of Keoladeo National Park (KNP), Bharatpur, India for a period of three years, i.e. 2003 through 2005. The study was undertaken to assess the temporal variation in the arrival pattern of fishes into the KNP, and explore the influence of the processes in the catchment on occurrence and abundance of fishes in the feeder canal. A total of 40,766 individuals belonging to 8 families, 18 genera and 24 species of fishes were recorded during the study. The species accumulation curve, however, suggests that there might be 30 species in the catchments of KNP. The present investigation recorded 7 new arrivals to the park, viz. *Brachydanio albolineatus, Cyprinus carpio communis, Clarias gariepinus, Danio dangila, Mystus tengara, Rasbora daniconius* and *Securicula gora*. The highest fish diversity was during 2004 (1.368) and the least in 2003 (0.934). The drift density and rate differed among the catches (ANOVA, $P < 0.05$) and were positively correlated with each other (r = 0.703, P

---

[*] Corresponding author. Email: anjaneia@gmail.com
[†] rachnaeia@gmail.com
[‡] mshahhussain@rediffmail.com
[§] azeezpa@gmail.com

< 0.05). The water speed (m/sec) was negatively correlated with the number of individuals per catch. Except *B. albolineatus, Catla catla, C. gariepinus* and *Mastacembelus armatus*, all other fish species were distinct and independent with respect to their abundance (χ^2, $P < 0.05$). During hierarchical cluster analysis of all the recorded species *Puntius sophore* evidently formed a distinct and separate cluster. *Chanda ranga* and *Puntius sarana* formed one cluster while the rest of the species together formed another cluster.

The quantity and duration of water release, a function of total rainfall in the catchment, influenced the composition of fishes and number entering the park. The decline in the fish species recruitment in comparison to the previous reports necessitates a review of the ongoing anthropogenic pressures on fishes, and habitat destruction for infrastructure development and urbanization in the catchments.

Keywords: Keoladeo National Park, fish diversity, fish recruitment, wetland, Bharatpur, Michaelis-Menton equation

INTRODUCTION

The Indian subcontinent harbors one of the richest fish fauna in the world. About 11% of the 20,000 fish species reported worldwide have been recorded from different ecosystems of India (Anonymous1992-1993, Dehadrai & Poniah 1997). The National Bureau of Fish Genetic Resources, Lucknow, and Zoo Outreach Organization, Coimbatore have recently evaluated 329 freshwater fish species (ZOO/CBSG 1998, Balasundaram et al. 1999) in terms of the available knowledge base on each species. The recent thrust on biodiversity conservation necessitates documentation of fauna and identification of threats associated with them. Study of the fish community in wetlands would be useful in assessing the influence of environmental conditions on those species. An understanding of the species that are annually recruited from outside and those that breed in the system is a prerequisite for rational, conservation oriented and successful wetland ecosystem management.

The most important factor for the fish variability and movement in aquatic systems is water availability (Kumar & Mittal 1993) and vegetation (Kumar & Vijayan 1988). The hydro period, the seasonal or monsoonal changes in water level because of variations in water input (Mitsch & Gosselink 2000), and the ecological and geographical setup of the wetland and its catchment have wide implications on the fish fauna, their recruitment and abundance. The Keoladeo National Park (KNP), a World Heritage site in India, is known to harbor about 40 species of piscivorous birds (Kumar et al. 1995) and several other water fowls. The status of birds in the KNP largely depend on prey availability; the availability of fishes. In India, Ali (1953) was the pioneer to accentuate the importance of fish to the piscivorous birds in this National Park. However, the first ecological research work on fish was started by Kumar & Vijayan (1988). Subsequently several studies were conducted in this national park to address specific ecological issues related to fishes and piscivorous birds (Kumar 1991, Vijayan 1991, Kumar & Mittal 1993, Kumar & Asthana 1994, Kumar et al. 1995). The present investigation was undertaken to assess and analyze the variations in the fish recruitment into the KNP during 2003 to 2005 and to compare the scenario with that of the mid eighties.

STUDY AREA

The KNP (27° 7.6' to 12.2' N and 77° 29.5' to 77° 33.9' E, Figure 9), located in the state of Rajasthan, India is almost 180 km from New Delhi. KNP, a monsoonal largely seasonal wetland, is situated in the flood plains of Banganga and Gambhir, two non-perennial river systems in the State of Rajasthan (Prusty et al. 2007b, Figure 10.). The area falling under semi-arid hot dry zone of India (Pal et al. 2000) experiences climatic extremes from hot dry summer (April to June) to a chilly cold winter (November to January) and a short monsoon (July to September) and post monsoon (September to October, Azeez et al. 1992). During the study period the minimum rainfall recorded in KNP was 1.8 mm (in April 2004) and the maximum 659.9 mm (in July 2005, Figure 11). Usually the temperature in the park varies from about 1°C to 49°C with strong diurnal and seasonal variations (Prusty et al. 2007a). The monthly variation in the minimum and maximum temperature experienced during the study is presented in Figure 12.

The more than 250 years old KNP is a world famous wetland and one of the early Ramsar sites (Mathur et al. 2005) and a world heritage site. It is unique with multiple habitats with congregation of thousands of birds on the onset of monsoon (Naturetrek 2005). The aquatic birds are attracted to the park mainly due to the availability of diversified food, including fishes. The park occupying an area of about 2900 ha, gently slopes to the centre forming depression, the wetland, of about 850 ha (the central dotted area in the map, Figure 9). The park is divided into 15 blocks / compartments, referred alphabetically from A to O, by means of dykes. The water level in the compartments is regulated by means of sluice gates placed at strategic locations. The central low lying blocks remain mostly inundated, while the others are terrestrial supporting woodland and grassland. Parts of terrestrial areas that border the wetland blocks also are partially inundated during monsoon and after monsoon depending upon the quantum of rainfall and water inflow. The water level in the wetland system starts falling towards the end of October and the wetland almost dries up by April leaving only certain patches of water, puddles and pools. During this period, the park management pumps out ground water as a stop gap arrangement to avoid mass kill of fish and to provide drinking water to other denizens of the park.

The park with almost a flat terrain (173 to 176 m ASL) has a thick alluvial clayey soil and scattered saline patches specially in the terrestrial areas. The geologic formation of the Bharatpur district, in which the Park falls, includes different sedimentary and meta-sedimentary rocks of Bhilwara, Delhi and Vindyan subgroups and quaternary alluvials (Azeez et al. 2000) and falls under the Aravalli Super group. The major subsurface geology is of quaternary alluvium. The major ground water reserves in this area are those occurring in unconsolidated formations of Quaternary sediments consisting of clay, silt, kankar and gravel (Bhushan & Sharma 1987). Kankar, nodules of calcium carbonate usually found in the older alluvium or stiff clay of the Indo-Gangetic plain, is found at the depth of 80-90 cm in certain locations in the park. The water in the phreatic aquifer in the area is predominantly saline and contributes to the salinity and chlorinity (Prusty & Azeez 2007) of the park water.

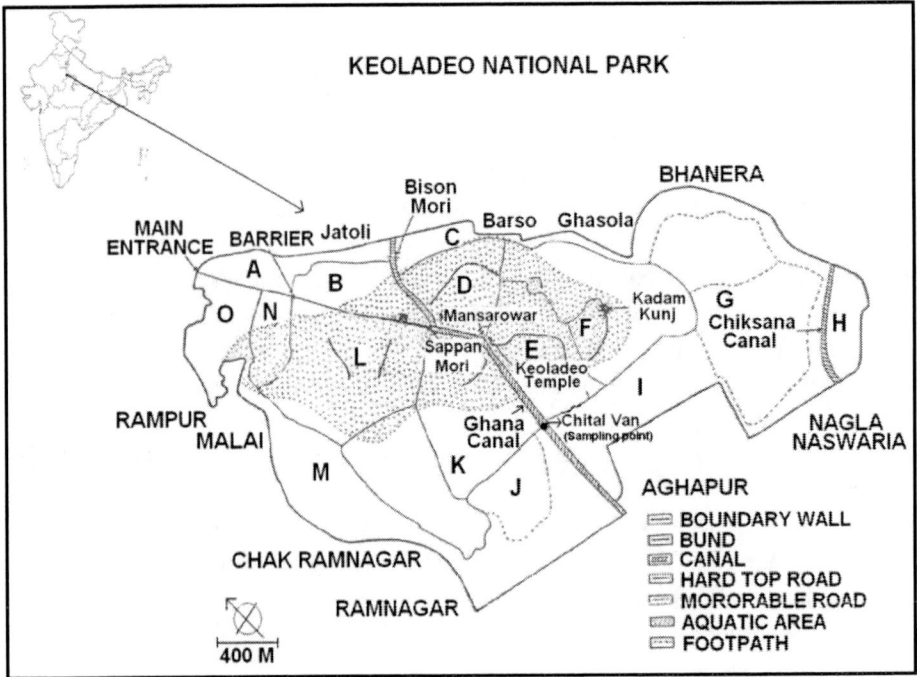

Figure 9. Study area, the Keoladeo National Park.

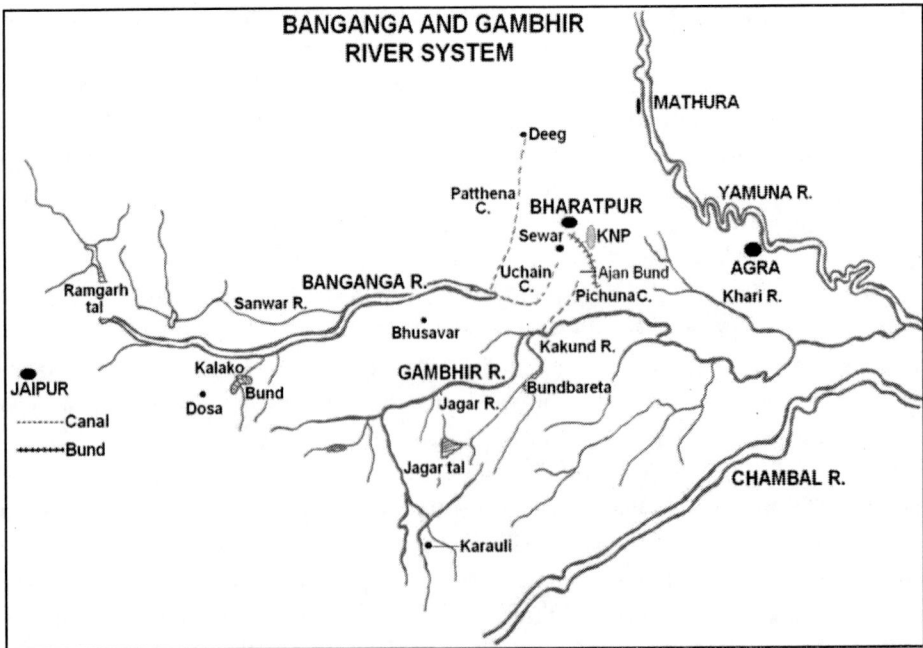

Figure 10. Banganga and Gambhir river system.

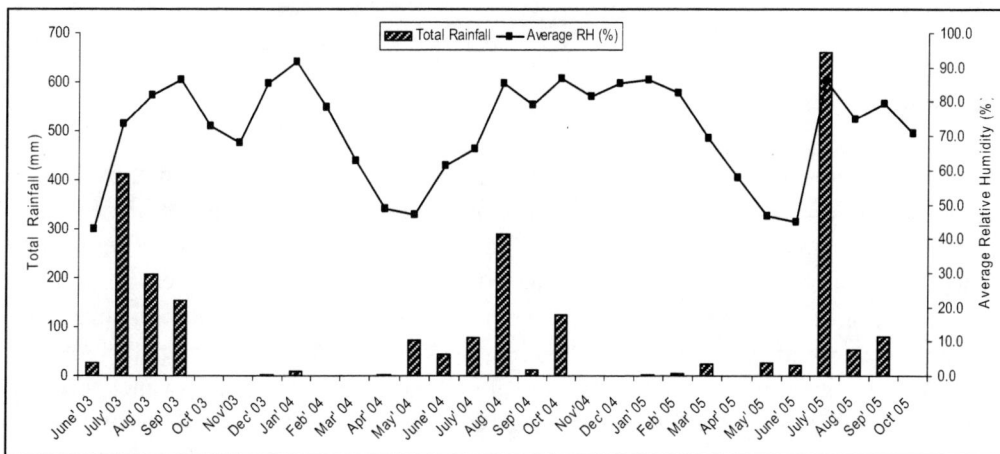

Figure 11. Monthly variation in total rainfall and average relative humidity during the study.

Being a monsoonal wetland, water inputs decide the very survival of KNP. The KNP supports almost 400 avian species, several of which are piscivorous arriving in thousands during breeding season. Therefore, for the park ecosystem to sustain in particular with its avian glory, recruitment of millions of fishes every year is vital. The park receives water mainly from the Ajan bund, an earthen dam constructed for the purpose, situated 500 m away, in addition to the seasonal rain that falls in the park area. The rain falling directly over the park area is negligible compared to the total quantity of water flowing from the Ajan bund. The Ajan bund is controlled by the Rajasthan State Irrigation Department and the water is released in a controlled manner through sluice gates and canals. Water from the catchment of the rivers Banganga and Gambhir is drained into the dam and this water carries millions of fry, micro and macro invertebrates and nutrients in to the park (Kumar & Vijayan 1988, Azeez 1991, Azeez et al. 2007] setting in motion a cycle of intense activities in the park such as congregation and breeding of piscivorous birds.

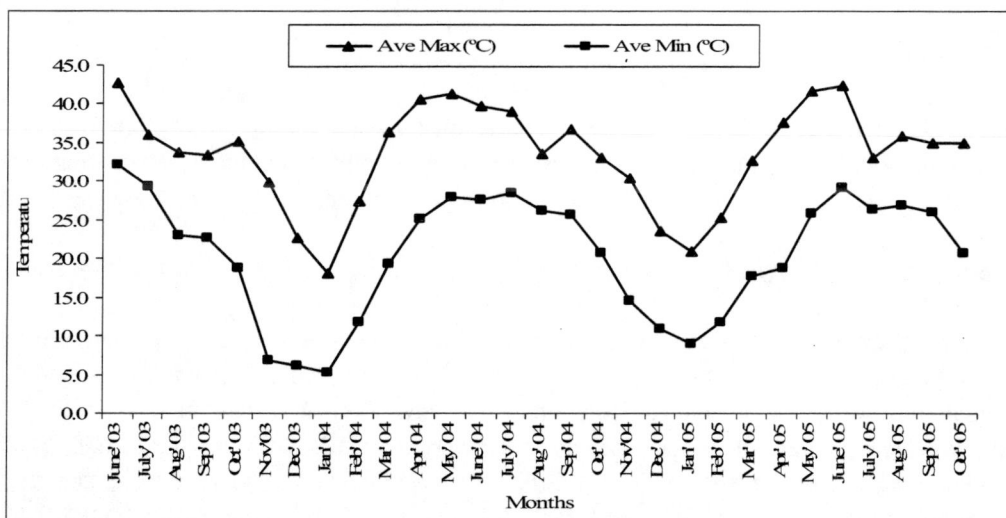

Figure 12. Monthly average maximum and minimum temperature during the study.

METHODS

Fish Sampling

A modified version of stow net (square framed net bag, pyramidal in shape), made up of mosquito netting was used in the present study. The net having a square mouth of 1.83 x 2.01 had a funnel like tapering extending up to 1.5 m. Fry was sampled from the feeder canal to the park at a point named "Chital Van" (Figure 9), about a kilometer downstream of the Ajan bund and a kilometer before the water is diverted to different blocks of the sanctuary. The bridge at this point has two pillars thus, dividing the channel into three flow paths. The area of the fishing net mouth was equal to the area of each compartment of the channel, to ensure that the metal frame of the net fits properly and remains in a stable and steady position in the running water. During sampling, the net was dipped in water so as to remain in firm contact with the pillars and pressed the bottom of the channel. The net was kept so in each outlet for a fixed duration of time (05 sec) and then lifted up. This exercise was repeated thrice in each outlet (Prusty et al. 2007b). However, the time of each operation was changed when the number of fry was very high, to reduce mortality and counting error. The fry trapped in the net were identified on the spot, the live weight (gm) for each catch was taken and the individuals counted. Subsequently the total live weight and number equivalent for 45 seconds (5 seconds x 3 replicates x 3 compartments) was estimated. Unidentified species were marked and coded on the spot; some specimens were preserved and brought to State Fisheries Department, Rajasthan for further identification and confirmation. Day (1986), Talwar and Jhingran (1991a, 1991b) were also referred for identification. Some of the fry being too small to handle live and to identify were considered under miscellaneous category. During 2003 and 2004, sampling was done during dawn and dusk. In 2005, sampling was done in two lots, i.e. during July and during September-October, since water release happened in two lots. During the first lot only dawn and dusk sampling was done, while during the second lot round the day sampling was done with 3 hrs intervals.

Data Analyses

The cumulative number of species encountered during each sampling was plotted against each sampling unit, considering one catch as one sampling unit, to obtain a scatter of points, to which we fitted the following Michaelis-Menton-like equation,

$$S = S_{max} N / (K_m + N)$$

where, S is the number of species, N the sampling unit, S_{max} the maximum number of species that could be present in the catchment and K_m the Michaelis-Menton constant. The species accumulation was also examined using the software EstimateS (Colwell 1994-2004). Shannon-Weiner index of species diversity (H') was calculated following Ludwig & Reynolds (1988) and Jaccard's index of association was calculated to assess the species-specific association. The Test of Association was performed following Schluter (1984) to assess the overall association among the fish species. Drift density, a measure of abundance throughout the study was calculated following Reichard et al. (2004) as the number of individuals per 100 m^3 of water entering the park (current velocity multiplied by the area of the net's mouth). Drift rate, the number of individuals drifted through the sampled section per unit time, was also calculated. Two-Way ANOVA was performed to assess variations in the number of fry across the species and years, and drift density and drift rate. Chi Square test (Goodness of Fit) was done to confirm the distinctness of the species in terms of their arrival during different years. Pearson's test of correlation (Two-tail) was performed to examine the interrelations of fish species among each other based on their occurrence and abundance. All the above cited tests were performed using MEGASTAT (Orris 2000). A hierarchical cluster analysis that depicts the clustering of fishes with respect to their abundance was performed using SPSS (Norusis 1990).

RESULTS

The total number of catch and the resultant species of fishes were found to vary widely on an annual basis (Table 7). In overall, 24 species of fish belonging to 8 families and 18 genera entered the park during the study period. However, the species accumulation curve (Figure 13.) shows that there might be 30 species in the catchment of KNP. The results obtained by analyzing with the EstimateS ranged between 24.75 (First order Chao) and 26.96 species (Second order Jackknife). Species richness was maximum during 2005 (19) followed by 2003 (12 species) and minimum during 2004 (5 species). The family Cyprinidae ranked highest with 15 species. Three species were common arrivals in all the 3 years, viz. *Chanda nama, Mystus vittatus* and *S. gora*. Seven species (*B. albolineatus, C. c. communis, C. gariepinus, D. dangila, M. tengara, R. daniconius* and *S. gora*) were new additions to the Park's list of fish species reported earlier (Kumar & Vijayan 1988). Of these new arrivals, *S. gora* was recorded common for all the three years. Shannon Wiener index of species diversity (H') was highest during 2004 (1.368) followed by 2005 (1.168) and lowest during 2003 (0.934).

The annual variation in the percentage composition of fish species that drifted through the bridge is presented in Table 8. The species composition of fry that entered the park varied drastically during the study (Figure 14.) although numerically the major fish species remained more or less the same. Overall for the whole period of the study, the fish species occurrence and abundance was significantly distinct and independent for all the three years (χ^2, $P < 0.05$). Moreover, the test performed to assess the fact with respect to individual species, showed that except *B. albolineatus, Catla catla, C. gariepinus* and *M. armatus*, all other species were distinct and independent in their occurrence and abundance (χ^2, $P < 0.05$, Table 9.). The recruited species were found to be significantly and positively associated among themselves (W = 103.26, $P < 0.05$). When association among all the recorded fish species was estimated using Jaccard's index (Table 10.), *Puntius sophore*, the most abundant species during 2003 and 2005, showed 100% association with *C. catla, Channa punctatus* and *M. armatus* and 67% association with *C. nama, M. vittatus* and *S. gora*. It showed 50% association with *B. albolineatus, C. ranga, Chela bacaila, Cirrhinus mrigala, Cirrhinus reba, C. gariepinus, Colisa fasciatus, C. c. communis, D. dangila, Esomus danricus, M. tengara, Salmostoma bacaila, Chela cachius, Rasbora daniconius* and *Wallago attu*. In the case of

Labeo rohita, the most dominant species during 2004, highest association was seen with *C. nama*, *M. vittatus* and *S. gora*. It showed 50% association with *B. albolineatus*, *C. fasciatus*, *C. communis* and *E. danricus*.

Table 7. Comparative details of fish sampling protocol

No	Parameters	2003	2004	2005
1	Total annual rainfall (mm) in KNP	849.8	630.0	873.8
2	Total water Park received from Ajan bund during the sampling period (mm^3)	4.876	0.51	13.617
3	Total number of species recorded	12	5	19
4	Number of first record (species)	3	1	3
5	Total number of catch	17	4	52
6	Total duration of catch	12 minutes 45 seconds	3 minutes	39 minutes
7	Total number of fry	1356	905	38505
8	Total biomass (live weight in kg)	1.75	1.31	28.09
9	Total duration of water release (hours)	266	51.5	418.5
10	Estimated No. (fry) that entered the Park (millions)	1.69	0.93	24.79
11	Estimated biomass (fry) that entered the Park (tons)	2.19	1.35	18.08
12	Estimated number (fry) per 100 hours (in millions)	0.635	1.805	5.924
13	Estimated number (fry) per 1 million m^3 (in millions)	0.346	1.823	1.821

Table 8. Order of species based on percentage composition

Year	Order of species with their percentage composition in parenthesis
2003	Pso (75.442) > Cn (11.283) > Lr (5.752) > Cf (3.024) > Sg (2.212) > Cc (1.032) > Cp (0.590) > Ed (0.221) > Ma = Mv (0.148) > Ccl = Ba (0.074)
2004	Lr (31.271) > Sg (29.945) > Psa (19.448) > Cn (19.227) > Mv (0.111)
2005	Pso (53.183) > Psa (26.150) > Cr (16.922) > Sb (1.356) > Cn (0.919) > Cb (0.883) > Mv (0.369) > Dd (0.044) > Cre (0.041) > Cm (0.034) > Sg (0.029) > Rd (0.018) > Cc (0.013) > Wa = Mt (0.010) > Ma (0.008) > Ccl (0.005) > Cp = Cg (0.003)

Ba: *Brachydanio albolineatus*, Cc: *Chela cachius*, Ccl: *Catla catla*, Cg: *Clarias gariepinus*, Cn: *Chanda nama*, Cr: *Chanda ranga*, Cb: *Chela bacaila*, Cm: *Cirrhinus mrigala*, Cp: *Channa punctatus*, Cre: *Cirrhinus reba*, Cf: *Colisa fasciatus*, Cc: *Cyprinus carpio communis*, Dd: *Danio dangila*, Ed: *Esomus danricus*, Lr: *Labeo rohita*, Ma: *Mastacembelus armatus*, Mt: *Mystus tengara*, Mv: *Mystus vittatus*, Psa: *Puntius sarana*, Pso: *Puntius sophore*, Rd: *Rasbora daniconius*, Sb: *Salmostoma bacaila*, Sg: *Securicula gora*, Wa: *Wallago attu*.

Table 9. Chi square values for fish species

No.	Species	Chi square (χ^2)
1	*Brachydanio albolineatus*	2.0*
2	*Catla catla*	2.0*
3	*Chanda nama*	107.5
4	*Chanda ranga*	13032.0
5	*Channa punctatus*	12.7
6	*Chela bacaila*	680.0
7	*Chela cachius*	10.0
8	*Cirrhinus mrigala*	26.0
9	*Cirrhinus reba*	32.0
10	*Clarias gariepinus*	2.0*
11	*Colisa fasciatus*	82.0
12	*Cyprinus carpio communis*	28.0
13	*Danio dangila*	34.0
14	*Esomus danricus*	6.0
15	*Labeo rohita*	355.1
16	*Mastacembelus armatus*	2.8*
17	*Mystus tengara*	8.0
18	*Mystus vittatus*	272.3
19	*Puntius sarana*	19452.1
20	*Puntius sophore*	37156.0
21	*Rasbora daniconius*	14.0
22	*Salmostoma bacaila*	1044.0
23	*Securicola gora*	403.9
24	*Wallago attu*	8.0

* Not significant at $P < 0.05$.

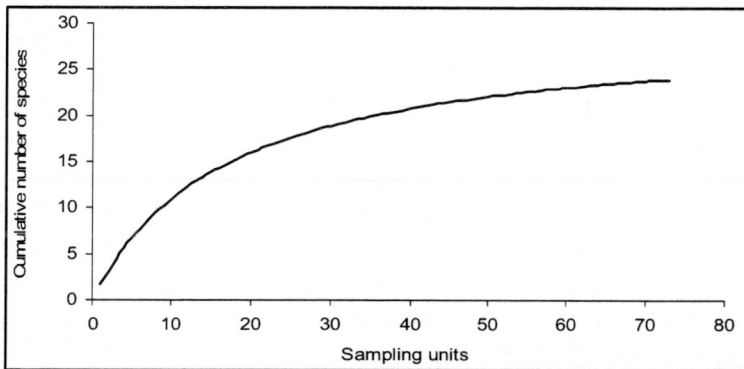

Figure 13. Species recruitment curve for the KNP. A Michaelis-Menton equation $S = S_{max} N / (Km + N)$, where S is the number of species, N the sampling unit, S_{max} the maximum number of species that could be present in the catchment and K_m the Michaelis-Menton constant, was fitted to the data. Estimated parameters for the best fit Michaelis-Menton equation in this case were: $S_{max} = 29.59$, $K_m = 17$ ($r = 0.971$, $P < 0.05$).

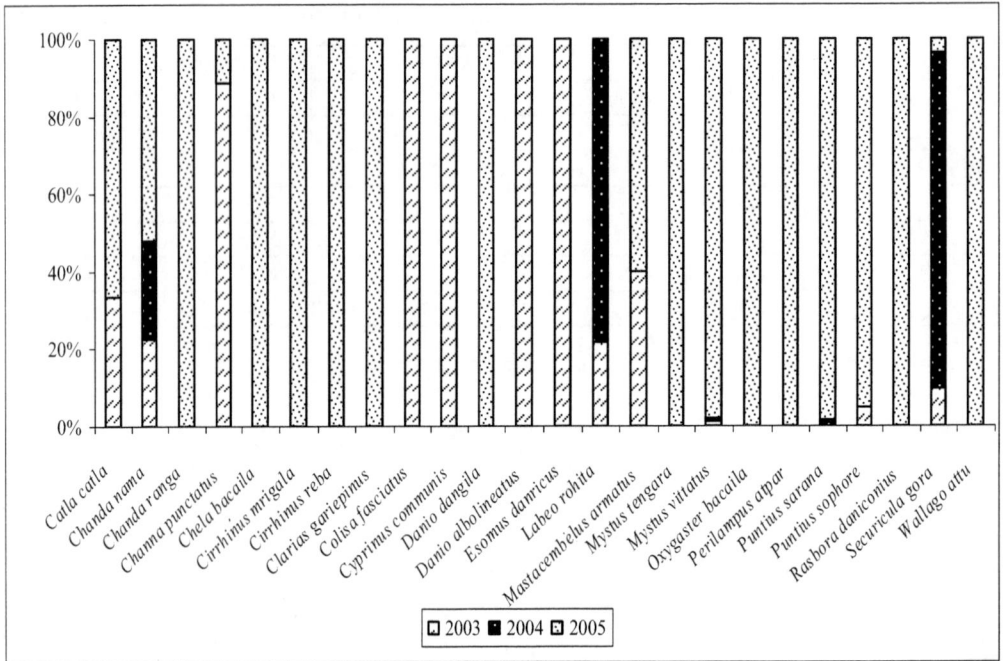

Figure 14. Species composition for the whole study duration.

The abundance of fish species varied with the duration and quantum of water release (Table 7). Water speed (m/sec) and total number of individuals per catch in the canal were negatively correlated indicating more number of individuals coming in towards the end period of water release, i.e. when water speed slows down. The drift density varied drastically among the catches and was in the range of 45.57 to 1.01×10^6 (individuals per 100 m^3 water), while the drift rate ranged from 0.2 to 93.8 (individuals entering per second). The variation in both the cases was significant among the catches (One Way ANOVA, $P < 0.05$).

The correlation matrix (Two-tail) depicting the interrelation of fish species showed that *C. nama* was positively correlated with *L. rohita* and *S. gora*. Similarly, *C. ranga* was positively correlated with *M. vittatus*, *O. bacaila* and *P. sophore* ($P < 0.05$, Table 11). *C. bacaila* and *C. fasciatus* were positively correlated with *P. sophore* and *C. c. communis* respectively. *D. dangila* and *L. rohita* were significantly correlated with *M. vittatus* and *S. gora*. Similarly, *M. armatus* and *M. tengara* were correlated with *P. atpar* and *R. daniconius*. *M. vittatus* and *O. bacaila* were correlated with *O. bacaila* and *P. sophore* (Two-tail, $P < 0.05$). The hierarchical cluster analysis and the resultant dendrogram showed three clusters in total (Figure 15.). *P. sophore* was apparently distinct forming a separate cluster. The second cluster comprised of *C. ranga* and *P. sarana,* while the third one comprised of the rest of the species (21 out of total 24 species).

Table 10. Species association (Jaccard's index X 100) among the recorded fish species

Species*	Ba	Ccl	Cn	Cr	Cp	Cb	Cm	Cre	Cg	Cf	Cc	Dd	Ed	Lr	Ma	Mt	Mv	Sb	Cc	Psa	Pso	Rd	Sg	Wa
Ba	100																							
Ccl	50	100																						
Cn	33	67	100																					
Cr	0	50	33	100																				
Cp	50	100	67	50	100																			
Cb	0	50	33	100	50	100																		
Cm	0	50	33	100	50	100	100																	
Cre	0	50	33	100	50	100	100	100																
Cg	0	50	33	100	50	100	100	100	100															
Cf	100	50	33	0	50	0	0	0	0	100														
Cc	100	50	33	0	50	0	0	0	0	100	100													
Dd	0	50	33	100	50	100	100	100	100	0	0	100												
Ed	100	50	33	0	50	0	0	0	0	100	100	0	100											
Lr	50	33	67	0	33	0	0	0	0	50	50	0	50	100										
Ma	50	100	67	50	100	50	50	50	50	50	50	50	50	33	100									
Mt	0	50	33	100	50	100	100	100	100	0	0	100	0	0	50	100								
Mv	33	67	100	33	67	33	33	33	33	33	33	33	33	67	67	33	100							
Sb	0	50	33	100	50	100	100	100	100	0	0	100	0	0	50	100	33	100						
Cc	0	50	33	100	50	100	100	100	100	0	0	100	0	0	50	100	33	100	100					
Psa	0	33	67	50	33	50	50	50	50	50	50	50	50	33	33	50	67	50	50	100				
Pso	50	100	67	50	100	50	50	50	50	50	50	50	50	33	100	50	67	50	50	33	100			
Rd	0	50	33	100	50	100	100	100	100	0	0	100	0	0	50	100	33	100	100	50	50	100		
Sg	33	67	100	33	67	33	33	33	33	33	33	33	33	67	67	33	100	33	33	67	67	33	100	
Wa	0	50	33	100	50	100	100	100	100	0	0	100	0	0	50	100	33	100	100	50	50	100	33	100

* The expansions of the abbreviations are given in Table 8.

Table 11. Correlation matrix of the recorded fish species

Species*	Cn	Cr	Cb	Cm	Cre	Cf	Cc	Dd	Lr	Ma	Mt	Mv	Sb	Cc	Psa	Pso	Rd	Sg
Cn	1.0																	
Cr	NS	1.0																
Cb	NS	NS	1.0															
Cm	NS	NS	NS	1.0														
Cre	NS	NS	NS	NS	.0													
Cf	NS	NS	NS	NS	NS	1.0												
Cc	NS	NS	NS	NS	NS	.324	.0											
Dd	NS	NS	NS	NS	NS	NS	NS	1.0										
Lr	.328	NS	NS	NS	NS	NS	NS	NS	1.0									
Ma	NS	NS	NS	NS	NS	NS	NS	NS	NS	1.0								
Mt	NS	NS	NS	NS	NS	NS	NS	NS	NS	NS	1.0							
Mv	NS	.253	NS	NS	NS	NS	NS	.475	NS	NS	NS	1.0						
Sb	NS	.657	NS	NS	NS	NS	NS	NS	NS	NS	NS	.314	1.0					
Cc	NS	NS	NS	NS	NS	NS	NS	NS	NS	.482	NS	NS	NS	1.0				
Psa	NS	NS	NS	NS	NS	NS	NS	NS	NS	NS	NS	NS	NS	NS	1.0			
Pso	NS	.754	.497	NS	NS	NS	NS	NS	NS	NS	NS	NS	.368	NS	NS	1.0		
Rd	NS	NS	NS	NS	NS	NS	NS	NS	NS	NS	.433	NS	NS	NS	NS	NS	1.0	
Sg	.390	NS	NS	NS	NS	NS	NS	NS	.861	NS	NS	NS	NS	NS	NS	NS	NS	1.0

73: Sample size, ± .230 critical value .05 (two-tail), NS: Not significant.
* The expansions of the abbreviations are given in Table 8.

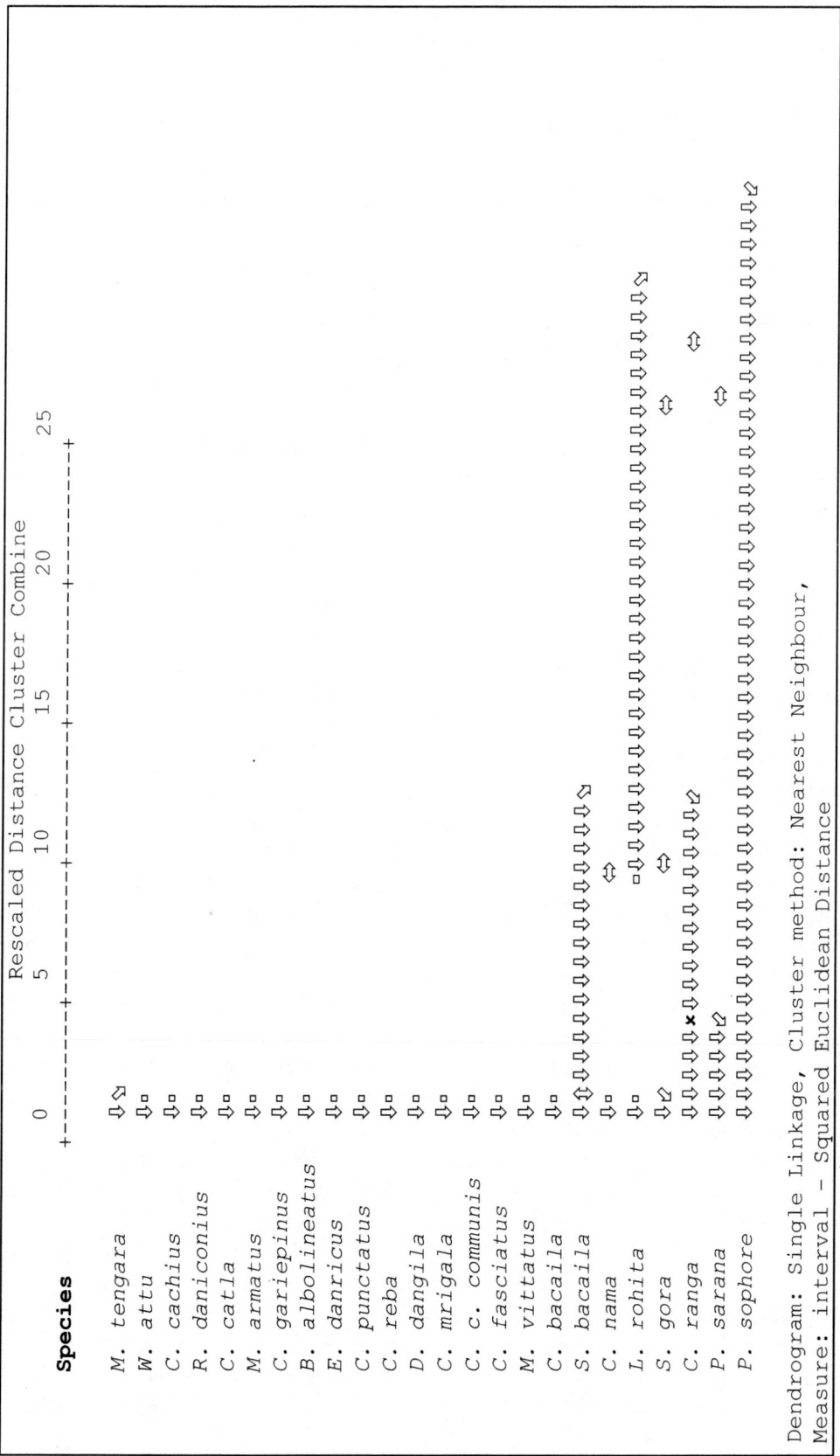

Figure 15. Hierarchical Cluster Analysis with respect to fish abundance.

DISCUSSION

The present investigation, reports a unique arrival pattern of fish on yearly basis. The total number of species recorded during the present investigation was found to be lesser than the earlier findings by Kumar & Vijayan (1988) who had reported 50 species. The species accumulation curve suggesting 30 species to be present in the catchment is notably lesser than the previously recorded number of 50, which imply that some of the earlier recorded species have become locally extinct or rare in the catchment. The complete absence of certain species such as *Notopterus* sp. (Kumar & Asthana 1994), *Puntius ticto*, *Channa gachua*, *Labeo calbasu* and *Ompok* sp. could be associated with several reasons, of which the major one may be primarily the local extinction of the species resulting from the changes in the catchment and ongoing habitat alterations. Kar et al. (2006) observed that alterations in the local limnological and hydrological conditions are likely to disturb the normal sustenance of certain species. The presence of fish in the canal may also be decided by the physico-chemical conditions, prey-predator availability, food availability (Lewin et al. 2004) and movements in water column relating to Ajan bund or its catchment. Changes in the habitat inside the park, alteration or destruction of habitats in the breeding area outside the park, over-exploitation, and displacement or competitive exclusion by the 'invaders' (Kumar & Vijayan 1988) are the possible causes for the reported variations. The two important factors that affect the survival of the fingerlings as well as adults in the summer pools are space availability and intense predation pressure. The studies performed by Miller et al. (1989) and Master (1990) in North America reveals that the aquatic species are more vulnerable to anthropogenic disturbances and become endangered and extinct at a much faster rate than other animals.

The present year wise variation in the occurrence and abundance of fishes is observed to be a function of the quantum of rainfall and amount of water released into the wetland system. The quantum of water released to the park varied strikingly from a low 0.51 million m^3 in 2004 to a high 13.6 million m^3 in 2005 (Table 7). Total number of species and individuals recorded were high during 2005, which could be associated with the high rainfall during the year (Table 7) resulting in wider flood in the catchment of the park that facilitated the movement of fishes from small water holes, pools and other water holding structures into the park. Although, the quantity of water released was highest during 2005, the estimated number of fry per million m^3 of water was highest during 2004. This fact could be associated with the uncommon release of fry in the dam for increasing the fish supply in the park by the park management. In the years with below normal rainfall, such practice is resorted by the management to augment the availability of fishes in the aquatic blocks. Such release also is likely to alter the species composition of the fry entering the park. In addition, water speed, up to a certain extent, decides the abundance of species in the canal, as total number of individuals per catch was negatively correlated with the water speed. The fry movement appears passive (drifting) during the initial period of water release when the speed of water was high and the fry arriving was younger. However, in due course of time, with maturity the fry become capable of moving more actively in the water current and this would have resulted in such negative correlation.

Factors such as the amount of water release, water speed, duration of water release, time of water release, and the growth stage of fry control the composition of fry entering the park. However, source of water and source of fry appears to be factors that are more fundamental.

The rivers Banganga and Gambhir from which the Ajan dam draws water play a critical role in the diversity and abundance of fish fauna of the park, by way of the above two major factors. Banganga, a mountain torrent with a coarse textured bed of sand mixed gravel (Kumar et al. 1995), has several pools and puddles. The river Banganga has less number of fish species probably because of its non-connectivity with the river Yamuna to which it was formerly connected. As it dries up on the way to Bharatpur, the fishes get restricted to the available small water bodies in the catchment. Moreover, intermittent droughts and the available flow being distributed for agricultural, domestic and industrial uses (Sebastian 2005, Varshney 2005) leaves the river dry. The river water not reaching the park causes further depletion of the aquatic fauna of the park including fish (Dehadrai & Poniah 1997, Saunders et al. 2002). The watershed of Gambhir river lies in Karauli Hills of Sawai Madhopur district (Rajasthan), running through Bayana and Rupbas tehsils and some parts of Uttar Pradesh to re-enter Rajakhera tehsil, Bharatpur district (Kumar 1993) before draining into river Yamuna (Figure 10.). Considerable changes have happened along the river course, including changes due to urbanization, agricultural practices and water usage.

During the last three decades, the catchment of KNP and Ajan bund has experienced severe alterations. Irrigation canals, construction of barrages on the rivers, and intensive agriculture and alteration along the course of rivers and channels are some such changes which are likely to affect the ichthyofaunal composition in the area (Johal et al. 1990). In lean months several of the fishes that reach the KNP are known to occupy small ponds and ditches in the catchment. These small water structures face problems of eutrophication and over exploitation and hence might result in the loss of certain species (Dehadrai & Poniah 1997, Kharat et al. 2003). As reported by Welcomme (1994) and Saunders et al. (2002) changes in fish species composition are commonly associated with flow regulation. Schmitz et al. (1991) and Saunders et al. (2002) have reported that introduction of *Eichhornia* and *Hydrilla* to fresh waters has resulted in changes in dissolved oxygen, pH, temperature, turbidity, and composition of native plant and animal communities. They restrict the movement of fish, act as silt traps (Sharma & Praveen 2002), and hence are known to influence fish availability and distributions (Miranda & Hodges 2000). Lewis (2003) was also of the view that growth of certain weed species in the aquatic systems reduces the fish populations. The temperature and light intensity also affects the availability of fish fauna in the canal. For the last 20 years, there has been notable change in environmental as well as climatic conditions (Sebastian 2005). With periods of prolonged drought and flood there has been a significant variation in the habitat pattern and preference of different species at local level.

ACKNOWLEDGEMENTS

The work is part of a multidisciplinary study on the ecology of the KNP undertaken by SACON. We are thankful for the helps to Prof. LL Sharma and Dr. D Devarshi (MSJ College, Bharatpur), Mr. MA Kadri (Rajasthan State Fisheries Department, Bharatpur), Dr. S Panigrahi and Mr. S Mohanty (Jharsuguda College, Sambalpur University, Orissa). The help and assistance by M/s Amlan Datta, M. Sudheer, Randhir Singh, Brijendra Singh and Rajesh Singh during the fieldwork is highly acknowledged. At KNP, we are thankful to Mr. KCA

Arun Prasad, IFS, Director and Mr. Bholu Abrar Khan, forester for their help throughout the study.

REFERENCES

Ali, S. (1953). The Keoladeo Ghana of Bharatpur (Rajasthan). *Journal of Bombay Natural History Society*, 51(3): 531-536.

Anonymous. (1992-1993). Annual report. National Bureau of Fish Genetic Resources, Lucknow, India.

Azeez, P.A. (1991). *Hydrology of Keoladeo National Park*. Seminar on Wetland Ecology and Management. Keoladeo National Park, Bharatpur, India, 1990.

Azeez, P. A., Nadarajan, N. R. & Mittal, D. D (2000). The impact of a monsoonal wetland on ground water chemistry. *Pollution Research*, 19 (2): 249-255.

Azeez, P. A., Nadarajan, N. R. & Prusty, B. A. K. (2007). Macrophyte decomposition and changes in water quality. In: K. K. Singh, A. Juwarkar & A. K. Singh (Eds.), *Environmental Degradation and Protection* (Volume – II, pp. 115-156.). New Delhi, MD Publications.

Azeez, P. A., Ramachandran, N. K. & Vijayan, V. S (1992). The socioeconomics of the villagers around Keoladeo National Park, Bharatpur, Rajasthan. *International Journal of Ecology and Environmental Sciences*, 18: 169-179.

Balasundaram, C., Dheepa, A. & Mariappan, P. (1999). Fish Diversity in grand Anicut, river Cauvery (Tiruchirapalli, Tamil Nadu). *Zoo's Print Journal*, 14 (8): 87-88.

Bhushan, B. & Sharma, R. C. (1987). *Ground water resources and development potential in Bharatpur district*. Central Ground Water Board, Government of India. 99 pp.

Colwell, R. K. (1994-2004). *EstimateS: Statistical estimation of species richness and shared species from samples*. Version 7. Persistent URL <purl.oclc.org/estimates>

Day, F. (1986). *The Fishes of India, being a natural history of fishes known to inhabit the seas and the fresh waters of India, Burma and Ceylon* (Text and plates in two parts, pp. 778, pls. 195). New Delhi, Today & Tomorrow's Book Agency.

Dehadrai, P. V. & Poniah, A. G. (1997). Conserving India's Fish Diversity. *International Journal of Ecology and Environmental Sciences*, 23: 315-326.

Johal, M. S., Chahal, J. S & Tandon, K. K. (1990). Ichthyofauna of Rajasthan State (India). *Journal of Bombay Natural History Society*, 90: 404-411.

Kar, D., Nagarathna, A. V., Ramachandra, T. V. & Dey, S. C. (2006). Fish diversity and conservation aspects in an aquatic ecosystem in northeastern India. *Zoo's Print Journal*, 21 (7): 2308-2315.

Kharat, S., Dahanukar, N., Raut, R. & Mahabaleshwarkar, M. (2003). Long-term changes in fresh water fish species composition in North Western Ghats, Pune district. *Current Science*, 84 (6): 816-820.

Kumar, C. R. A. (1991). *Habitat segregation of fishes in Keoladeo National Park, Bharatpur, Rajasthan*. Ph D thesis, Kanpur University.

Kumar, C. R. A. (1993). *Aplocheilus panchax* (HAM.) - An addition to the fish fauna of Rajasthan. *Journal of Bombay Natural History Society*, 90 (1): 115.

Kumar, C. R. A. & Asthana, A. (1994). Circadian variation in the movement of fry in a feeder canal. *Journal of Bombay Natural History Society*, 91: 194-202.

Kumar, C. R. A. & Mittal, D. D. (1993). Habitat preference of fishes in wetlands in relation to aquatic vegetation and water chemistry. *Journal of Bombay Natural History Society*, 90 (2): 181-192.

Kumar, C. R. A. & Vijayan, V. S. (1988). On the fish fauna of Keoladeo national Park, Bharatpur (Rajasthan). *Journal of Bombay Natural History Society*, 85 (1): 44-49.

Kumar, C. R. A., Ramachandran, N. K. & Asthana, A. (1995). Composition, abundance and distribution of fish in Banaganga-Gambhir river system and source of fish to the Keoladeo National Park, Bharatpur. *Journal of Bombay Natural History Society*, 92 (1): 30-39.

Lewin, W. C., Okun, N. & Mehner, T. (2004). Determinants of the distribution of juvenile fish in the littoral area of a shallow lake. *Freshwater Biology*, 49: 410-424.

Lewis, M. (2003). Cattle and conservation at Bharatpur: A case study in science and advocacy. *Conservation and Society*, 1: 1. New Delhi, SAGE Publications.

Ludwig, J. A. & Reynolds, J. F. (1988). *Statistical Ecology: A primer on methods and computing* (337 pp.). New York, John Wiley and Sons.

Master, L. M. (1990). The imperiled status of North America animals. *Biodiversity Network New*, 3: 7-8.

Mathur, V. B., Sinha, P. R. & Mishra, M. (2005). *Keoladeo National Park World Heritage Site. Technical Report No. 5*, UNESCO-IUCN-Wild Life Institute of India, Dehradun, India.

Miller, R. S., Williams, J. D. & Williams, J. E. (1989). Extinction of North American fishes during the past century. *Fisheries*, 14: 22-38.

Miranda, L. E. & Hodges, K. B. (2000). Role of aquatic vegetation coverage on hypoxia and sunfish abundance in bays of a eutrophic reservoir. *Hydrobiologia*, 427: 51-57.

Mitsch, W. J. & Gosselink, J. G. (2000). *Wetlands* (Third Edition, 920 pp.) New York, John Wiley and Sons.

Naturetrek. (2005). *India – Bharatpur & Chambal*. Naturetrek Tour Dossier 2006. August 05, pp. 06.

Norusis, M. J. (1990). *SPSS/PC+4.0 Base Manual – Statistical Data Analysis*. SPSS Inc.

Orris J. B. (2000). MEGASTAT, version 8.8. http://www.mhhe.com/bstat.

Pal, D. K., Bhattacharyya, T., Deshpande, S. B., Sarma, V. A. K. & Velayutham, M. (2000). *Significance of minerals in soil environment of India (NBSS Review Series 1)*. National Bureau of Soil Science & Land Use Planning, Nagpur, India, pp. 68.

Prusty, B. A. K. & Azeez, P. A. (2007). Alkali and alkaline earth metals in the soil profile of a wetland-terrestrial ecosystem complex in India. *Australian Journal of Soil Research*, 45 (7): 533-542.

Prusty, B. A. K., Azeez, P. A. & Jagadeesh, E. P. (2007a). Alkali and transition metals in macrophytes of a wetland system. *Bulletin of Environmental Contamination and Toxicology*, 78 (5): 405-410.

Prusty, B. A. K., Chandra, R., Azeez, P. A. & Sharma, L. L. (2007b). New additions to the ichthyofauna of Keoladeo National Park, a World Heritage Site in India. *Zoo's Print Journal*, 22 (10): 2848-2852.

Reichard, M., Jurajda, P. & Smith, C. (2004). Spatial distribution of drifting cyprinid fishes in a shallow lowland river. *Archives of Hydrobiology*, 159 (3): 395-407.

Saunders, D. L., Meeuwig, J. J. & Vincent, C. J. (2002). Freshwater Protected Areas: Strategies for Conservation. *Conservation Biology*, 16 (1): 30-41.

Schluter, D. (1984). A variance test for detecting species associations, with some example applications. *Ecology*, 65: 998-1005.

Schmitz, D. C., Schardt, J. D., Lesile, A. J., Dray Jr., F. A., Osborne, J. A. & Nelson, B. V. (1991). The ecological impact and management history of three invasive alien aquatic plant species in Florida. In: B. N. Mcknight (Ed.), *Biological Pollution: the control and impact of invasive exotic species* (pp. 173-194.). Indianpolis, Indiana Academy of Science.

Sebastian, S. (2005). Keoladeo Ghana: Last gasp for survival. *The Hindu Survey of The Environment 2005*, pp. 55-59.

Sharma, S. & Praveen, B. (2002). *Management Plan, Keoladeo National Park, Bharatpur*, 142 pp..

Talwar, P. K. & Jhingran, A. G. K. (1991a). *Inland fishes of India and adjacent countries* (Volume 1, pp. 01-541). New Delhi, Oxford and IBH Publishing Co. PVT Ltd.

Talwar, P. K. & Jhingran, A. G. K. (1991b). *Inland fishes of India and adjacent countries* (Volume 1, pp. 542-1158). New Delhi, Oxford and IBH Publishing Co. PVT Ltd.

Varshney, V. (2005). Gripe Water: Bird-brained politics mars Bharatpur wetlands. *Down To Earth*. March 31, 7-8.

Vijayan, V. S. (1991). *Keoladeo National Park Ecology Study (1980 – 1990), final report.* Bombay Natural History Society, India, 337 pp.

Welcomme, R. L. (1994). The status of large river habitats. In: I. G. Cowx (Ed.), *Rehabilitation of freshwater fisheries* (pp. 11-20). Oxford, United Kingdom, Fishing New Books.

Zoo Outreach Organization / CBSG India (1998). *Report of the workshop "Conservation Assessment and Management Plan of Freshwater fishes of India"*, 156 pp.

Chapter 5

METAL AND NUTRIENT REMOVAL IN A CONSTRUCTED WETLAND FOR INDUSTRIAL WASTEWATER TREATMENT IN ARGENTINA

M. A. Maine[1,2,], H. R. Hadad[1,2], N .L. Suñé[1], G. Sanchez[1], S. Caffaratti[1] and C. Bonetto[3]*

[1]Química Analítica, Facultad de Ingeniería Química, Universidad Nacional del Litoral.
Santiago del Estero 2829 (3000) Santa Fe, Argentina
[2]Consejo Nacional de Investigaciones Científicas y Técnicas
[3]Instituto de Limnología "Dr. Ringuelet", Av. Calchaquí Km. 23.5, (1888) Florencio
Varela, Buenos Aires, Argentina. Consejo Nacional de Investigaciones
Científicas y Técnicas

ABSTRACT

This chapter summarizes the nutrient and metal removal of a free water surface constructed wetland and compares it with a previous small-scale prototype. The wetland was built to treat wastewater containing metals (Cr, Ni, Zn) and nutrients (P and N) from a tool factory in Santo Tomé, Santa Fe, Argentina. Water, sediment and macrophytes were sampled at the inlet and outlet areas of the constructed wetland during three years. Both wetlands showed high efficiencies in contaminant removal. The development of the vegetation showed the same pattern in both the small-scale prototype and the large-scale wetland. Three successive phases of vegetation dominance were developed and three different patterns of contaminant retention were observed. A first phase of floating macrophyte colonization with dominance of *Eichhornia crassipes* was followed by a combined floating-emergent phase ending in an long, on-going stand of emergent macrophyte phase, with dominance of *Typha domingensis*. During the *E. crassipes* dominance, contaminants were retained in the macrophyte biomass; during the *E. crassipes + Typha domingensis* stage, contaminants were retained in the sediment and in

* Corresponding author: Química Analítica, Facultad de Ingeniería Química, Universidad Nacional del Litoral. Santiago del Estero 2829 (3000) Santa Fe, Argentina. Tel.: 54-0342-4571164 Int. 2515. E-mail: amaine@fiq.unl.edu.ar

the *T. domingensis* dominance stage, contaminants were retained in sediment and in the macrophyte biomass. Even though retention mechanisms were different, removal efficiencies did not show significant differences among the three vegetation stages, except for NH_4^+ and soluble reactive phosphorus (SRP). Greenhouse experiments were carried out to explain the early disappearance of floating macrophytes. Thresholds for conductivity, pH and metal damage were determined. High conductivity and pH of the incoming wastewater were the cause of the disappearance of the floating species. Because of its highest tolerance, *T. domingensis* is the best choice to treat wastewater of high pH and conductivity with heavy metals, a common result from many industrial processes.

Keywords: constructed wetland; sediment; macrophytes; wastewater; Argentina

1. INTRODUCTION

Constructed wetlands (CWs) are mainly used for nutrient and organic matter retention in domestic and municipal sewage, storm water and agricultural runoff [Hammer, 1989; Moshiri, 1993; Kadlec and Knight, 1996; Vymazal et al., 1998; Kadlec et al., 2000]. The application of CW for industrial wastewater treatment is a promising alternative in Argentina, since there is a large availability of marginal land around most cities together with a low population density. The central and northern areas of the country have mild winters, allowing extended growing periods for plants.

The choice of plants is an important issue in CWs, as they must survive the potentially toxic effects of the wastewater and its variability. Common reeds (*Phragmites australis* (Cav.) Trin.), cattails (*Typha* spp.), bulrushes (*Scirpus* spp.) and reed canary grass (*Phalaris arundinacea* L.) have been used for both domestic and industrial wastewater treatment [Shepherd et al., 2001; Mbuligwe, 2005; Vymazal, 2005; Vymazal and Krópfelová, 2005; Maine et al., 2007a]. Although regionally abundant macrophyte species are adapted to the local climatic and edaphic conditions, their performance under the environmental conditions imposed by wastewater was unknown. Bahco metallurgic industry constructed a small-scale experimental wetland to assess the feasibility of treating wastewater at its tool factory in Santo Tomé, Santa Fe (Argentina). Metals, biological and chemical oxygen demand and nutrients (except SRP and NH_4^+) largely decreased in the wetland effluent [Maine et al., 2005]. The success in improving wastewater quality led this industrial plant to develop a large scale CW to treat the wastewater of the whole factory. The CW has been in operation since 2002. The wastewater received a primary treatment (precipitation, sieving and decantation); however it contained Cr, Ni and Zn and showed high pH and conductivity. An assemblage of locally common macrophytes was transplanted to the studied wetland. Plant growth showed three different periods, the first one dominated by *Eichhornia crassipes* (Mart.) Solms., followed by a decline of *E. crassipes* and increased *Typha domingensis* Pers. cover period, and finally an on-going *T. domingensis* dominance period [Maine et al., 2007a]. This chapter summarizes the removal efficiency of the previous small-scale prototype and the large scale CW, and compares it during the different macrophyte dominance stages.

2. DESIGN OF THE SMALL AND LARGE-SCALE CONSTRUCTED WETLANDS

The wetlands were constructed at Bahco Argentina S. A. metallurgic plant, located in Santo Tomé, Argentina (S 31° 40' 01,9"; W 60° 47' 06,9").

The small-scale experimental wetland was 6 m length, 3 m wide and 0.4 m deep (Figure 1a). A polyethylene impermeable film was placed at the bottom and a soil layer of 30 cm was added. The influent entered the wetland through a PVC tube (diameter: 63-mm) with a perpendicular drip dispersion tube with aligned holes to produce a laminar flow. The inflow discharge was 1000 l d^{-1} and the approximate hydraulic residence time was 7 days. Three floating (*Pistia stratiotes* L., *E. crassipes* and *Salvinia rotundifolia* L.) and eight emergent (*Cyperus alternifolius* L., *Panicum elephantipes* Ness. ex Trin., *Thalia geniculata* L., *Polygonum punctatum* Elliott., *Pontederia cordata* L., *Pontederia rotundifolia* L. f., *T. domingensis* and *Aechmea distichantia* L.) macrophytes were transplanted. The experimental wetland was studied during 12 months.

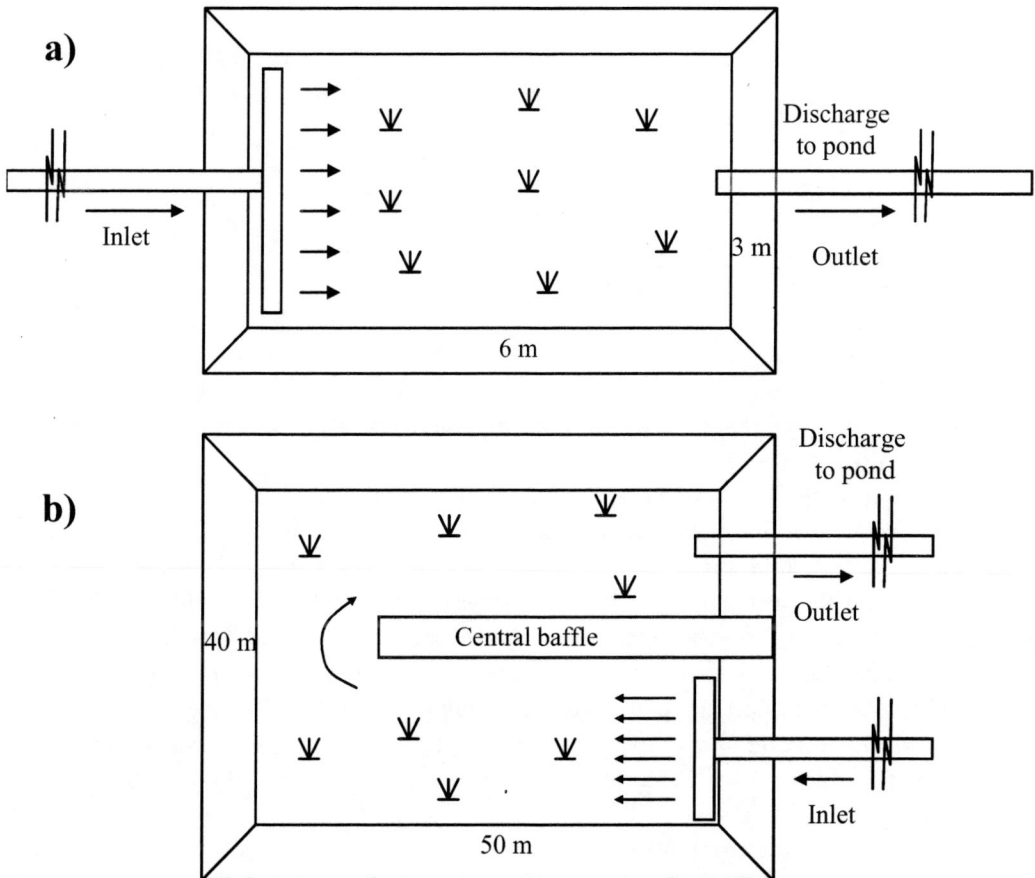

Figure 1. Scheme of the small-scale wetland (a) and the large-scale wetland (b).

The definitive, large-scale free water surface wetland was 50 m length by 40 m wide and 0.5-0.8 m deep, with a central baffle (Figure 1b). The baffle doubled the flow-path and resulted in a 5:1 length-wide ratio. The wetland received wastewater through a PVC pipe provided with a perpendicular distribution pipe with holes at regular distances in order to allow uniform distribution of flow. Mean wastewater discharge was 100 $m^3 d^{-1}$ and the hydraulic residence time ranged from 7 to 12 days. The wetland was rendered impermeable by means of a bentonite layer (5 compacted layers of bentonite - approximate total depth: 0.6 m). A layer of 1 m of soil was placed on top of the bentonite layer. Several locally available macrophytes were transplanted into the wetland, being *E. crassipes*, *T. domingensis* and *P. cordata* those of the larger initial cover. The effluent, after passing through the wetland, is led by an excavated channel into a 1.5 ha pond (3-5 m depth). Both wastewater from the industrial processes and sewage from the factory were treated together. The idea of treating sewage was based on the fact that sewage composition is rich in organic matter and nutrients. High nutrient concentrations could improve macrophyte tolerance to metals. This hypothesis was confirmed by a greenhouse experiment. Wastewater reached the wetland after a primary treatment. During the first stage of the wetland operation (from October 2002 to February 2003) only diluted sewage of the factory was treated (the composition of the influent was 25 $m^3 d^{-1}$ of sewage + 75 $m^3 d^{-1}$ of pond water). Sewage consisted of wastewater from a staff of 750 employees of the factory. Later (March 2003), industrial wastewater and sewage were treated together (25 $m^3 d^{-1}$ of sewage + 75 $m^3 d^{-1}$ of industrial wastewater).

3. WETLAND STUDIES

3.1. Experimental Wetland Study

Table 1 summarizes the variables measured at the inlet and outlet, and the removal percentages corresponding to the small-scale wetland prototype. Concentrations were significantly lower ($p<0.05$) at the outlet in all measured variables except for HCO_3^-, CO_3^{2-} and NH_4 [Maine et al., 2005].

Dissolved oxygen (DO) concentration showed large variations in the influent (mean 6.2 mg l^{-1}) and was generally lower at the effluent (mean 3.2 mg l^{-1}). In two occasions the inlet water was anoxic, while the outlet attained 2-4 mg l^{-1}. Therefore, there was oxygen production within the wetland. The outlet water was anoxic only once. Both, biochemical oxygen demand (BOD) and chemical oxygen demand (COD) were an order of magnitude lower at the outlet than at the inlet water. Since most often the DO concentrations were lower in the outlet, biologically produced DO was likely consumed by the COD and BOD. The pH at the inlet water was strongly alkaline (8.0-12.5) and decreased through the wetland resulting in a pH that ranged from 7.0 to 9.0 at the outlet. The decrease in pH was likely caused by the mineralization of organic matter as shown by the observed decrease in the COD and BOD. Ca^{2+} was retained at a rate of 40%. HCO_3^- concentration was the only single measured variable that was larger in the outlet, probably due to pH changes.

Table 1. Measured variables and removal percentages in the small-scale constructed wetland (n=16)

Parameter	Inlet			Outlet			% Removal		
	Mean	Max.	Min.	Mean	Max.	Min.	Mean	Max	Min
Temperature (^{o}C)	18.4	26.0	11.0	18.4	26.0	11.0	---	---	---
Conductivity (μS cm^{-1})	5130	8500	3300	3120	5400	750	35.7	62.1	3.57
DO(mg l^{-1})	6.18	10.6	0.0	3.15	5.80	0.0	---	---	---
pH	10.1	12.5	8.03	7.98	9.03	7.03	19.5	39.2	4.40
Solids (mg l^{-1})	4180	8050	2350	2690	4260	610	33.0	92.4	1.23
HCO_3^- (mg l^{-1})	218	615	0.0	399	713	203	---	---	---
CO_3^{2-} (mg l^{-1})	97.1	403	0.0	10.5	48.4	0.0	91.3	100	72.2
Cl^- (mg l^{-1})	752	1840	387	500	875	70	31.6	86.4	4.54
SO_4^{2-} (mg l^{-1})	1440	2320	298	929	1620	204	35.2	72.6	0.0
Ca^{2+} (mg l^{-1})	171	292	43.2	97	238	22.3	40.5	79.3	0.982
Mg^{2+} (mg l^{-1})	25.2	95.9	0.0	11.1	19.3	4.5	38.6	85.5	0.0
Na^+ (mg l^{-1})	1070	1710	544	781	1280	180	26.6	66.9	0.0
K^+ (mg l^{-1})	20.6	24.7	12.8	18.0	23.0	11.6	12.9	35.8	0.0
SRP (mg l^{-1})	0.549	2.56	0.026	0.141	0.572	0.010	75.0	98.4	0.0
TP (mg l^{-1})	0.972	4.66	0.082	0.228	0.689	0.025	73.6	97.8	0.0
$N-NO_3^-$ (mg l^{-1})	20.3	86.1	4.93	1.52	3.98	0.214	88.1	97.6	54.1
$N-NO_2^-$ (mg l^{-1})	2.25	12.0	0.027	0.150	0.890	0.001	84.9	99.8	53.0
$N-NH_4^+$ (mg l^{-1})	2.21	4.86	0.320	2.00	9.55	0.006	36.3	99.4	0.0
Inorg-N (mg l^{-1})	24.8	95.1	9.54	3.91	14.4	0.310	79.1	96.8	24.8
Fe (mg l^{-1})	9.09	32.0	0.053	0.212	0.7	0.048	81.7	99.8	9.10
Cr (mg l^{-1})	0.127	0.589	0.005	0.013	0.105	0.001	81.1	99.8	23.5
Ni (mg l^{-1})	0.181	0.750	0.003	0.051	0.190	0.004	66.1	96.2	18.2
Zn (mg l^{-1})	0.060	0.210	0.04	0.040	0.069	0.010	59.0	81.0	0.0
COD (mg l^{-1})	276	583	57	40	85	22	78.4	94.7	7.18
BOD (mg l^{-1})	136	400	17	15	38	2	78.2	98.8	52.9

NO_3^- and NO_2^- were depleted at the outlet attaining a mean removal of 88 and 85%, respectively (Tables 1 and 2). SO_4^{2-} decreased to a smaller extent (35%) and NH_4^+ sometimes decreased and sometimes increased at the outlet without a net change over the sampling period. Since the amounts stored in the macrophyte biomass and in the sediment pool represented a small proportion of the removed NO_3^- and NO_2^-, intensive denitrification must have occurred. Minzoni et al. (1988) measured large N losses through denitrification in enclosures installed in rice fields, and Golterman et al. (1988) confirmed the results by mass balances performed in experimental plots. Reddy et al. (1989) showed large denitrification rates in the rhizosphere of emergent macrophytes of deltaic wetlands. D'angelo and Reddy (1993) determined that most of the ^{15}N-nitrate (roughly 90%) applied to sediment-water cores was lost by denitrification. Precise N[15] balances performed by Matheson et al. (2002) in wetland microcosms showed that 61% of the added NO_3^- was lost through denitrification, 25% immobilized in the soil and 14% assimilated by plants. Our results suggest that diffusion from the water into the anoxic sediment followed by denitrification was the main N pathway in the wetland. Diffusion from water followed by reduction within sediment would also have accounted for the decreased in SO_4^{2-} concentration at the oultlet. SO_4^{2-} can be either retained as FeS or released as H_2S to the atmosphere. Strong smell denoted sulfide presence only in the period in which DO was exhausted in the outlet water. Golterman (1995) reported that adding FeS to sediment increased denitrification rates considerably, proposing the following reaction:

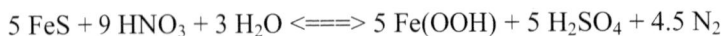

$$5 \text{ FeS} + 9 \text{ HNO}_3 + 3 \text{ H}_2\text{O} \Longleftrightarrow 5 \text{ Fe(OOH)} + 5 \text{ H}_2\text{SO}_4 + 4.5 \text{ N}_2$$

Thus, the high concentration of Fe and SO_4^{2-} in the inlet water would have increased the N-removal by simultaneously enhancing deniftrification and promoting adsorption of P onto FeOOH.

Maine et al. (2005) reported that both total phosphorus (TP) and soluble reactive phosphorus (SRP) showed the same trend (Table 2), attaining a 75 and 74% average retention, respectively. In three successive samplings during Sept. 2001, TP and SRP were higher at the outlet than at the inlet water ($p<0.05$). DO at the outlet water was not exhausted. In the sample in which the outlet water was anoxic, both TP and SRP concentrations were reduced.

Table 2. Nutrient and metal concentrations (in mg l^{-1}) in the inlet and outlet of the small-scale experimental wetland along the studied period

Date	TP		SRP		N-NO$_2^-$		N-NO$_3^-$		Ni		Cr	
	Inlet	Outlet	Inlet	Outlet	Inlet	Outlet	Inlet	Outlet	Inlet	Outlet	Inlet	Outlet
(2001) Jul. 26	4.66	0.610	2.56	0.370	2.67	0.089	13.7	1.0	0.359	0.016	0.478	0.005
Aug. 1	3.42	0.340	1.80	0.120	7.03	0.089	86.1	1.24	0.043	0.021	0.005	0.002
Aug. 21	1.80	0.040	0.530	0.040	1.27	0.045	11.6	1.29	0.026	0.010	0.011	0.006
Aug. 28	0.720	0.220	0.640	0.040	3.72	0.026	11.1	1.50	0.003	0.001	0.009	0.002
Sept. 3	0.833	0.055	0.056	0.026	0.027	0.487	12.2	2.12	0.003	0.001	0.028	0.002
Sept. 11	0.271	0.689	0.110	0.572	12.0	0.031	24.3	2.10	0.149	0.006	0.016	0.012
Sept. 19	0.082	0.283	0.043	0.155	1.22	0.310	4.93	2.03	0.750	0.190	0.069	0.020
Sept. 27	0.183	0.312	0.164	0.249	1.90	0.890	35.7	3.98	0.470	0.150	0.458	0.013
Oct. 5	0.489	0.040	0.374	0.040	0.30	0.003	6.43	1.55	0.150	0.120	0.356	0.105
Oct. 11	0.892	0.040	0.765	0.040	0.788	0.023	24.2	1.17	0.170	0.130	0.107	0.013
Oct. 16	0.890	0.040	0.808	0.040	2.21	0.329	13.1	1.19	0.110	0.090	0.167	0.017
Oct. 23	0.699	0.025	0.608	0.010	0.179	0.044	35.2	1.14	0.079	0.017	0.026	0.003
Nov. 1	0.218	0.167	0.124	0.116	1.46	0.005	10.9	2.12	0.419	0.027	0.013	0.002
Dec. 1	0.218	0.137	0.124	0.087	1.05	0.002	13.2	0.691	0.042	0.018	0.013	0.001
(2002) Feb. 22	0.255	0.133	0.119	0.026	0.058	0.009	13.6	0.690	0.063	0.004	0.589	0.001
Sept. 2	0.258	0.178	0.186	0.095	0.121	0.001	8.76	0.214	0.060	0.005	0.150	0.004

Table 3. Concentrations of sediment of the small-scale experimental wetland

Parameter	Initial	Final	
		Inlet	Outlet
Cr (μg g^{-1})	6.3	239	96
Ni (μg g^{-1})	5.2	149	65
Zn (μg g^{-1})	30.3	189	75
Ca (mg g^{-1})	2.37	3.14	2.24
Fe (mg g^{-1})	18.3	24.5	17.4
Organic matter (%)	7.5	8.9	7.8
N (mg g^{-1})	0.33	0.37	0.32
TP (mg g^{-1})	0.45	0.78	0.38
Fe(OOH)-P (mg g^{-1})	0.127	0.296	0.121
CaCO$_3$-P (mg g^{-1})	0.082	0.147	0.087
Org-P (mg g^{-1})	0.245	0.341	0.171

The concentration of organic matter, nutrients and metal concentrations in bottom sediment increased at the inlet through the studied period (Table 3). Organic matter and N-Kjeldhal increased by 19 and 12%, respectively; Ca and Fe by 32 and 34%, respectively, while TP roughly doubled the initial concentration. All P-fractions increased, especially Fe(OOH)$\approx$P (2.3 times), followed by CaCO$_3$$\approx$P (1.8 times) and org-P (1.4 times). Since both Fe and Ca were also retained within the wetland (82 and 40%, respectively) it seems likely that P might have coprecipitated with them. Phosphate adsorption to iron colloids decreases with increasing pH, while adsorption to calcareous mater shows the opposite trend [Golterman, 1995]. Golterman (1998) developed a phase diagram to determine SRP adsorption at different pH, Ca and Fe concentrations. The high pH, Ca and, in most samplings, CO$_3^{-2}$ concentrations at the inlet water, suggest that P would co-precipitate together with CaCO$_3$ at the inlet of the wetland. However, Fe(OOH)$\approx$P was the fraction that increased to a larger extent within the sediment. Mineralization of organic matter maintained the sediment at a lower pH range than the high values prevailing at inlet wastewater. CO$_3^{-2}$ would undergo partial dissolution and the released SRP be readsorbed onto the Fe(OOH)$\approx$P fraction. Golterman (1995) showed that an increase in Ca, Na and Mg concentrations increased the adsorption of SRP to the Fe(OOH)$\approx$P fraction. The high ionic concentration prevailing at the inlet wastewater therefore contributed to the enhanced SRP retention in this fraction. Similarly, Golterman and Booman (1987) observed that Fe(OOH)$\approx$P was the dominant fraction in wetlands from Camargue (France) in spite of containing large amounts of calcareous material. It seems therefore that CaCO$_3$$\approx$P represents the main precipitating mechanism while, through the equilibrium with interstitial SRP and adsorption onto Fe(OOH), an equilibrium will be established with Fe(OOH)$\approx$P within the sediment. No significant differences ($p<0.05$) were observed between initial and final sediment concentrations at the outlet. When the capacity of a wetland to retain P depends solely on the fixation capacity of the parental material, this fixation capacity will eventually be attained and the wetland will have a limited time life. Once the sediment becomes saturated, P retention will be strongly decreased. However, since the environment for P retention (high pH and Fe, Ca and ionic concentrations) is largely provided by the inlet wastewater it will be expected that the small-scale experimental wetland would continue to retain P as far as the composition of the inlet solution is maintained.

After a year of wetland operation, Cr and Ni concentrations at the inlet sediment were two orders of magnitude higher, and Zn concentration was 6 times higher, than the initial sediment concentration (Table 3). At the outlet, final sediment concentrations were lower than at the inlet, but larger than the initial ones ($p<0.05$). In spite of the high pH and alkalinity of the inlet wastewater, the sediment pH remained at a comparatively low pH range (7.4-7.6) throughout the sampling period. Metals concentrations in water were significantly lower in the outlet with respect to the inlet ($p<0.05$) (Table 2). The largest average retention was attained by Fe (82%), followed by Cr (81%), Ni (66%) and lastly Zn (59%). Zn concentrations in the inlet water were the lowest, and were often below the analytical method detection limits in both the inlet and outlet water. High metal retention is a common feature and has repeatedly been reported in constructed wetlands; metals are mainly stored in the sediment compartment while accumulation in the macrophyte biomass is comparatively small [Mays and Edwards, 2001; Dombeck et al., 1998]. Dombeck et al. (1998) considered that settling was the main retention mechanism and reported Ni as being retained at lower rate due to the formation of strong soluble complexes with organic matter that led to a poor sedimentation. Guo et al. (1997a) studied the Ni speciation in wetland soil and water slurries at different redox potentials. At high redox potentials Ni was mainly adsorb to Fe and Mn oxides while remaining in solution as much as 15% of the total amount. As redox potential decreased (-140 mV) most of the Ni was bound to carbonates (50% of the total) and soluble Ni was strongly depleted. At lower redox potentials (-170 mV) Ni was bound to sulfides and large molecular humic compounds represented the second largest important fraction attaining 30% of the total Ni pool. High alkalinity and pH conditions prevailing in our studied wetland enhanced Ni association with carbonates and increased retention.

Cr^{6+} is easily reduced to Cr^{3+} by organic matter [Bodeck et al., 1988; Losi et al., 1994; Fendorf, 1995]. The high organic matter and low DO concentration attained in the wetland explains that Cr^{6+} was not detected throughout the study period. The low solubility of Cr^{3+} makes it precipitates as $Cr(OH)_3$ [Guo et al., 1997b]. Cr^{6+} has also been reported as being adsorbed to Fe colloids under oxic conditions in slurries while Cr^{3+} was bound to insoluble humic substances of large molecular weight at low redox conditions [Guo et al., 1997b]. In an attempt to simulate the Cr behavior during flooding of a wetland soil, Masscheleyn et al. (1992) made a stepwise reduction of water and soil slurries and observed Cr^{6+} reduction to Cr^{3+} to proceed in the same redox potentials in which NO_3^- disappeared, being completely exhausted before NO_3^- was depleted. The high Cr retention in our experimental wetland was consistent with the high NO_3^- removal as a result of the steep redox gradients in the superficial sediment boundary.

Guo et al. (1997b) reported that Zn bound to Fe and Mn oxides was replaced by Zn associated with CO_3^{2-} when the redox potential decreased, while Zn associated with insoluble large humic substances and sulfides became the main fraction when redox was further reduced to about -170 mv. Nuttall and Younger (2000) studied the process of Zn removal in alkaline wastewater from abandoned Zn and Pb mines and suggest that the observed precipitation corresponded to smithsonite ($ZnCO_3$) in the pH range of 7.8-8.5, and to amorphous Zn oxides (ZnO) at higher pH levels such as those presently reported. As surface sediment progressively becomes buried by settling matter they reach lower redox potentials. Then, Zn associated with sulfides seems the likely final fate of Zn within the sediment. Our results are consistent with the literature showing that Zn is effectively retained in constructed wetlands. Cheng et al. (2002) showed extremely efficient Zn removal from inlet water containing high Zn concentrations (4 mg l^{-1}) in *C. alternifolius* constructed cells. After 114 days, 5% of the inlet Zn was present in the plant biomass, similar to our present results, but attained a higher Zn concentration in plant tissue and sediment than in the present work.

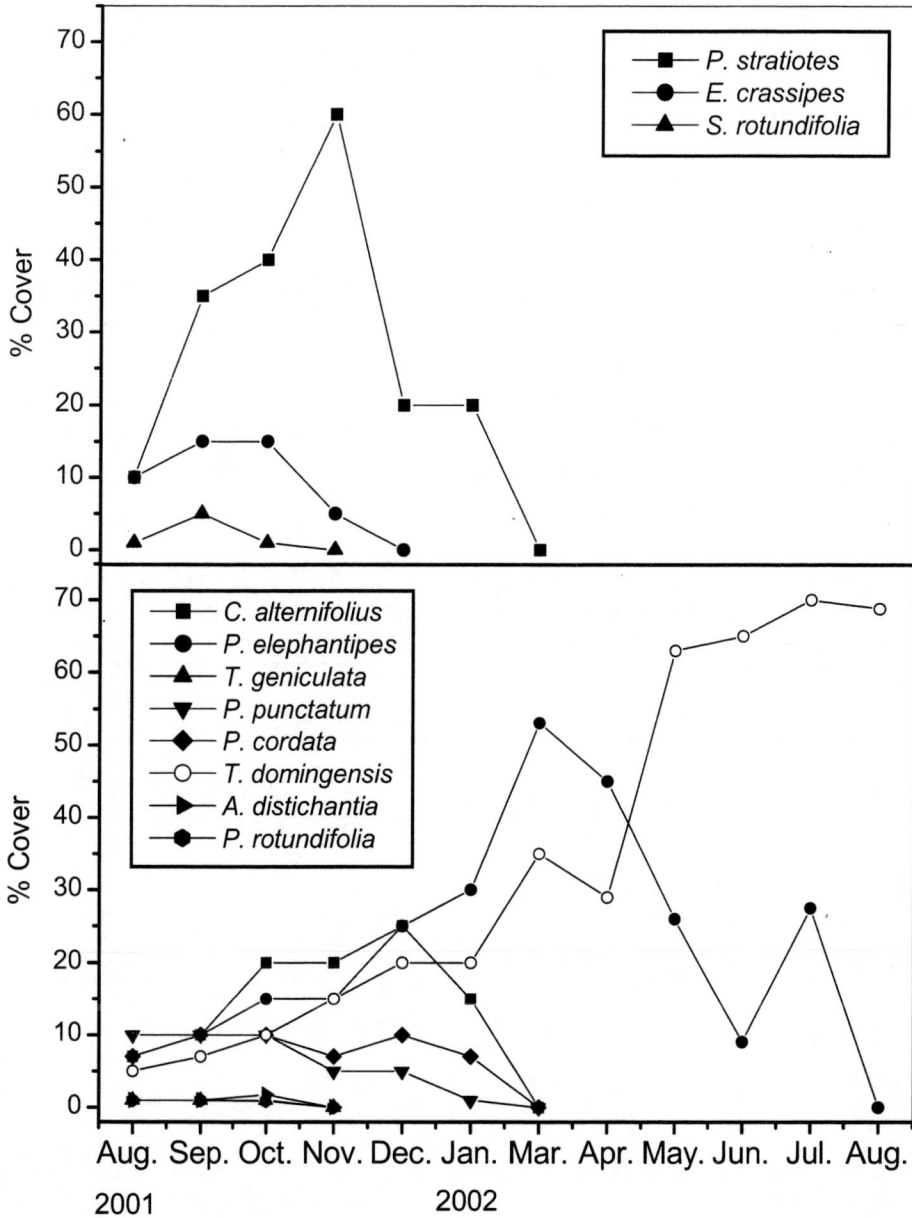

Figure 2. Macrophyte cover in the experimental wetland throughout the study period.

Figure 2 shows the growth of the macrophyte species throughout the sampling period. Some of the transplanted macrophytes (*T. geniculata*, *P. rotundifolia* and *A. distichantia*) did not grow at all and disappeared. Floating macrophytes showed some initial growth, noticeably *P. stratiotes*, attaining 60% cover in November 2001, but later decreased until disappearance. *S. rotundifolia* disappeared in November 2001, *E. crassipes* in December 2001 and *P. stratiotes* in March 2002, together with the emergent *P. punctatum* and *P. cordata*. Only two emergent macrophytes, *P. elephantipes* and *T. domingensis*, remained in the wetland on March 2002. The former attained its maximum cover in March 2002, and decreased since then, until it disappeared in August 2002. *T. domingensis* increased its cover and became the only single species at the end of the sampling period. The biomass attained by *P. elephantipes* at its maximum cover resembles the biomass attained in natural wetlands close to the study site. Similarly, the biomass attained by *T. domingensis* at the end of the study period was similar to that naturally attained in nearby undisturbed environments [Hadad et al., 2006].

The floating macrophytes and the emergent *P. cordata* and *C. alternifolius* increased their P concentrations in plant tissue along time but decreased in the last sample before disappearing (Table 4). *P. elephantipes* attained maximum P concentrations in December 2001, earlier than attaining its maximum cover though it later decreased. *T. domingensis* attained its maximum P concentration in November 2001 and latter decreased to remain relatively uniform within a range of 1.2-1.4 $\mu g\ g^{-1}$ from March till September 2001 when it attained its maximum biomass.

Table 4. P concentration in plant tissue ($\mu g\ g^{-1}$) in the small-scale experimental wetland (*a* stands for above and *b* for below-ground biomass)

Species	2001				2002							
	Jul.	Aug.	Sept.	Oct.	Nov.	Feb.	Mar.	Apr.	May	Jun.	Jul.	Sept.
E. crassipes (a)	1.02	1.36	2.13	1.85								
E. crassipes (b)	0.98	1.48	2.09	2.02								
S. herzogii (a)	1.23	2.18	4.34	6.34								
S. herzogii (b)	1.34	2.37	3.99	2.75								
P. stratiotes (a)	0.96	1.44	2.87	2.83	1.27							
P. stratiotes (b)	1.15	1.55	2.36	2.37	1.44							
C. alternifolius (a)	0.49	0.80	1.5	1.94	0.95							
C. alternifolius (b)	1.12				0.92							
P. cordata (a)	1.18	1.64	2.03	3.25	1.01							
P. cordata (b)	1.87				0.91							
P. elephantipes (a)	1.51				1.05	1.26	0.77	0.76	0.44	0.86	0.98	0.56
P. elephantipes (b)	1.23				0.96		0.86	0.85	0.82	1.18	0.57	1.25
T. domingensis (a)	0.70	0.85	1.6	2.02	0.86	1.87	1.32	1.20	1.27	1.25	1.44	1.46
T. domingensis (b)	1.24				0.99		1.39	1.34	1.47	1.30	1.38	2.04

Table 5. Cr concentration in plant tissue (µg g⁻¹) in the small-scale constructed wetland. (*a* stands for aboveground and *b* for below-ground biomass)

Species	2001					2002	
	Jul.	Aug.	Sept.	Oct.	Nov.	Feb.	Sept.
E. crassipes (a)	0.2	0.7	12.4	27.1			
E. crassipes (b)	0.8	1.1	20.4	101.6			
S. herzogii (a)	0.1	1.0	7.4	21.9			
S. herzogii (b)	0.2	2.0	20.7	57.9			
P. stratiotes (a)	0.1	1.8	6.3	24.1	21.4		
P. stratiotes (b)	0.9	1.5	11.9	55.1	39.2		
C. alternifolius (a)	0.1	0.8	15.8	11.3	28.0		
C. alternifolius (a)	0.2				26.5		
P. cordata (a)	1.0	1.6	5.7	15.0	20.2		
P. cordata (b)	1.2				20.9		
P. elephantipes (a)	8.8				10.8	24.4	128.8
P. elephantipes (b)	11.2				55.6		318.1
T. domingensis (a)	0.1	0.4	4.8	11.1	19.8	18.3	35.8
T. domingensis (b)	0.2				24.0		109.5

Table 6. Ni concentration in plant tissue (µg g⁻¹) in the small-scale constructed wetland (*a* stands for aboveground and *b* for belowground biomass)

Species	2001					2002	
	Jul.	Aug.	Sept.	Oct.	Nov.	Feb.	Sept.
E. crassipes (a)	1.2	2.6	6.6	93.2			
E. crassipes (b)	9.2	19.7	18.0	130.2			
S. herzogii (a)	1.2	4.5	4.8	50.0			
S. herzogii (b)	6.5	18.9	20.3	119.9			
P. stratiotes (a)	1.9	3.6	4.7	5.5	53.7		
P. stratiotes (b)	10.1	18.2	16.4	167.4	212.0		
C. alternifolius (a)	1.4	3.4	3.2	25.8	25.7		
C. alternifolius (b)	10.2				11.5		
P. cordata (a)	3.1	5.2	2.4	3.1	3.3		
P. cordata (b)	7.6				9.1		
P. elephantipes (a)	1.1				1.1	14.7	66.4
P. elephantipes (b)	10.0				32.2		169.4
T. domingensis (a)	1.1	2.1	4.8	18.9	14.6	14.4	32.0
T. domingensis (b)	1.3				39.8		70.0

Cr, Ni and Zn concentrations in plant tissue increased with time throughout the study period (Tables 5, 6 and 7, respectively). Macrophytes generally attained its maximum metal concentration in the last samples before disappearance. Floating macrophytes showed generally faster increases than emergent macrophytes. Metal in the roots were higher than in stems, most often 2 to 3 times higher. Faster metal accumulation in floating than emergent macrophytes, can be attained because their roots are immersed in the wastewater solution while the emergent macrophytes developed their roots in the reduced soil matrix where metals are largely immobilized. A higher metal accumulation in floating than emergent macrophytes was also reported by Mays and Edwards (2001). Dushenkov et al. (1995) grew *Brassica juncea* in hydroponic cultures exposed to high Ni and Zn concentrations (10 and 100 mg l^{-1}, respectively) and attained higher plant tissue concentrations (1 and 13 mg g^{-1}, respectively) than here reported. Sen and Battacharyya (1994) reported tolerance levels for *Salvinia natans* up to 10 mg l^{-1} of Ni. MacCabe and Otte (1997) reported higher Zn concentrations in *Glyceria fluitans* in a natural wetland receiving mine wastewater than in the present study.

At the end of the study period the macrophyte biomass represented 15% of the TP and 3% of the inorganic nitrogen removed from the inlet water. Similarly, a small fraction of the metal retention in the wetland (8, 4, and 3% of the Zn, Cr, and Ni, respectively) was stored in the vegetation tissue. Manios et al. (2003) studied the effect of metal accumulation in *Typha latifolia* in a 10 week experiment. Similarly to our findings, the metal concentration in plant tissue increased as the experiment progressed. According to the authors each plant has a threshold of tolerance, once the threshold is surpassed toxicity takes its toll. In the experiment performed by Manios et al. (2003), toxicity was early indicated by a chlorophyll decrease and a change in the chlorophyll *a* to chlorophyll *b* ratio. In our study, a P decrease in plant tissue was observed in the last sampling before disappearance suggesting that P concentration represented an equally suitable early indicator of toxicity. Manios et al. (2003) applied high metal concentrations together with nutrients and observed a decrease in chlorophyll concentrations only at the highest dose (40 mg Ni l^{-1} and 80 mg Zn l^{-1}). They interpreted this result as an effect of increased tolerance by the nutrient addition. In our study, Ni concentration in the plant tissue of *T. dominguensis* was larger than that recorded by Manios et al. (2003) while no symptom of toxicity was observed, and being its biomass similar to that recorded in nearby non-disturbed environments. Treating the industrial wastewater together with the sewage discharge likely enhanced the development of *T. domingensis* in the wetland and resulted in high removal efficiencies of both metals and nutrients. Because its high tolerance to the wastewater toxicity *T. dominguensis* is therefore a promising alternative to use in wetlands for the treatment of metal contaminated wastewater.

3.2. Greenhouse Experiments

3.2.1. Effect of Nutrient Addition on Metal Tolerance of Floating Macrophytes

We hypothesized that nutrient enrichment enhances the metal tolerance of floating macrophytes and would therefore enable the development of floating vegetation in constructed wetlands at metal concentrations which would otherwise inhibit plant growth. Increased nutrient concentrations might be attained by mixing the sewage of the factory with the industrial wastewater once the appropriate primary treatment has been undertaken.

Hadad et al. (2007) studied in three range-finding experiments the relative growth rate of *S. herzogii* exposed to concentrations of Zn, Ni and Cr of 4, 6, 8 and 10 mg l^{-1}. The aim was to establish approximate toxicity thresholds for each metal in order to carry out the subsequent definitive toxicity experiments.

Table 7. Zn concentration in plant tissue ($\mu g\ g^{-1}$) in the small-scale constructed wetland (*a* stands for above and *b* for below-ground biomass)

Species	2001					2002	
	Jul.	Aug.	Sept.	Oct.	Nov.	Feb.	Sept.
E. crassipes (a)	10.2	44.7	41.7	91.3			
E. crassipes (b)	1.1	6.5	64.7	182.9			
S. herzogii (a)	3.3	35.3	51.0	66.3			
S. herzogii (b)	12.8	55.3	77.2	122.8			
P. stratiotes (a)	10.3	41.3	70.4	105.8	748.9		
P. stratiotes (b)	25.9	57.2	55.9	185.1	971.1		
C. alternifolius (a)	12.5	22.9	30.7	46.7	97.9		
C. alternifolius (b)	24.5				74.0		
P. cordata (a)	22.5	26.0	30.8	22.1	20.8		
P. cordata (b)	60.8				331.2		
P. elephantipes (a)	23.2				84.6	34.2	40.6
P. elephantipes (b)	41.6				35.9		70.7
T. domingensis (a)	6.1	15.7	24.3	22.7	21.8	23.1	38.7
T. domingensis (b)	56.7				45.5		66.7

In definitive experiments, plant growth rates in response to metal exposures were compared with exposures enriched with 5 mg l^{-1} of P (as H_2KPO_4), 5 mg l^{-1} of NO_3^--N (as KNO_3) and 5 mg l^{-1} NH_4^+-N (as NH_4Cl). Three concentrations for each metal were studied as follows:

Treatments 1, 2 and 3:	exposure to metal concentrations 1, 2 and 3
Treatment 4:	control without metal addition
Treatments 5, 6 and 7:	exposure to metal concentrations 1, 2 and 3 and simultaneous nutrient enrichment
Treatment 8:	xtreatment with nutrient enrichment, without metal exposition

Metal concentrations 1, 2 and 3 were chosen according to the results previously obtained in the range-finding experiments and were different for each metal.

The range-finding experiment showed differences in the toxicity thresholds for *S. herzogii* of the assayed metals. Relative growth rates were significantly different from the control at 6 mg Cr l^{-1}, 10 mg Zn l^{-1} and all assayed Ni concentrations (minimum 4 mg l^{-1}) (Figure 3). The range-finding experiments showed that relative growth rates of *S. herzogii* were negatively correlated with metal concentration for Ni, Cr and Zn at the assayed concentrations, being the linear regression coefficients significantly different from zero ($p<0.05$) (Table 8). Relative growth rates of *S. herzogii* were also inversely correlated with metal concentrations in the definitive experiments (Table 8). On the contrary, the treatments enriched with nutrients did not show significant correlation. Therefore, nutrient addition eliminated the observed decrease of the relative growth rate in response to metal exposition.

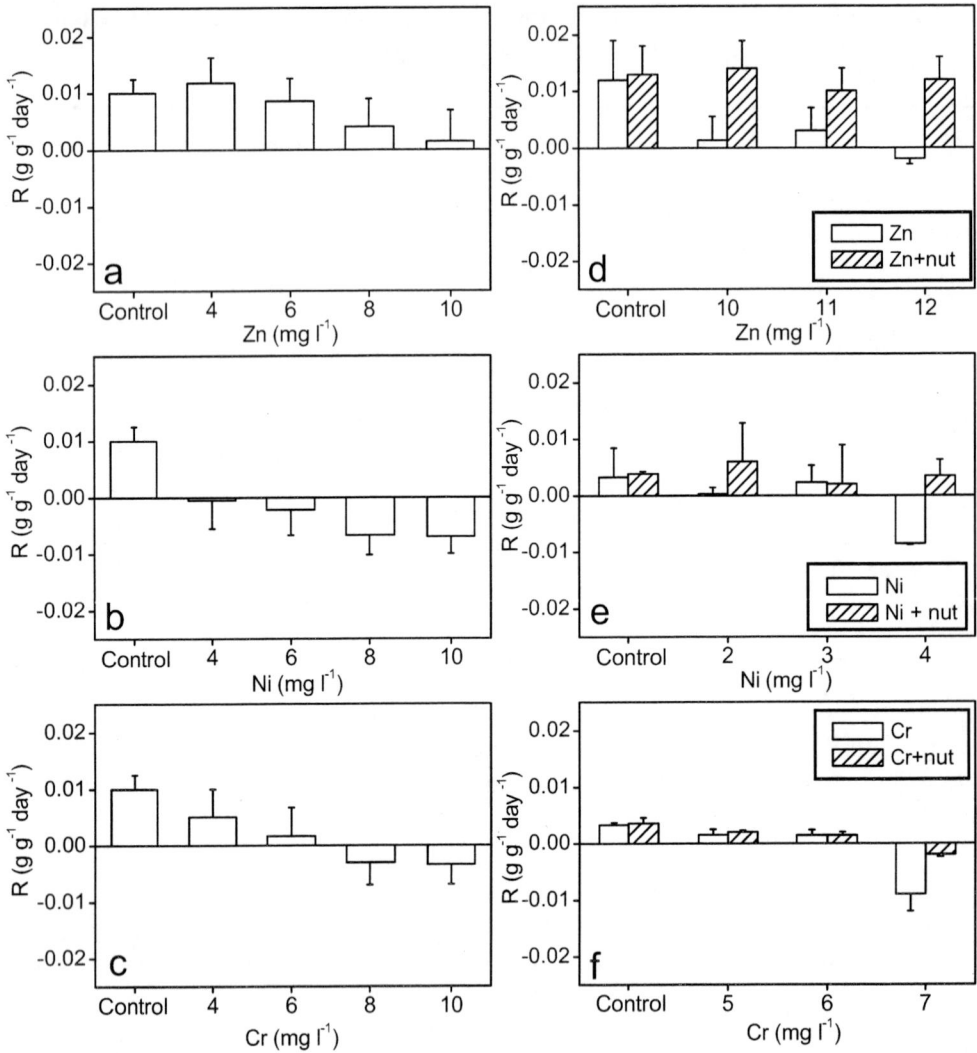

Figure 3. Relative growth rates (R) obtained in the Zn, Ni and Cr range-finding (a, b and c, respectively) and definitive metal exposure experiments (d, e and f, respectively). The bars represent standard deviation.

Nutrient enrichment enabled *S. herzogii* growth at Zn and Ni exposures that impaired growth in treatments without nutrient addition. The trend to increase Zn concentrations in leaves with increasing metal exposure was reduced by nutrient enrichment. Chaney (1993) reported that plants exhibiting Zn toxicity had lower P levels in shoots, and argued that it might result either from inhibited root growth or from insoluble Zn-phosphate formation and immobilization in the roots. Loneragan and Webb (1993) reported that P depresses either Zn absorption by roots or tanslocation from roots to shoots and commented that under conditions of high Zn supply, immobilization in the roots through the formation of Zn-phytate has been shown to occur. Simultaneous P and Zn immobilization in the roots is likely the mechanism for the increased tolerance of *S. herzogii* to Zn exposure.

Table 8. Linear regression analyses used to examine relationships among relative growth rate and metal exposures

Parameter	Zn	Zn + nut.	Ni	Ni + nut.	Cr	Cr + nut.
Relative growth rate (range-finding experiments)	-0.822 (0.0035) -0.0009	—	-0.967 (<0.0001) -0.002	—	-0.977 (<0.0001) -0.002	—
Relative growth rate (definitive experiments)	-0.932 (0.0007) -0.001	—	-0.757 (0.030) -0.002	—	-0.734 (0.038) -0.0009	—
Root biomass	-0.936 (0.0006) -0.036	-0.929 (0.0009) -0.020	-0.756 (0.030) -0.040	—	-0.860 (0.006) -0.033	-0.876 (0.0044) -0.011
Chlorophyll concentrations	—	-0.966 (<0.0001) -0.309	-0.873 (0.005) -0.430	-0.855 (0.007) -0.258	—	—

Correlation coefficients, p values in parenthesis, and slopes, are given at the top, middle and bottom, respectively. Non significant correlations are not presented.

Ni exposure decreased relative growth rate and chlorophyll concentration simultaneously (Table 8), suggesting that both variables represent equally suitable indicators of Ni toxicity. Monni et al. (2000) reported growth inhibition in *Empetrum nigrum*, an ericaceous shrub, with increasing Ni concentration in solution. The authors suggested that growth supression is the cost of tolerance because, although growth was affected, plant survival did not decrease during the experiments.

Root growth inhibition by Cr exposure observed in the present study has previously been reported and represents a sensitive indicator of Cr toxicity [Shanker et al., 2005; Scoccianti et al., 2006]. Reductions in root biomass due to Cr exposure were attenuated by nutrient enrichment, suggesting an improving effect of nutrient enrichment on the Cr tolerance of *S. herzogii* (Table 8). Since no effect on chlorophyll concentration was observed, Cr toxicity may not be mediated through the photosynthetic metabolism for *S. herzogii* under these experimental conditions. Maine et al. (2004) reported that the chlorophyll concentration of *P. stratiotes* decreased at Cr concentrations in water of over 1 mg l^{-1}, while *S. herzogii* did not show significant changes in chlorophyll concentrations up to 6 mg Cr l^{-1}.

Göthberg et al. (2004) found high metal concentrations in *I. aquatica* cultivated for human consumption in freshwater courses near Bangkok receiving variable amounts of cultural nutrient loads. The authors proposed fertilization as a means to attenuate metal accumulation. Their experimental work, in agreement with our findings, showed that nutrient enrichment increased *I. aquatica* tolerance to Cd, Pb and Hg. Different patterns of metal concentration in leaves of *I. aquatica* by nutrient addition were reported. Cadmium concentrations did not change with nutrient enrichment, as was observed with Ni and Cr in the present study, while Hg and Pb concentrations decresead with increasing nutrient concentrations, as was observed with Zn in the present study. Göthberg et al. (2004) suggested the competition between metals and nutrients in the root uptake in connection with the translocation to the shoots as the probable mechanism governing metal concentrations in leaves and roots. Higher metal concentrations in roots than in leaves in macrophytes exposed to metals have been observed in the present study as has been shown in other previously reported research [Sen and Bhattacharyya, 1994; Banerjee and Sarker, 1997; Manios et al., 2003, Göthberg et al., 2004; Paris et al., 2005]. Therefore, a higher tolerance of roots than shoots together with a trend to decrease translocation with increasing metal concentration in the roots represents a common feature for the different metals and plants studied. Binding positively charged toxic metal ions to negative charges in the cell walls of the roots, metal-phosphate and metal-phytate formation, and chelation to phytochelatins followed by accumulation in vacuoles have been invoked as mechanisms to reduce metal transport and increase metal tolerance [Chaney, 1993; Loneragan and Webb, 1993; Göthberg et al., 2004]. Nutrient enrichment increased the tolerance of S. herzogii to metals. Nutrient enrichment will improve metal removal by increasing macrophyte production, leading to a higher metal uptake by the macrophyte biomass, and also by enhancing the overall biological activity, attaining a higher retention in the detrital fractions.

3.2.2. Effect of pH and Conductivity on Floating Species

Hadad et al. (2006) carried out greenhouse experiments in order to explain the disappearance of the floating macrophytes in the studied wetland. Taking into account that metal concentrations in the treated wastewater were always below the toxicity thresholds determined in the previous experiment, the effects of pH and conductivity of wastewater on the free floating species (S. herzogii, P. stratiotes and E. crassipes) were studied. Four treatments at pH 8, 9, 10 and 11 were performed. Such values were obtained adding NaOH to water until the various values were adjusted. As for conductivity, five treatments with values of 2, 3, 4, 6 and 8 mS cm^{-1} were done. Such values were reached preparing solutions of Na_2SO_4. This salt was chosen since ions Na^+ and SO_4^{2-} presented the highest concentrations at the influent of the pilot-scale wetland. Conductivity and pH were kept constant through time with the addition of NaOH and Na_2SO_4 when necessary.

Conductivity presented statistically significant effects on the relative growth rates of E. crassipes (Figure 4). Plants subjected to conductivities of 4 mS cm^{-1} or higher showed significant differences from control treatments, whereas at values of 2 and 3 mS cm^{-1} there were no significant differences ($p < 0.05$). The plants subjected to conductivities of 6 and 8 mS cm^{-1} showed saline secretions on their petioles at the end of the experiment. The damage threshold would be in the range of 3-4 mS cm^{-1}. At pH 10 and 11, growth was significantly lower than that of the control ($p < 0.05$). (Figure 4). The other assayed pH values (8 and 9) did not cause significant effects on growth ($p < 0.05$). At a pH of 10 and 11, and conductivity of 4, 6 and 8 mS cm^{-1} E. crassipes showed visible signs of chlorosis and necrosis.

For P. stratiotes, there were statistically significant differences on growth rate between control treatment and the treatment of 8 mS cm^{-1} and pH = 11 ($p < 0.05$). Signs of chlorosis and necrosis were observed in these treatments.

S. herzogii did not show differences between control treatment and the several conductivity treatments applied ($p < 0.05$). However, it did show differences in growth rate between treatment of pH = 11 and the treatments of other pH. As to conductivity treatment, saline secretion was observed on the edges of leaves of S. herzogii at the end of the experiment.

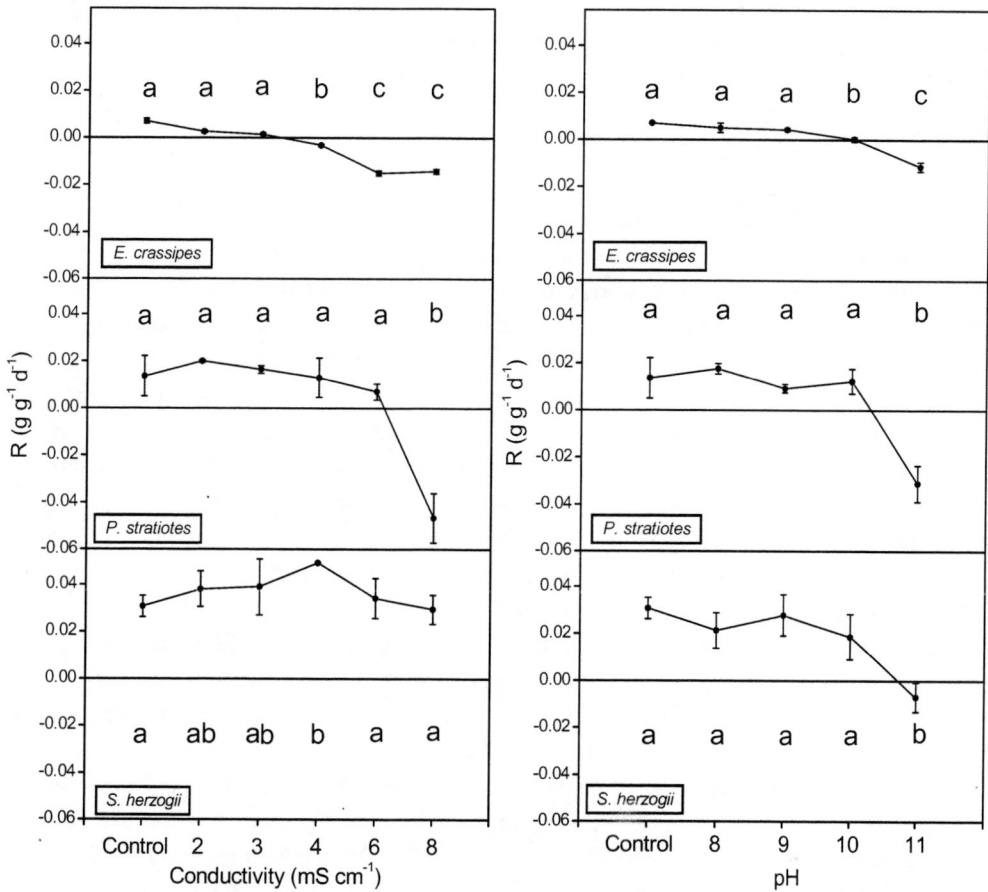

Figure 4. Conductivity and pH effects on the relative growth rate (R) of *E. crassipes*, *P. stratiotes* and *S. herzogii* in greenhouse experiments. Different letters represent significant statistical differences.

Conductivity in the wetland incoming wastewater was higher than the tolerance threshold determined experimentally for *E. crassipes* in 14 out of 16 samplings, and pH in approximately half the samplings. *P. stratiotes* and *S. herzogii* showed a higher tolerance range than *E. crassipes*. Consistently, *P. stratiotes* persistence within the wetland was longer than that of *E. crassipes*. Water pH in the incoming wastewater was 12.5 and 11.9 in Oct. 2001 and Feb. 2002, the samplings previous to the disappearance of *S. herzogii* and *P. stratiotes*, respectively. *S. herzogii* was transplanted into the constructed wetland at a lower density than the others because of a low availability at the undisturbed environment at the time when the macrophytes were transplanted into de constructed wetland. As previously stated, metal concentration in both, incoming wastewater and plant tissues, were lower than the metal thresholds determined in the previous experiment and reported in literature [Gibson and Pollard, 1988; Delgado et al., 1993; Sen and Bhattacharyya, 1994; Selvapathy et al., 1997; Cardwell et al., 2002; Fakayode and Onianwa, 2002; Ingole and Bhole, 2003; Manios et al., 2003; Soltan and Rashed, 2003; Maine et al., 2004]. Although high pH and conductivity might shift the toxicity thresholds for metals, the greenhouse experiments showed that floating macrophytes could not develop within the pH and conductivity prevailing in the incoming wastewater even in the absence of metals.

3.3. Large-scale Wetland

3.3.1. First Stage: Sewage Treatment

During the first stage of the wetland operation (from October 2002 to February 2003) only diluted sewage of the factory was treated. The main chemical characteristics of the wastewater at the inlet and outlet zone during the five sampled months are shown in Table 9. According to the values of concentrations of the measured variables in the water at the inlet and outlet areas, it was determined that the wetland carried out an efficient removal of nutrients until anoxic conditions developed in the outlet zone. As water flows along the wetland, the concentrations of SRP, NO_2^-, NH_4^+ and NO_3^- decreased significantly ($p<0.05$). In aquatic systems, NH_4^+ is removed by sedimentation, nitrification followed by denitrification, plants uptake or volatilization [Reddy and Sutton, 1984]. NO_3^- decrease in water was mainly due to the process of denitrification and macrophyte uptake. The denitrification process needs energy, which is derived from the oxidation of organic matter possibly present in the anaerobic sediment. Macrophytes may improve conditions for denitrification in nitrate-rich waters by supplying org-C, which can be used directly by denitrifying bacteria or can stimulate denitrification indirectly by contributing to a lower redox potential [Weisner et al., 1994]. Although SRP was efficiently retained by the wetland, anaerobiosis occurred at the outlet area in March (last sampling), causing an increase of SRP in this area with respect to the middle area. The presence of N-species in the wastewater probably favored P sorption. Panigatti and Maine (2003) showed that the presence of N-species increased SRP removal from water, NH_4^+ being the most effective species. BOD and COD also decreased significantly between the inlet and the outlet ($p<0.05$). $SO_4^=$, Ca^{+2} and Mg^{+2} decreased to a smaller extent. pH decreased through the wetland, while Fe decreased largely.

The initial mean concentration of organic matter in the sediment was 3.5% (Table 10), the mean concentration of Ca^{2+} was 1.16 mg g^{-1} (d.w.) and that of Fe was 19.7 mg g^{-1} (d.w.). Mean Eh was 95 mV and pH was 7.70. The mean concentration of TP was 0.383 mg g^{-1} (d.w.). The largest fraction was Fe(OOH)≈P representing 28% of the TP, followed by alkaline extractable organic P (21%) and CaCO$_3$≈P (20%). No statistically significant differences were found in the chemical composition of sediment between inlet and outlet areas ($p<0.05$), probably due to the fact that during the construction of the artificial wetland, the sediment was removed, mixed and deposited on the bentonite layer.

Regarding the final conditions of the sediment the concentration of organic matter in the sediment was found to be 3.50-4.44% (Table 10). The concentration of Ca^{2+} was between 0.839-2.41 mg g^{-1} (d.w.) and that of Fe between 16.65-22.45 mg g^{-1} (d.w.). Eh values ranged between 90-162 mV and pH between 6.32-7.86. The sediment of the inlet zone showed a significant increase in the concentration of TP in relation to initial conditions ($p<0.05$). This fact is due to the high concentration of SRP in this zone, in agreement with the good correlation found between TP and SRP at the different sampling points ($r= 0.9010$, $p<0,001$, $n= 15$) [Maine et al., 2007b]. Both Fe(OOH)≈P and org-P$_{\rightarrow alk}$ fractions of the sediment increased significantly ($p<0.05$). The increase of the Fe(OOH)≈P fraction is really remarkable (aproximately 30%, Table 10) demonstrating that adsorption onto Fe(OOH) is the process that governed the phosphate binding onto this sediment. The P-adsorption onto the Fe(OOH)≈P fraction can be probably due to the high Fe content of the sediment and the low concentration of $CaCO_3$. The importance of Fe(OOH) as an adsorbent in sediments was understood long time ago [Mortimer, 1941, 1942; Olsen, 1964; Golterman, 1977; Lijklema, 1977, 1980].

At the outlet, anoxic conditions led to a release of P from the sediment. This was observed not only as an increase of SRP concentration, but also in the morphological characteristics and amount of P in tissues of *E. crassipes*, mainly in the aerial parts (Table 11). Maine et al. (2007b) proposed that probably the plants sorbed the P released from the sediments and transported it to the aerial parts.

Table 9. Main characteristics of the large-scale CW at the stage in which only sewage was added

Parameter	Inlet area			Outlet area		
	Mean	Min.	Max.	Mean	Min.	Max.
pH	7.63*	7.22	8.33	7.17	6.92	7.52
Water Temperature (°C)	21	17	25	21	17	25
Conductivity (μmhos cm^{-1})	645*	480	1250	571	470	700
SRP (mg l^{-1})	0.346*	0.205	0.512	0.113	< 0.01	0.274
N-NO$_2^-$ (mg l^{-1})	0.079*	0.073	0.121	0.007	<0.005	0.026
N-NH$_4^+$ (mg l^{-1})	1.11*	0.271	1.981	0.476	< 0.05	0.633
N-NO$_3^-$ (mg l^{-1})	6.72*	1.62	10.7	2.52	1.22	3.45
DO (mg l^{-1})	7.0*	6.5	8.0	4.5	0.0	8.0
Ca^{2+}(mg l^{-1})	20.5	19.6	26.8	17.9	13.9	22.7
Mg^{2+} (mg l^{-1})	4.7	3.7	13.6	3.5	3.5	13.1
Fe (mg l^{-1})	0.458*	0.265	0.587	0.153	0.053	0.167
Alkalinity (mg l^{-1})	221	199	246	211	189	245
Cl$^-$ (mg l^{-1})	71.2	65.2	74.7	71.5	63.5	73.6
SO$_4^{2-}$ (mg l^{-1})	126*	86.5	125.1	89.8	82.8	99.7
COD (mg l^{-1})	49*	29	68	14	<5	23
BOD (mg l^{-1})	14*	16	29	<2	<2	12

* Significant differences between inlet and outlet concentrations ($p<0.05$).

In consequence, it can be suggested that the observed SRP decrease from the inlet to the outlet of the wetland was due to the absorption of aquatic plants as well as the adsorption to the sediment. This is observed by the increase of macrophyte biomass together with its TP tissue amount and the TP retention in the sediment. In order to compare the removal rates in sediment and plants, P concentrations were converted into P amount (mg m^{-2}). Thus, we have considered not only the concentration but also each compartment mass, i.e. the amount. Initial and final amounts of P in the sediment (considering a depth of 2 cm) and in the three main species *E. crassipes*, *P. cordata* and *T. domingensis* were estimated. The P retained during the experiment (Table 11) is the difference of P amount between the initial and final conditions. According to the amounts of P in the different compartments of the system, it can be concluded that the sediment is the main P compartment, in agreement with the results obtained in natural systems [Maine et al., 1998]. However, taking into account the P retained in each compartment during the 5 months of experiment, it can be seen that the floating macrophyte *E. crassipes* carried out the higher P removal, probably due to its high productivity (Table 11). The advantage of macrophytes is the possibility of being harvested, which leads to important P removal rate in a short time. Adler et al. (1996) proposed a wetland system that better exploits the nutrient uptake mechanisms of plants. P stored within the plant biomass was periodically harvested. Because of the mechanism of P uptake was not based on the soil matrix, P could not be reversible desorbed and sites for P sorption could not be saturated.

This study shows the high P removal rate of *E. crassipes*. However, Panigatti and Maine (2003), using laboratory reactors to simulate a wetland, demonstrated that when macrophytes are not present, sediment would increase its removal rate reaching almost the same values as in reactors with macrophytes plus sediment.

Table 10. Mean Sediment concentrations of P fractions, organic matter, Fe, Ca (expressed in mg g^{-1} d.s.)

Zone	Fe(OOH)≈P	CaCO3≈P	P$_{\to acid}$	P$_{\to alk.}$	TP	Organic Matter	Fe	Ca	pH	Eh
Initial	0.109	0.080	0.063	0.084	0.383	3.5	19.70	1.16	7.70	95
Final Inlet	0.138	0.078	0.069	0.098	0.42	3.86	22.45	2.41	7.86	162
Outlet	0.099	0.088	0.068	0.054	0.349	4.24	19.05	1.07	7.75	90

Table 11. P mass balance in the small-scale wetland. P amount are expressed in mg m^{-2}

	E. crassipes		P. cordata		T. domingensis		TP Macrophyte	TP Sediment	Total
Inlet	Aerial	Roots	Aerial	Roots	Aerial	Roots			
Initial P amount	1,285.8	294.9	125.5	53.5	79.4	5.7	1,844.9	9,031.1	10,875.9
Final P amount	3,125.9	489.4	148.2	52.5	976.8	356.1	5,148.9	9,903.6	15,032.5
P retained	1,840.1	180.1	22.7	-1	897.4	350.4	**3,304.0**	**872.5**	**4,086.6**
Outlet									
Initial P amount	542.1	311.7	221.6	44.7	49.3	21.4	1,190.8	8,583.1	9,773.9
Final P amount	3,378.4	374.0	513.6	112.0	668.5	162.3	5,208.8	8,229.4	13,438.2
P retained	2,836.3	62.3	292.0	67.3	619.2	140.9	**4,038.0**	**-353.7**	**3,664.3**

3.3.2. Second Stage: Sewage and Industrial Wastewater Treatment

3.3.2.1. Wetland Efficiency

Maine et al. (2007b) reported that the composition of the incoming wastewater showed important variations through time (Table 12). The concentrations of the parameters measured at the outcoming effluent of the wetland were found to be below the guide levels for industrial wastewater according to the national regulation. The effluent showed not only lower concentrations but also lower variability of the measured parameters regarding the influent, proving the buffer capacity of the wetland.

Table 12. Mean loading rates (g m^{-2} d^{-1}), mean and range of temperature (°C), conductivity (µS cm^{-1}), concentrations (mg l^{-1}) of the inlet and outlet, and estimated removal efficiency in the large scale constructed wetland along the studied period. Different letters represent statistical significant differences (p<0.05)

Parameter	Inlet			Outlet			Mean removal (%)
	Loading rate (g m^{-2} d^{-1})	Concentration Mean	Range	Loading rate (g m^{-2} d^{-1})	Concentration Mean	Range	
Temperature	---	19.1	10.2-28	---	16.7	6.6-26	---
pH	---	9.01[a]	6.5-12.3	---	7.66[b]	6.9-9.1	---
Conductivity	---	3249[a]	480-8500	---	1799[b]	470-5000	---
DO	---	1.53[a]	0-7.1	---	0.898[b]	0-7.5	---
Suspended Solids	137.9	2758.4[a]	699-8550	72.5	1450.9[b]	524-3693	36
Ca^{2+}	7.90	157.9[a]	24-651	3.33	66.7[b]	17.3-268	34
Mg^{2+}	0.81	16.2[a]	0.5-59	0.77	15.5[a]	3.3-62	5
Alkalinity (CaCO$_3$)	18.5	369.1[a]	71.2-1187	14.0	280.1[b]	95.2-475	33
SO$_4^{2-}$	62.4	1247.1[a]	98.1-3598	30.5	609.1[b]	158-2238	34
Cl$^-$	13.4	268.4[a]	70.4-778	7.70	154.0[b]	38.6-320	34
Na$^+$	35.4	708.0[a]	200.2-1680	20.3	405.4[b]	135-1136	34
K$^+$	0.878	17.6[a]	7.3-38	0.854	17.1[a]	2.4-39	5
Fe	0.387	7.73[a]	0.05-73.9	0.012	0.237[b]	0.05-1.22	74
Cr	0.0009	0.018[a]	0.001-0.164	0.0002	0.004[b]	0.001-0.015	53
Ni	0.001	0.028[a]	0.002-0.2	0.0006	0.013[b]	0.003-0.10	39
SRP	0.005	0.097[a]	0.001-0.51	0.005	0.104[a]	0.005-0.43	-3
TP	0.020	0.405[a]	0.046-1.39	0.015	0.302[a]	0.032-1.51	6
N-NO$_2^-$	0.014	0.285[a]	0.001-1.6	0.001	0.016[b]	0.001-0.30	75
N-NO$_3^-$	0.225	4.53[a]	0.018-16	0.044	0877[b]	0.07-7.0	68
N-NH$_4^+$	0.101	2.02[a]	0.05-9.65	0.097	1.94[a]	0.04-13.8	7
BOD	3.53	70.7[a]	6.5-360	1.07	21.4[b]	5.0-83.4	66
COD	10.6	211.7[a]	21.8-1082	2.87	57.5[b]	11.2-172.5	72

The DO concentration at the inlet and outlet showed a high variability, being anoxic at several samplings. At the outlet, DO concentrations were different at the surface layer than at the bottom, being DO lower at the bottom. Differences in DO concentrations induced differences between surface and bottom concentrations in most other parameters. Since the outlet pipe was placed at the bottom, it was decided to build a wall at the outlet to force the effluent into forming a small waterfall, which in turn favoured the increase in the DO concentration. Slight modifications in design may remarkably improve the efficiency of contaminant removal.

NO_3^- and NO_2^- were removed (Table 12), while NH_4^+ was not removed. SRP concentration of the effluent had an overall mean concentration 3% greater than in the influent. COD and BOD reductions suggested a large mineralization of the incoming organic matter. Organic matter mineralization increased the CO_2 concentration in water, which, in turn, reduced the water pH range from 6.5-12.3 in the influent to 6.9-9.1 in the effluent. Mean calcium concentrations decreased 34% and alkalinity 33%. Removal percentages of calcium and alkalinity were greater when pH of the influent was higher (9.2-12.3), suggesting that calcium carbonate precipitation within the wetland represents an important pathway governed by the pH of the incoming water.

Metal concentrations were significantly lower in the effluent than in the influent ($p<0.05$). The overall mean retention throughout the study period was 74%, 53% and 39% for Fe, Cr and Ni, respectively. The removal percentages of each metal remained almost constant during the experimental period. Consequently, the higher the incoming loads, the larger the removed metal amounts. Zn concentration was below 50 ug l^{-1} (detection limit of the analytical method) both in influent and effluent throughout the study period. Simultaneous SO_4^{2-} and Fe removal and DO depletion in the water column suggest insoluble FeS formation. Because of the high SO_4^{2-} concentration in the incoming wastewater, most of the organic matter mineralization took place at the expense of biological SO_4^{2-} reduction as observed in coastal marine sediment where SO_4^{2-} reduction is responsible for 25-79% of the total organic matter mineralization [Giblin, 1988]. S^{2-} released by SO_4^{2-} reduction subsequently reacts with iron to form FeS minerals [Giblin, 1988]. Several monosulphide minerals early precipitated are later converted to pyrite [Giblin, 1988]. FeS formation depends on the rate of Fe(II) and S^{2-} supply. The lower concentrations of Fe than SO_4^{2-} in the incoming wastewater and the almost complete Fe retention within the wetland (Table 12) suggest that Fe availability limits FeS formation in the CW.

3.3.2.2. Macrophyte Dominance Stages

E. crassipes became dominant and covered about 80% of the surface from March 2003 to January 2004, decreasing progressively until its disappearance over the following six months (Figure 5). In September 2003 roughly 20% of the biomass was harvested. In January 2004 the wetland was emptied for a few days and the plants remained anchored in the mud, without apparent damage. *E. crassipes* decreased progressively since then to attain a very small cover in July 2004. In August 2004, the wetland was emptied again, the few remaining floating macrophytes were harvested and soil was added in strips, perpendicularly to the water circulation, where *T. domingensis* was transplanted. Once the wetland was refilled, water depth was reduced to 0.5 m and 0.3 m at the strips with added soil in order to favour *T. domingensis* growth and the oxygenation of water. *E. crassipes* was transplanted again in August 2004. Although it showed some initial growth, it soon decreased its cover. Metal concentrations in water were not sufficiently high to cause its disappearance. As it was demonstrated in the greenhouse experiment, *E. crassipes* tolerance thresholds for pH and conductivity lay between 9-10 and 3-4 mS cm[-1], respectively. In many samplings, conductivity and pH were higher than the tolerance thresholds, possibly being the cause of *E. crassipes* disappearance. *T. domingensis* developed along the banks, covering roughly 4-13% of the wetland surface during 2003 (Figure 5). After soil strips were added in August 2004 plant cover steadily increased to attain roughly 30% of the surface by the end of 2004. In October 2005 the wetland was emptied again and the organic matter remaining on the bottom was extracted. As floating macrophytes showed low tolerance to the conditions and caused DO depletion, only new specimens of *T. domingensis* were planted. *T. domingensis* covered 65% of the surface by the end of the study.

Maine et al. (2007a) divided the experimental period in the three successive stages described for the vegetation development: *E. crassipes* dominance stage (March 2003-June 2004); *E. crassipes* + *T. domingensis* dominance stage (September 2004-February 2005); and *T. domingensis* dominance stage (March 2005-March 2006).

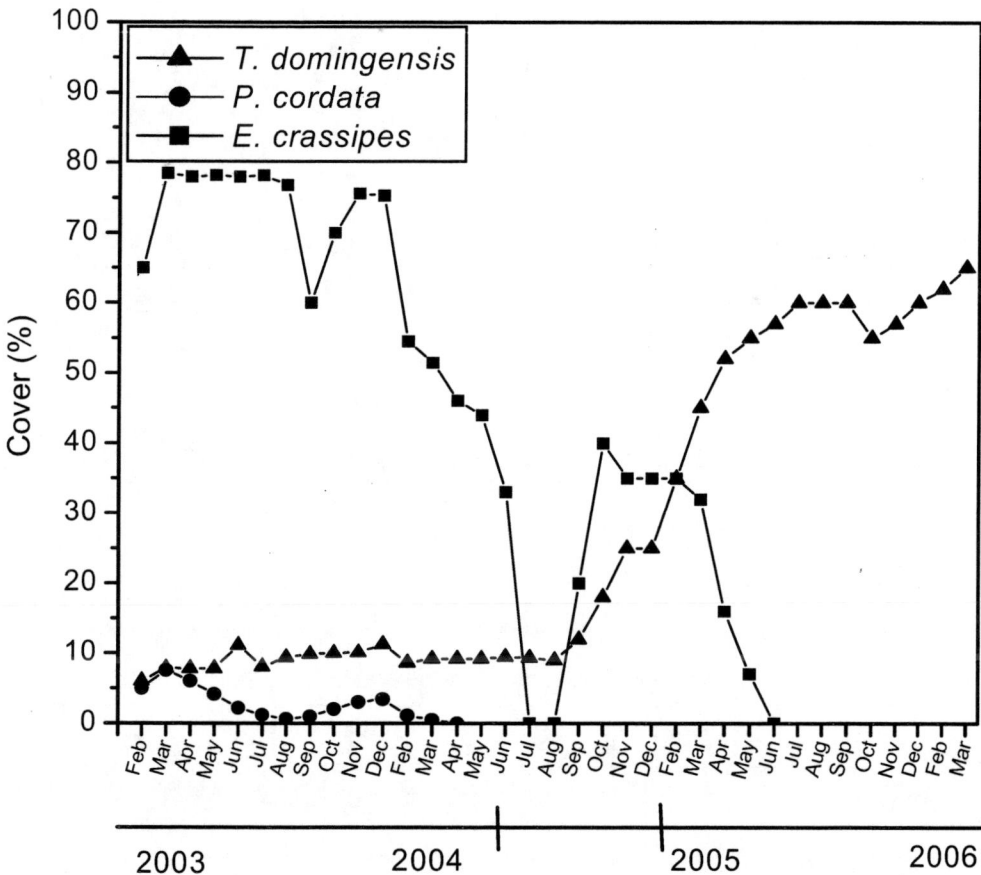

Figure 5. Macrophyte cover in the CW throughout the study period. The line below the figure indicates the different vegetation dominance stages.

3.3.2.3. Conthaminant Concentrations in Macrophyte and Sediment

Large temporal variations were observed in nutrient and metal plant tissue concentrations (Figure 6 and 7, respectively) along the studied period, probably due to biomass temporal variation. Metal concentration in plant tissue was higher in roots than in leaves (Figure 7). TP, Cr, Ni and Zn concentrations in sediment (Figure 8) did not increase significantly through the *E. crassipes* stage in spite of the fact that the concentrations between influent and effluent revealed wetland retention ($p<0.05$). *E. crassipes* was responsible for the P, Cr, Ni and Zn removal; mass balance (Table 13) confirmed these results. Permanent DO depletion in the *E. crassipes* stage prevented metal co-precipitation with iron oxy hydroxides and metals were sorbed by macrophytes.

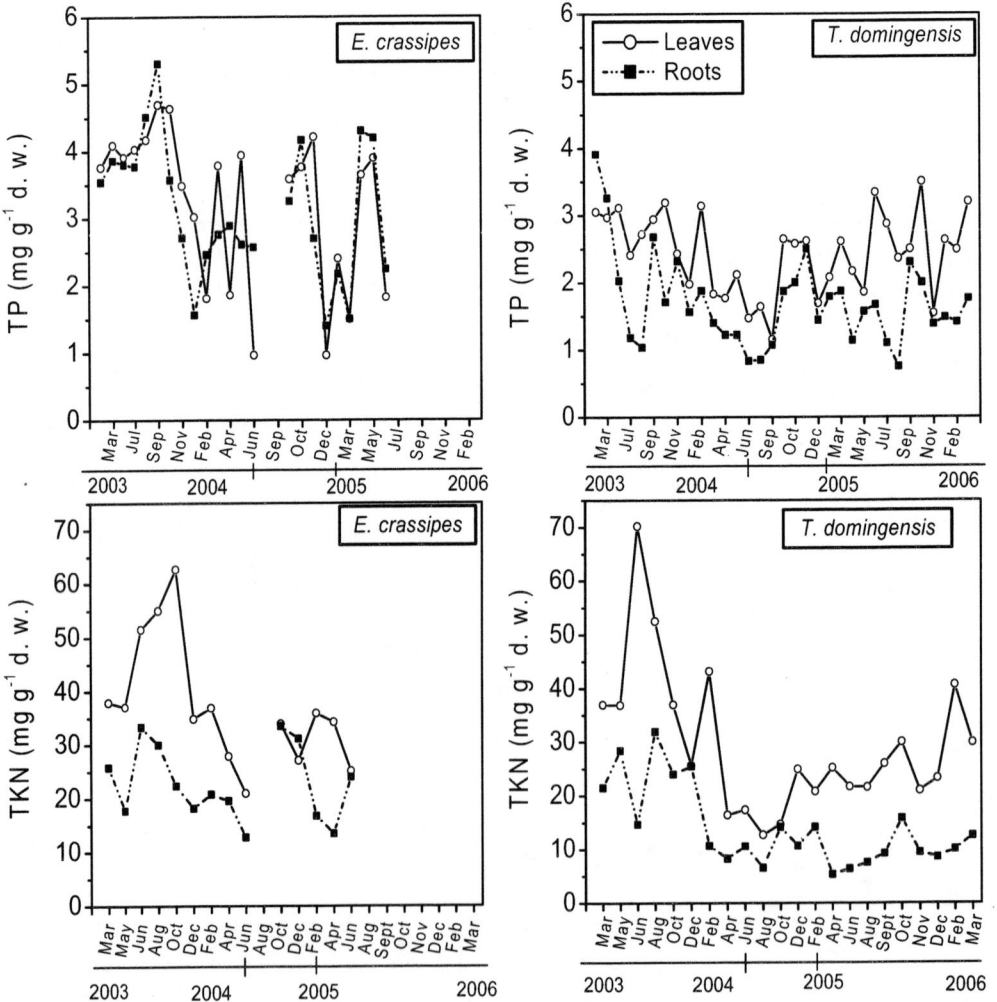

Figure 6. Mean TP and TKN concentrations in tissues (*E. crassipes* and *T. domingensis*). The line below the figure indicates the different vegetation dominance stages.

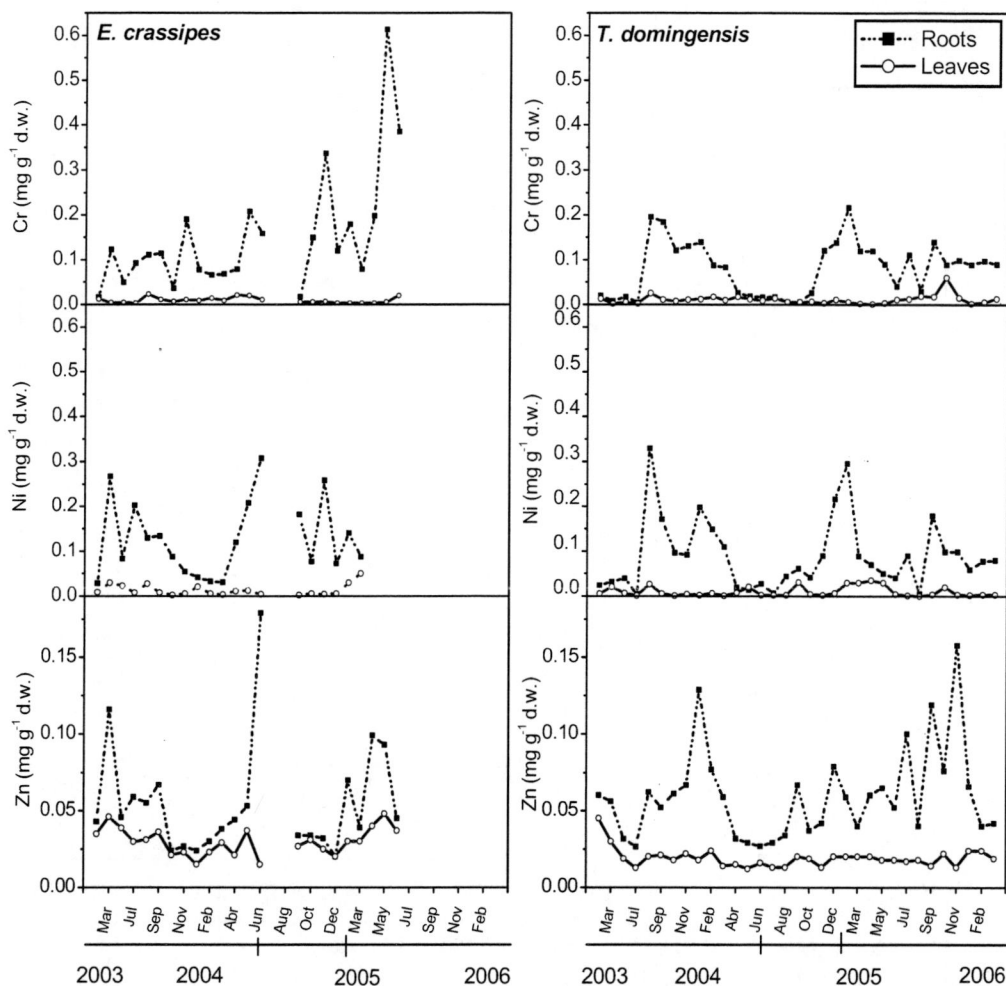

Figure 7. Mean Cr, Ni and Zn concentrations in tissues (*E. crassipes* and *T. domingensis*). The line below the figure indicates the different vegetation dominance stages.

Table 13. Distribution of the removed metals and nutrients between macrophytes and sediment during the three stages (%)

Stages	Cr		Ni		Zn		P		N	
	Sed.	Macr.	Sed.	Macr	Sed.	Macr.	Sed.	Macr.	Sed.	Macr.
E. crassipes	12	88	8	93	2	98	2	98	--	25
E. c. + T. d.	93	7	95	5	89	11	79	21	--	16
T. domingensis	70	30	87	13	59	41	62	38	--	23

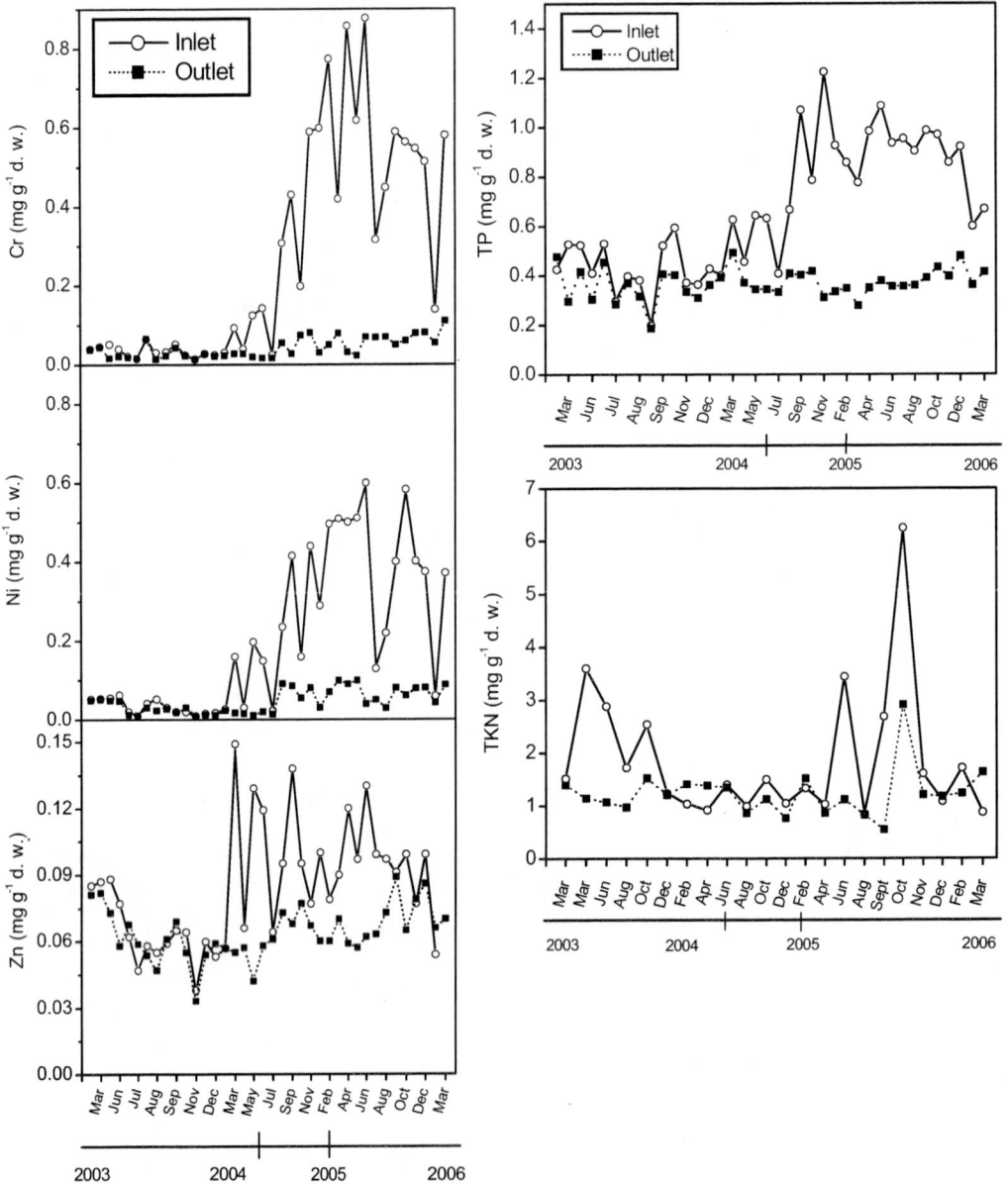

Figure 8. Mean TP, TKN, Cr, Ni and Zn concentrations in sediment in the inlet and outlet areas. The line below the figure indicates the different vegetation dominance stages.

Since June 2004 (*E. crassipes* + *T. domingensis* and *T. domingensis* dominance stages) significant increases in TP, Cr, Ni and Zn sediment concentration at the inlet area were registered. During the *E. crassipes* + *T. domingensis* stage, Cr, Ni and Zn were retained by the sediment (Table 13 and Figure 8). During the *T. domingensis* dominance stage, Cr, Ni, Zn and P were accumulated both in sediment and macrophytes, but mainly in the sediment. Due to the fact that its roots are not in contact with residual water, it retains metals in a lower proportion than the free floating species. Nevertheless, emergent macrophytes contribute to wastewater treatment processes in a number of ways, such as favouring the settlement of suspended solids, providing surface area for bacterial growth, carrying oxygen from the aerial parts to the roots, creating the proper environment in the rhizosphere for the proliferation of microorganisms and promoting a variety of chemical and biochemical reactions which enhance metal retention by the sediment [Brix, 1994, 1997; Kadlec et al., 2000].

Although Zn concentration in the influent and effluent remained below detection limits in most samplings, the increase in plant tissues and in sediment concentrations during the last two vegetation stages suggested that the wetland effectively retained Zn.

There were no significant differences between the initial and final TKN concentrations in sediment, both in the inlet and the outlet areas. Mass balance of N suggested that the biomass N pool accounted for 16-23% of the N removed from the incoming wastewater (Table 13). Given the observed DO depletion it could be proposed that denitrification is the major removal process.

3.3.2.4. Comparison of the Removal Efficiencies during the Three Macrophyte Dominance Stages

As the retention mechanisms at the three vegetation stages were different [Maine et al., 2007a], the removal efficiency at each stage of vegetation dominance defined from plant cover was evaluated. It was observed that there were no significant differences in the removal efficiency of the parameters measured among the three stages, except for NH_4^+, TP and SRP. NH_4^+ showed a different behaviour at the different phases of vegetation development (Table 14). $E.$ $crassipes$ produces a large amount of detritus, which decomposes in the anoxic bottom thus reducing the redox potential in the water column. Organic matter mineralization represents a source of NH_4^+, which is not nitrified because of DO depletion and therefore NH_4^+ was often higher in the effluent than in the influent. During the $E.$ $crassipes$ + $T.$ $domingensis$ period, decreased $E.$ $crassipes$ cover likely allowed a higher oxygen contribution from the atmosphere to the water column. Besides, emergent macrophytes release oxygen from the roots increasing aerobic degradation of organic matter and producing a strong positive effect on nitrifying bacteria in the rhizosphere [Bodelier et al., 1996; Brix, 1997]. The simultaneous occurrence of partial oxidation of ammonium and denitrification account for the ammonium removal observed during the last two stages. Ammonium removal related to the increase of $T.$ $domingensis$ cover is consistent with the removal of ammonium observed in the constructed small-scale wetland in which $T.$ $domingensis$ was the dominant species.

In the first stage, SRP concentration of the effluent had an overall mean concentration 14% higher than at the influent. The low SRP concentrations observed at the inlet in coincidence with the samplings of high pH, calcium and carbonate might be caused by phosphorous co-precipitation with calcium carbonate. The analysis of P fractionation in sediment (Figure 9) showed that the $CaCO_3 \approx P$ fraction increased significantly at the inlet. The concentrations of SRP, higher at the outlet than at the inlet may be due to the mineralization of organic matter. As pH decreases, P sorption to carbonates decreases, while the adsorption to Fe^{+3} oxy hydroxides increases [Golterman, 1995]. However, DO depletion prevented adsorption to iron, resulting in often higher SRP concentrations in the effluent throughout the period of $E.$ $crassipes$ dominance. During the second and third stages, SRP removal percentage was 15% and 22%, respectively. During these stages, SRP continued co-precipitating with $CaCO_3$ at the wetland inlet. The increase of this fraction was remarkable in the inlet sediment (Figure 9). At the outlet there was a significant increase of the $Fe(OOH) \approx P$ fraction at the expense of $CaCO_3 \approx P$ fraction (Figure 9). Therefore, it seems that $CaCO_3 \approx P$

represents the main precipitating mechanism. However, mineralization of organic matter maintained the outlet sediment at a pH range lower than the high values prevailing in the influent. CO_3^{-2} could undergo partial dissolution and the released SRP could be readsorbed onto the $Fe(OOH) \approx P$ fraction [Maine et al., 2007b]. Emergent macrophytes influence the biogeochemical cycles of the sediment through the effects on the redox status of the sediment, due to their capacity to transport oxygen from roots into the rhizosphere [Barko et al., 1991; Sorrel and Boon, 1992]. Qualitatively this is easily visualized by the reddish colour associated with the oxidized forms of iron on the surface of the roots and the surrounding sediment.

Table 14. Mean water composition of the influent and the effluent of the wetland and removal efficiency during the three vegetation stages at the large scale constructed wetland. Different letters represent statistically significant differences in the removal percentages among the periods ($p<0.05$)

Stages	*E. crassipes*			*E. c. + T. d.*			*T. domingensis*		
Parameter	Influent	Effluent	% removal	Influent	Effluent	% removal	Influent	Effluent	% removal
pH	8.8	7.3	-	8.8	7.7	-	9.60	8.13	-
Conductivity (μS cm^{-1})	2803	1372	-	2371	1664	-	3949	2352	-
DO (mg l^{-1})	2.6	0.2	-	0.5	2.3	-	0.8	4.4	-
Ca^{2+} (mg l^{-1})	144.1	50.4	41[a]	154.8	92.2	22[a]	179.8	77.1	44[a]
Mg^{2+} (mg l^{-1})	17.4	12.6	11[a]	17.2	19.3	5[a]	11.3	10.6	10[a]
Alkalin. (mg l^{-1})	422.4	263.2	32[a]	325.9	305.7	12[a]	332.5	233.6	34[a]
SO_4^{2-} (mg l^{-1})	904.0	423.6	37[a]	1338.5	755.4	25[a]	1624.0	812.3	42[a]
NO_3^- (mg l^{-1})	3.80	0.72	65[a]	2.51	0.43	86[a]	6.43	1.20	81[a]
NO_2^- (mg l^{-1})	0.240	0.023	78[a]	0.098	0.008	63[a]	0.413	0.006	91[a]
NH_4^- (mg l^{-1})	1.51	1.76	-24[a]	6.65	2.41	19[b]	2.75	2.59	11[b]
SRP (mg l^{-1})	0.162	0.177	-14[a]	0.058	0.037	15[b]	0.042	0.032	22[b]
TP (mg l^{-1})	0.251	0.269	-8[a]	0.182	0.128	17[b]	0.429	0.321	14[b]
Fe (mg l^{-1})	12.2	0.305	72[a]	0.432	0.096	65[a]	4.76	0.206	73[a]
Cr (μg l^{-1})	12.0	2.33	66[a]	3.6	3.0	55[a]	32	6.0	65[a]
Ni (μg l^{-1})	13.4	7.20	48[a]	47.0	30.4	40[a]	44	15	52[a]
BOD (mg l^{-1})	45.7	12.8	62[a]	40.5	15.1	55[a]	111.4	33.1	68[a]
COD (mg l^{-1})	194.9	38.1	68[a]	95.9	33.3	54[a]	272.9	88.9	70[a]

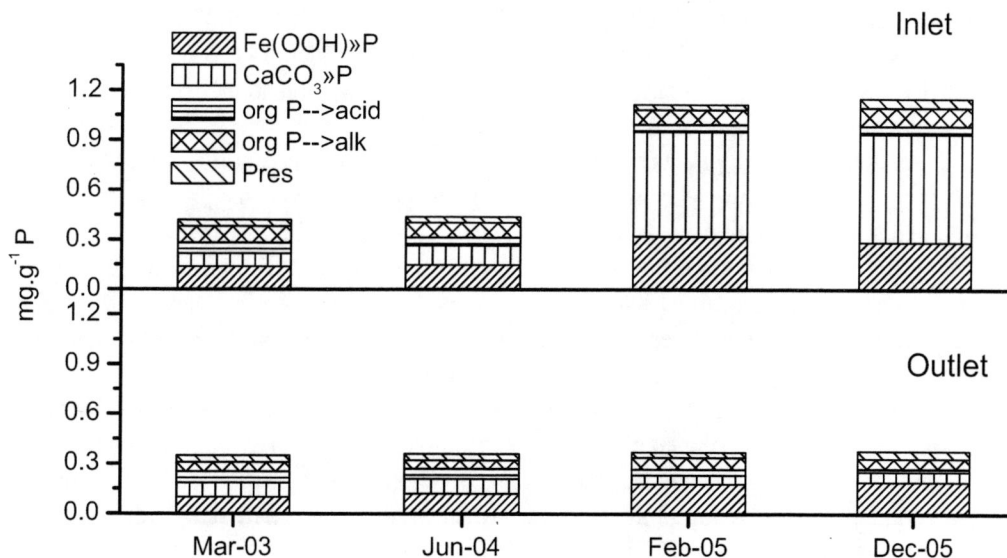

Figure 9. P fractions in sediment at the inlet and outlet at the different vegetation dominance stages.

A complete root-rhizome development for a newly CW may require 3-5 years. The CW performance improves with wetland maturity [Kadlec et al., 2000]. Vymazal and Krópfelová (2005) reported that for *Phragmites* spp., three to four seasons are usually needed to reach maximum standing crop but in some systems it may take even longer. This CW has been dominated by *T. domingensis* for only about a year. Its performance is expected to improve with time.

Plant harvest and extraction of organic matter from the bottom is periodically carried out. These materials are used in the manufacture of compost for growing ornamental plant species in a greenhouse located on the same grounds. Metal concentration in compost and tissues of the ornamental species are analysed on a regular basis, concentrations remain below the permitted levels of national regulations.

3.4. Comparison of the Constructed Wetlands

Maine et al. (2007a) reported that the large-scale wetland showed a retention efficiency for metals similar to that of the small-scale prototype studied earlier. In both wetlands, vegetation development showed a similar pattern, being *T. domingensis* the dominant species after a year of operation. The incoming wastewater was different; the influent of the small-scale wetland presented greater pH, conductivity, nutrient and metal concentrations than the larger wetland, causing the earlier disappearance of the floating macrophytes. Three floating and eight emergent macrophytes were transplanted to the small-scale wetland. After some initial growth, the small-scale wetland became a monospecific stand of *T. domingensis* with a biomass similar to undisturbed environments. Metal concentration in both, incoming wastewater and plant tissues, were lower than the metal thresholds reported in literature [Gibson and Pollard, 1988; Delgado et al., 1993; Sen and Bhattacharyya, 1994; Selvapathy et al., 1997; Cardwell et al., 2002; Fakayode and Onianwa 2002; Ingole and Bhole, 2003;

Manios et al., 2003; Soltan and Rashed, 2003; Maine et al., 2004]. In greenhouse experiments, we found that pH toxicity thresholds of 10, 11 and 11 for *E. crassipes, P. stratiotes* y *S. herzogii*, respectively, and conductivity toxicity thresholds of 4 and 8 mS cm^{-1} for *E. crassipes* and *P. stratiotes*, respectively. In most of the samplings in the small-scale wetland conductivity and pH were higher than the tolerance thresholds. In consequence, conductivity and pH were the main cause of the progressive disappearance of most transplanted species. As regards emergent species, Klomjek and Nitisoravut (2005) studied the response given by eight species used in a constructed wetland, similar to the one studied at a pilot scale, which received saline wastewaters (14-16 mS cm^{-1}). *Typha angustifolia*, among others, proved to be tolerant and showed a satisfactory growth. These conductivity values are higher than those recorded for the studied wetlands, being the reason why *T. domingensis* adapted to the conditions of the system, presenting a positive growth rate. It seems therefore not surprising that floating macrophytes disappeared comparatively earlier than emergent ones. In consequence, the floating macrophytes could not develop within the pH and conductivity prevailing in the incoming wastewater of the small-scale wetland even in the absence of metals. Increased water depth and decreased pH, conductivity and metal concentration in the large-scale wetland resulted in the early dominance of water hyacinth. Then water level was lowered and *T. domingensis* became the dominant species.

At present, the large scale constructed wetland has been in operation for almost 6 years, about 80% of the surface is covered by *T. domingensis* , which attains a high biomass, and retention rates continue in the same magnitude as previously recorded.

4. CONCLUSION

- The CWs efficiently decreased mean concentrations and variability of the parameters analyzed in water, except for the case of SRP and NH_4^+ due to anoxia. The regulating capacity demonstrated by the CWs implies an important advantage if the primary treatment failed and there would be an accidental loading of high concentrations of metals, in which case the CW would retain them.
- Nutrient enrichment improve metal removal by increasing macrophyte tolerance to metals and production, leading to a higher metal uptake by the macrophyte biomass, and also by enhancing the overall biological activity, attaining a higher retention in the detrital fractions.
- Conductivity and pH of the influent were the cause of the disappearance of the floating macrophytes in the wetlands.
- During the *E. crassipes* dominance, contaminants were retained in the macrophyte biomass; during *E. crassipes* + *T. domingensis* stage, contaminants were retained in the sediment and in the *T. domingensis* dominance stage, contaminants were retained in sediment and in the macrophyte biomass.
- Even though retention mechanisms were different, removal efficiencies did not show significant differences among the three vegetation stages. So, the choice of the most suitable species to use should depend on the tolerance of the macrophytes to the conditions of the wastewater to be treated.

- *T. domingensis* was the best adapted species to the studied wastewater in terms of growth and propagation. This species produced high biomass which could be used after harvesting in the manufacture of compost. As a consequence it is the best choice to treat wastewater of high pH and conductivity with heavy metals, a common result from many industrial processes. Its development was favoured by regulating the level of water, attaining its best growth at approximately 0.3 m depth.

- Since the conditions for SRP removal (high pH, Fe, Ca and ionic concentrations) are largely provided by the influent, the CW would be expected to continue retaining SRP as far as the composition of the influent remains the same. Sediment will continue retaining metals, favoured by the presence of *T. domingensis*, while there are available adsorption sites in the sediment. Denitrification is the principal process for N removal.

REFERENCES

Adler, P. R., Summerfelt, S. T., Glenn, D. M. & Takeda, F. (1996). Evaluation of a wetland system designed to meet stringent phosphorus discharge requirements. *Water Environ. Res.*, 68(5), 836-840.

Banerjee, G. & Sarker, S. (1997). The role of *Salvinia rotundifolia* in scavenging aquatic Pb(II) pollution: a case study. *Bioprocess Eng.*, 17, 295-300.

Barko, J. W., Gunnison, D. & Carpenter, S. R. (1991). Sediment interactions with submersed macrophyte growth and community dynamics. *Aquat. Bot.*, 41, 41-65.

Bodeck, I., Lyman, J. W., Rechl, W. F. & Rosenblat, D. (1988). *Environmental Inorganic Chemistry*. New York, Pergamon Press.

Bodelier, P., Libochant, A. J., Blom, C. & Laanbroek, H. (1996). Dynamics of nitrification and denitrification in root-oxygenated sediment and adaptation of ammonia oxidizing bacteria to low-oxygen or anoxic habitats. *Appl. Environ. Microbiol.*, 62, 4100-4107.

Brix, H. (1994). Functions of macrophytes in constructed wetlands. *Water Sci. Technol.*, 29(4), 71-78.

Brix, H. (1997). Do macrophytes play a role in constructed treatment wetlands? *Water Sci. Technol.*, 35(5), 11-17.

Cardwell, A. J., Hawker, D. W. & Greenway, M. (2002). Metal accumulation in aquatic macrophytes from southeast Queensland, Australia. *Chemosphere*, 48, 653-663.

Chaney, R. (1993). Zinc phytotoxicity. In: A. Robson (Ed.), *Zinc in Soils and Plants* (pp. 135-150). Dordercht, Kluwer.

Cheng, S., Grosse, W., Karrenbrock, F. & Thoennessen, M. (2002). Efficiency of constructed wetlands in decontamination of water polluted by heavy metals. *Ecol. Eng.*, 18, 317-325.

D'angelo, E. & Reddy, K. (1993). Ammonium oxidation and nitrate reduction in sediment of a hypereutrophic lake. *Soil Sc. Soc. Am. J.*, 57(4), 1156-1163.

Delgado, M., Bigeriego, M. & Guardiola, E. (1993). Uptake of Zn, Cr and Cd by water hyacinths. *Water Res.*, 27, 269-272.

Dombeck, G., Perry, M. & Phinney, J. (1998). Mass balance on water column trace metals in a free-surface-flow-constructed wetlands in Sacramento, California. *Ecol. Eng.*, 10, 313-339.

Dushenkov, V. P., Nanda Kumar, B. A., Motto, H. & Raskin, Y. (1995). Rhizofiltration: The use of plants to remove heavy metals from aqueous streams. *Environ. Sci. Technol.,* 29, 1239-1245.

Fakayode, S. O. & Onianwa, P. C. (2002). Heavy metal contamination of soil, and bioaccumulation in Guinea grass (*Panicum maximum*) around Ikeja Industrial Estate, Lagos, Nigeria. *Environ. Geol.,* 43, 145-150.

Fendorf, S. (1995). Surface reactions of chromium in soils and waters. Geoderma, 67, 55-71.

Giblin, A. (1988). Pyrite formation in marshes during early diagenesis. *Geomicrobiol. J.,* 6, 77-97.

Gibson, J. P. & Pollard, A. J. (1988). Zinc tolerance in *Panicum virgatum* L. (Switchgrass) from the Picher Mine area. *Proceedings of the Oklahoma Academy of Science.* 68, 45-49.

Golterman, H. L. (1977). Sediments as a source of phosphate for algal growth. In: H. L. Golterman (Ed), *Interactions Between Sediments and Freshwater* (pp. 286-293). The Hague, Dr. W. Junk Publishers.

Golterman, H. L. (1995). The labyrinth of nutrient cycles and buffers in wetlands: results based on research in the Camargue (Southern France). *Hydrobiologia,* 315, 39-58.

Golterman, H. L. (1998). The distribution of phosphate over iron-bound and calcium-bound phosphate in stratified sediment. *Hydrobiologia,* 364, 75-81.

Golterman, H. L. & Booman, A. (1987). The sequential extraction of Ca and Fe-bound phosphates. *Verh. Inter. Verein. Limnol.,* 23, 904-909.

Golterman, H., Bonetto, C., & Minzoni, F. (1988). The nitrogen cycle in shallow water sediment systems of rice fields. Part III: The influence of N-application on the yield of rice. *Hydrobiologia,* 159, 211-217.

Göthberg, A., Greger, M., Holm, K. & Bengtsson, B. E. (2004). Influence of nutrient levels on uptake and effects of mercury, cadmium and lead in water spinach. *J. Environ. Qual.,* 33, 1247

Guo, T., Delaune, R. & Patrick, W. (1997a). The effect of sediment redox chemistry on solubility/chemically active forms of selected metals in bottom sediment receiving produced water discharge. *Spill Sci. & Technol. Bull.,* 4, 165-175.

Guo, T., Delaune, R. & Patrick, W. (1997b). The influence of sediment redox chemistry on chemically active forms of arsenic, cadmium, chromium and zinc in estuarine sediment. *Environ. Internat.,* 23, 305-316.

Hadad, H. R., Maine, M. A. & Bonetto, C. (2006). Macrophyte growth in a pilot-scale constructed wetland for industrial wastewater treatment. *Chemosphere,* 63(10), 1744-1753.

Hadad, H. R., Maine, M. A., Natale, G. S. & Bonetto, C. (2007). The effect of nutrient addition on metal tolerance in *Salvinia herzogii. Ecol. Eng.,* 31(2), 122-131.

Hammer, D. A. (1989). *Constructed Wetlands for Wastewater Treatment.* Chelsea, MI., Lewis Publishers Inc.

Ingole, N. W. & Bhòle, A. G. (2003). Removal of heavy metals from aqueous solution by water hyacinth (*Eichhornia crassipes*). *J. Water Serv. Res. Tec.,* 52, 119-128.

Kadlec, R. H. & Knight, R.L. (1996). *Treatment Wetlands.* Boca Raton, FL., Lewis Publishers.

Kadlec, R. H., Knight, R. L., Vymazal, J., Brix, H., Cooper, P. & Haberl, R. (2000). Constructed Wetlands for Pollution Control: Processes, Performance, Design and Operation. *IWA Specialist Group on use of Macrophytes in Water Pollution Control,* IWA Publishing, pp. 156.

Klomjek, P. & Nitisoravut, S. 2005. Constructed treatment wetland: a study of eight plant species under saline conditions. *Chemosphere,* 58, 585-593.

Lijklema, L. (1977). The role of iron in the exchange of phosphate between water and sediment. In: H. L. Golterman (Ed), *Interactions Between Sediments and Freshwater* (pp. 313-317). The Hague, Dr. W. Junk Publishers.

Lijklema, L. (1980). Interaction of ortho-phosphate with iron (III) and aluminum hydroxides. *Environ. Sci. Technol.,* 14, 537-541.

Loneragan, J. & Webb, M. (1993). Interactions between Zinc and other nutrients affecting the growth of plants. In: A. Robson (Ed.), *Zinc in Soils and Plants* (pp. 119-131). Dordercht, Kluwer.

Losi, M. E., Amrhein, C. & Frankenberger, W. (1994). Environmental biochemistry of chromium. *Rev. Environ. Contam. Toxicol.,* 136, 91-121.

MacCabe, O. & Otte, M. (1997). Revegetation of metal mine tailings under wetland conditions. *Proceedings of the 14th Annual National Meeting-Vision 2000:* An Environmental Commitment, American Society for Surface Mining and Reclamation. Texas, USA.

Maine, M., Panigatti, M. & Pizarro, M. (1998). Role of macrophytes in phosphorus removal in Paraná medio wetlands. *Pol. Arch. Hydrobiol.,* 45(1), 23-34.

Maine, M. A., Suñé, N. L. & Lagger, S. C. (2004). Chromium bioaccumulation: comparison of the capacity of two floating aquatic macrophytes. *Water Res.* 38, 1494-1501.

Maine, M. A., Suñé, N., Hadad, H. R., Sánchez, G. & Bonetto, C. (2005). Phosphate and metal retention in a small-scale constructed wetland for waste-water treatment. In: H. L. Golterman & L. Serrano (Eds.), Phosphates in Sediments. *Proceedings of the 4th International Symposium* (pp. 21-31). Leiden, Backhuys Publishers.

Maine, M. A., Suñé, N., Hadad, H. R., Sánchez, G. & Bonetto, C. (2007a). Removal efficiency of a constructed wetland for wastewater treatment according to vegetation dominance. *Chemosphere,* 68, 1105-1113.

Maine, M. A., Suñé, N., Hadad, H. R. & Sánchez, G. (2007b). Temporal and spatial variation of phosphate distribution in the sediment of a free surface water constructed wetland. *Sci. Tot. Environ.,* 380, 75-83.

Manios, T., Stentiford, E. & Millner, P. (2003). The effect of heavy metals accumulation on the chlorophyll concentration of *Typha latifolia* plants, growing in a substrate containing sewage sludge compost and watered with metaliferus water. *Ecol. Eng.,* 20, 65-74.

Masscheleyn, P., Pardue, J., Delaune, R. & Patrick, W. (1992). Chromium redox chemistry in a Lower Mississippi valley bottom land hardwood wetland. *Evriron. Sci. Technol.,* 26, 1217-1226.

Matheson, F., Nguyen, M., Cooper, A., Burt, T. & Bull, E. (2002). Fate of ^{15}N-nitrate in unplanted, planted and harvested riparian wetland soil microcosms. *Ecol. Eng.,* 19, 249-264.

Mays, P. & Edwards, G. (2001). Comparison of heavy metal accumulation in a natural wetland and constructed wetlands receiving acid mine discharge. *Ecol. Eng.,* 16, 487-500.

Mbuligwe, S. E. (2005). Comparative treatment of dye-rich wastewater in engineered wetland systems (EWSs) vegetated with different plants. *Water Res.,* 39, 271-280.

Minzoni, F., Bonetto, C. & Golterman, H. (1988). The nitrogen cycle in shallow water sediment systems of rice fields. Part I: The denitrification process. *Hydrobiologia,* 159, 189-202.

Monni, S., Salemaa, M. & Millar, N. (2000). The tolerance of *Empetrum nigrum* to copper and nickel. *Environ. Poll.,* 109, 221-229.

Mortimer, C. H. (1941). The exchange of dissolved substances between mud and water in lakes. *J. Ecol.,* 29, 280-329.

Mortimer, C. H. (1942). The exchange of dissolved substances between mud and water in lakes. *J. Ecol.,* 30, 147-201.

Moshiri, G. A. (1993). *Constructed Wetlands for Water Quality Improvement.* Boca Raton, Florida, CRC Press.

Nuttall, C. & Younger, P. (2000). Zinc removal from hard, circum-neutral mine waters using a novel closed-bed limestone reactor. *Wat. Res.,* 34, 1262-1268.

Olsen, S. (1964). Phosphate equilibrium between reduced sediments and water, laboratory experiments with radioactive phosphorus. *Verh Int Ver Limnol.,* 13, 915-922.

Panigatti, M. C. & Maine, M. A. (2003). Influence of nitrogen species (NH_4^+ and NO_3^-) on the dynamics of P in water-sediment-*Salvinia herzogii* systems. *Hydrobiologia,* 492, 151-157.

Paris, C., Hadad, H. R., Maine, M. A. & Suñé, N. (2005). Eficiencia de dos macrófitas flotantes libres en la absorción de metales pesados. *Limnetica,* 24(3-4), 237-244.

Reddy, K. R. & Sutton, D. L. (1984). Water hyacinths for water quality improvement and biomass production. *J. Environ. Qual.,* 13, 1-7.

Reddy, K., Patrick, W. & Lindau, C. (1989). Nitrification-denitrification at the plant root-sediment interface in wetlands. *Limnol. Oceanogr.,* 34, 1004-1013.

Scoccianti, V., Crinelli, R., Tirillini, B., Mancinelli, V. & Speranza, A. (2006). Uptake and toxicity of Cr(III) in celery seedlings. *Chemosphere,* 64(10), 1695-1703.

Selvaphathy, P., Juliet Jesline, J. & Prebha, S. (1997). Heavy metal removal from wastewater by Water Lettuce. *Indian J. Environ. Prot.,* 18, 1-6.

Sen, A. K. & Bhattacharyya, M. (1994). Studies of uptake and toxic effects of Ni (II) on *Salvinia natans*. *Wat. Air Soil Pollut.,* 78, 141-152.

Shanker, A., Cervantes, C., Loxa-Tavera, H. & Avudainayagam, S. (2005). Chromium toxicity in plants. *Environ. Int.,* 31(5), 739-753.

Shepherd, H. L., Grismer, M. E. & Tchobanoglous, G. (2001). Treatment of high-strength winery wastewater using a subsurface-flow constructed wetland. *Water Environ. Res.,* 73(4), 394-403.

Soltan, M. E. & Rashed, M. N. (2003). Laboratory study on the survival of water hyacinth under several conditions of heavy metal concentrations. *Adv. Environ. Res.,* 7, 321-224.

Sorrell, B. K. & Boon, P. L. (1992). Biogeochemistry of billabong sediments. II Seasonal variations in methane production. *Freshwater Biol.,* 27, 435-445.

Vymazal, J. (2005). Horizontal sub-surface flow and hybrid constructed wetlands systems for wastewater treatment. *Ecol. Eng.,* 25, 478-490.

Vymazal, J., Brix, H., Cooper, P. F., Green, M. B. & Haberl, R. (1998). *Constructed Wetlands for Wastewater Treatment in Europe.* Leiden, Backhuys Publishers, pp. 366.

Vymazal, J. & Krópfelová, L. (2005). Growth of *Phragmites australis* and *Phalaris arundinacea* in constructed wetlands for wastewater treatment in the Czech Republic. *Ecol. Eng.*, 25, 606-621.

Weisner, S., Eriksson, P., Granéli, W. & Leonardson, L. (1994). Influence of macrophytes on nitrate removal in wetlands. *Ambio.*, 23(6), 363-366.

Reviewed by Dra. Argelia Lenardón. Facultad de Ingeniería y Ciencias Hídricas, Universidad Nacional del Litoral. Paraje "El Pozo" s/n (3000) Santa Fe, Argentina. Consejo Nacional de Investigaciones científicas y Técnicas (CONICET). E-mail: lenardon@ceride.gov.ar

In: International Wetlands: Ecology, Conservation…   ISBN: 978-1-60456-999-5
Editor: Jose R. Herrara   © 2009 Nova Science Publishers, Inc.

Chapter 6

PLANKTONIC TROPHIC INTERACTIONS IN A TEMPERATE SHALLOW LAKE FROM A SOUTH AMERICAN WETLAND

Rodrigo Sinistro and Irina Izaguirre

Lab. Limnología, Depto. Ecología, Genética y Evolución, Facultad de Ciencias
Exactas y Naturales, Universidad de Buenos Aires, Argentina

ABSTRACT

We have analysed the planktonic communities and their trophic interactions in a vegetated wetland from the Lower Paraná River (Otamendi Natural Reserve, Argentina). The studies combined field surveys, and *in situ* experiments at microcosms and mesocosms scales.

Fields studies carried out in the main shallow lake of the wetland and in abandoned meanders of the river (relict oxbow lakes) show that the free-floating plants cover was one of the driving forces in the regulation of the structure of planktonic communities. Later on, the effect of the light attenuation due to floating macrophytes was simulated at microcosms scale, which showed changes in the structure of the microbial planktonic communities in response to the light penetration. In particular we found changes in the relationship autotrophs/heterotrophs + mixotrophs, depending on light conditions inside the enclosures. Further experiments with fluorescent-labeled bacteria (FLB), revealed the presence of mixotrophic phytoplankton species, which are frequent components of the phytoplankton in this wetland.

The zooplankton predation impact on phytoplankton was assessed using experiments at mesocosms scale in terms of abundance, size fraction structure and species composition. Likewise, the cascading responses of other components the microbial chains were also explored (heterotrophic nanoflagellates, ciliates and picoplankton)

On the other hand, the effect of nutrients (bottom up) and predation (top down) on the phytoplankton communities were simultaneously assessed. Regarding the top down effect, the impact of planktivorous fish predation over zooplankton was analyzed, as well as the consecutive cascading effect on phytoplankton. Moreover, the effect of nutrient release from natural lake sediments, was assessed.

In this communication we propose a first model of functioning of the plankton communities and their main interactions for this South American wetland, based on field and experimental data, which have been obtained along a six- years study period.

INTRODUCTION

South American wetlands were defined by Neiff et al. (1994) as "systems of sub-regional extent in which the spatial and temporal presence of variable water cover causes characteristic biogeochemical fluxes, soils of accentuated hydromorphism, and a biota whose structure and dynamics are well adapted to a wide range of variability". In South America the higher proportion of surface occupied by wetlands corresponds to the large rivers floodplains, most of them (80%) occurring in warm climates (Neiff & Malvares, 2004). Floodplains are complex wetlands which generally include a great number of shallow lakes with different limnological features. Junk et al. (1989) have described the floodplains wetlands as systems with regular flooding pulses, which together with the habitat heterogeneity, favor a high biodiversity of plants and animals.

The researches involved in this chapter were conducted in a floodplain wetland from the Lower Paraná River, located in the Otamendi Natural Reserve, between the Luján and the Paraná de las Palmas Rivers, Buenos Aires, Argentina (34° 10' - 34° 17' S ; 58 ° 48 ' - 58° 53' W). It is included in a region called "Bajo Delta del Paraná" (Malvares, 1999). The reserve has two permanent shallow lakes and several semi-permanent relict oxbow lakes (ROLs), which are surrounded by marshy vegetation, and temporarily covered by floating plants. The region has a temperate Pampean humid climate, and the hydrological cycles are controlled by rainfall. Precipitations occur during the whole year, with a mean annual value of 950 mm.

The water bodies have relatively high concentrations of phosphates, which are within the ranges reported for eutrophic systems, although nitrogen may be limiting for phytoplankton during periods of active algal growth (de Tezanos Pinto et al., 2007; Sinistro et al., 2006; Unrein, 2001). Waters are highly coloured, indicating high humic content (Rodríguez & Pizarro, 2007). According to the descriptions given by Williamson et al. (1999) this aquatic system can be defined as an ecosystem of mixotrophic lakes, with high concentrations of dissolved organic carbon (DOC) and total phosphorus.

Previous studies conducted in this wetland (Izaguirre et al., 2001; 2004; O'Farrell et al., 2003) have shown that there are important differences between the main shallow lake and the relic oxbow lakes, mainly triggered by the presence of free floating plants (Figure 1). There is a increase in light, dissolved oxygen and autotrophic/heterotrophic ratio from the ROLs to the shallow lakes, whereas the inverse trend occurs for nutrients, dissolved inorganic carbon and stability. The dense cover of floating plants provokes a drastic decrease in the light penetration in the water column, which in turn generates an imbalance between photosynthesis and respiration. This mainly accounts for the frequent anoxic conditions of the completely vegetated ROLs. The profuse vegetation in these abandoned meanders also accounts for their high concentrations of DOC. In general, relict water bodies exhibit a higher stability in their physico-chemical features and algal assemblages, due to the permanence of the floating plants (O'Farrell et al., 2003). During periods of complete high development of floating macrophytes, these water bodies generally show an algal flora well adapted to low light and anoxia (Izaguirre et al., 2001), with many algae species being either mixotrophic or heterotrophic. Conversely, when ROLs lack the dense macrophyte cover, the phytoplankton community undergoes compositional changes which are related to the improvement in the underwater light climate (O´Farrell et al., 2007). In permanent shallow lakes, that are generally free of floating plants, more pronounced environmental fluctuations occur along the year and the algal assemblages vary concordantly.

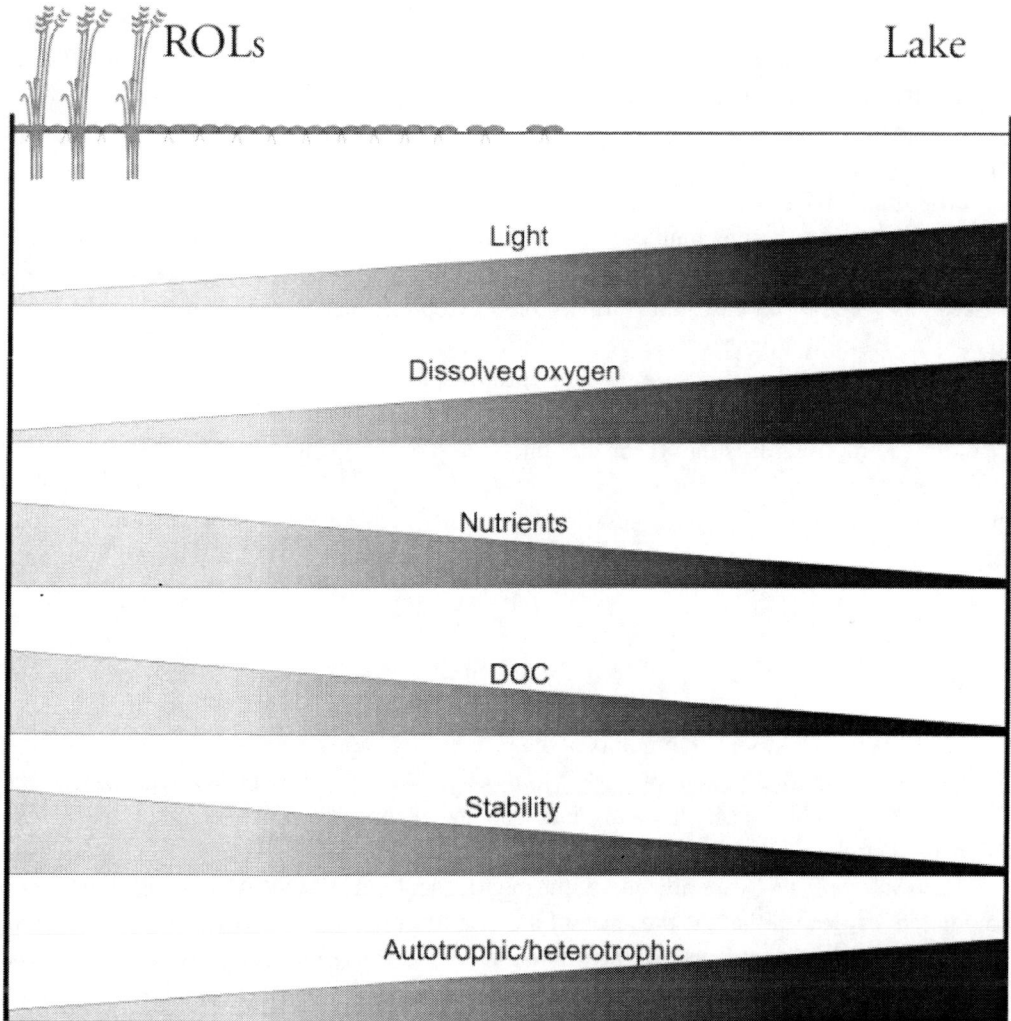

Figure 1. Characteristics of the floodplain wetland (Natural Reserve Otamendi).

METHODS

Different field experiments were conducted in the main shallow lake of the Otamendi Reserve with the aim to assess the main trophic interactions in the microbial plankton communities. Experimental studies were carried out in situ at microcosms and mesocosms scales (Sinistro et al., 2006; 2007). The hypotheses tested in the two first microcosms scale experiments were related to the shading effect of free floating plants on the microbial plankton communities; particularly focusing in changes in the relationship autotrophs to heterotrophs + mixotrophs. Microcosm scale experimental study were employed to explore the mixotrophic behavior of two *Cryptomonas* species (frequently occuring in this wetland), using fluorescent-labelled bacteria (FLB). Later, a further mesocosm scale experimentation was undergone with the aim to analyse the predation impact of zooplankton on phytoplankton, evaluating algae abundance, size structure and community composition. The cascading response of zooplankton on other components of the microbial chain (heterotrophic nanoflagellates, ciliates and picoplankton) was further explored (heterotrophic nanoflagellates, ciliates and picoplankton). The full description of the experimental designs, as well as the corresponding methodology of physical, chemical and biological analyses were given in Sinistro et al., (2006; 2007)

Furthermore, the effect of nutrients (bottom-up control) and predation (top-down control) on the phytoplankton communities were simultaneously analyzed at mesocosms scale. Regarding the top-down control, the grazing impact of planktivorous fish on zooplankton was assessed, as well as its cascading effects on phytoplankton. The bottom-up control was explored assessing nutrient release from natural sediments of the shallow lakes, by placing natural sediments in dialysis bags within mesocosms. The details corresponding to this experiment were described by Sinistro (2007).

We here propose a first model for the functioning of the plankton communities and their main interactions for this South American wetland. The model is based on field and experimental data from the above described studies, which have been obtained along a six year period.

CONCLUSION

The wetland planktonic community is conformed by many algal species of small size, among which some are mixotrophic, a wide variety of heterotrophic nanoflagellates and ciliates, heterotrophic bacteria and autotrophic picoplankton (dominated by pico-cyanobacteria rich in phycocianine)

The results of the experiments with microcosms, in which light conditions were manipulated, evidenced that the presence of a dense macrophyte cover, that reduced the light penetration, affected the structure of the microbial community. One of the most important and evident effects of the macrophytes on the water column is the decrease in light penetration and the concomitant decrease of the photosynthetic activities, which in turn leads to a decrease in oxygen levels. In such situations, the reduced chemical forms prevail in the lake. In the natural environment, as nutrients are more concentrated near the bottom sediments, and anoxia generally occurs, an ammonium vertical gradient in the water column is frequently observed, because it is uptaken by algae and bacteria. These characteristics are typical in all the vegetated water bodies in the wetland, as described in previous studies (Izaguirre et al., 2004; O'Farrell et al., 2003).

In general, due to the abundant aquatic vegetation, the water bodies in the wetland are characterized by a high concentration of humic substances, which are reflected in the very high dissolved organic carbon (DOC) values. Franco and Heath (1983) found that the availability of nutrients decreases in lakes with high concentration of DOC because they bond into complexes (humic substance-metal-phosphorus). Nevertheless, these complexes may also constitute potential phosphate reservoirs, which may be released from the complexes into the water column in situations of phosphorus deficit (Jones, 1998). The high levels of DOC in the aquatic environments in this wetland, together with good phosphorus availability, build optimal conditions for the development of the microbial pathways, and thus an important fraction of the bacterial production would be canalized towards the superior trophic levels through the "microbial loop". In this way, the trophic chains in the wetland profit the energy source provided by the organic matter resulting from the aquatic vegetation. In these situations, microzooplankton plays a very important role because they are the fraction that canalizes the supply of the organic matter towards the trophic cascade, by predating on picoplankton.

In our experiments performed with microcosms (Sinistro, 2006) where treatments with different light penetration were compared, we observed that the planktonic components analysed (autotrophic algae, mixotrophic algae, ciliates and HNF) evidenced changes associated to the light availability. In particular, it was observed that a decrease in the light intensity favored the replacement of obligate autotrophic species into mixorophic algae, which are probably better competitors under conditions of light deficiency, especially when a progressive decrease in nutrients occurs. Moreover, mixotrophic algae may be limited by nitrogen, whereas the HNF may be limited by carbon thus, the latter group would be better adapted than the former when the macrophyte cover is very profuse (light is limiting but DOC is high). In this sense, Jansson et al. (1996) explained the possible relations among proportions of mixotrophic to strictly heterotrophic organisms within a water body. When the biomass of HNF in a water body is relatively low compared to mixotrophic organisms, this may indicate that there is enough light for photosynthesis, and thus mixotrophy overruns heterotrophy.

Even if in our experiments the changes in the phytoplankton composition related to the different light conditions were not explored in detail, the relationship eucaryotes/procaryotes, as well as the proportion of the dominant clases, did show changes under different light conditions. In general, chlorophytes were favored under higher light conditions. Huissman *et al.* (1999) have shown that the optimal light intensity varies in a species specific way. The experiments of Flöder et al. (2002) showed that light fluctuations affected the diversity of the phytoplankton communities and that chlorophytes dominated in the treatments with permanent high light, whereas the cyanobacteria and the diatoms dominated under low light intensities, in agreement with our experiments.

Throughout our experiments, the dominant taxa within the group of mixotrophic algae were several species of the genus *Cryptomonas*. Following Jones (2000), these algae may be included in the protist group whose main nutrition is autotrophy but that can shift into phagotrophy whenever the periods of darkness are prolonged. Mixotrophy is more energetically costly than obligate autotrophy, because they should keep the "machinery" of both systems (photosynthetic and phagotrophic) resulting in higher energy requirements than maintaining one of the systems (to equal cellular sizes) (Raven, 1997). Stoecker (1998) and Jansson et al. (1996) proposed that the *Cryptomonas* spp. are within the group of mixotrophic algae that are mainly phototrophic but that they may feed by phagotrophy to obtain organic traces needed for their growth. In our bacterivory experiments, performed with fluorescent-labelled bacteria (FLB), we directly assayed the bacteria ingestion by alga that were assumed as potentially mixotrophic (*Cryptomonas erosa* and *C. marssonii*) in the first phase of the researches, and that are frequent in this wetland. The predation rates we calculated are within the highest values observed for cryptophytes (Panuska & Robertson, 1999; Tranvik et al., 1989). Nevertheless, they are within the ranges calculated by Domaizon et al. (2003).

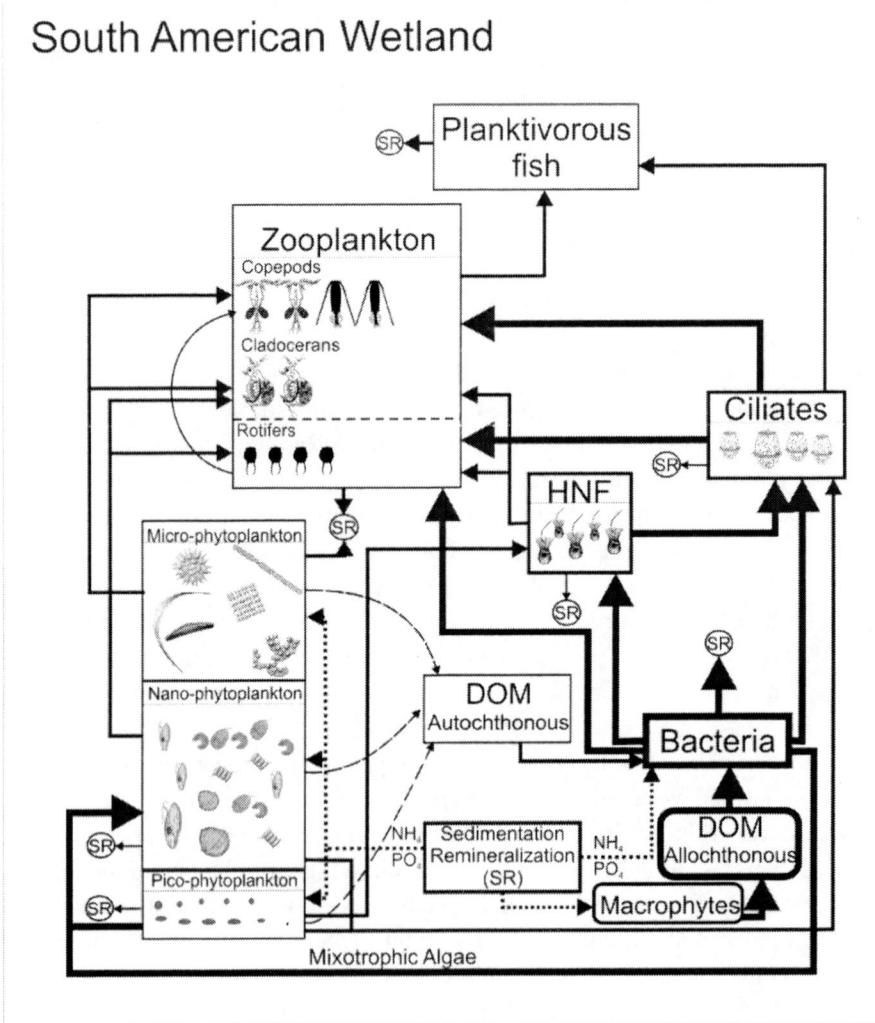

Figure 2. Scheme of the trophic ways in a South American floodplain wetland, where the boxes represent the principal components assayed, based on the microcosm and mesocosm scale experiments. The width of the arrows indicate the importance of the interactions among components.

The different experiments performed, both at microcosm and mesocosm scale, allowed to analyze the trophic interactions among the different planktonic components. In Figure 2, a model of the interactions between different planktonic components encountered in this wetland is proposed. In this scheme the arrows show the experimentally explored interactions among components, which present different width, depending on the importance of a given process in this wetland.

As regards trophic interactions at the level of the microbial chains, the experiments in microcosms evidenced straightforwardly, the effect of the predation of the potential picoplankton (heterotrophic nanoflagellates, ciliates and mixotrophic algae) predators. The occurrence of such predators produced a marked decline in the picoplankton abundance. This became evidently when comparing the treatments that included all the planktonic components with those were the predators were not present. Nevertheless, no significant differences in predation were recorded among the treatments with and without light. It is worth recalling that the HNF are the main bacteria predators in aquatic systems (Hahn & Hofle, 2001; Porter et al., 1985; Sanders et al., 1989), even if ciliates (Langenheder & Jürgens, 2001; Sherr & Sherr, 1987), and flagellated mixotrophs (Jansson et al., 1999; Jones, 1994) are capable of feeding on picoplankton.

On the other hand, experiments performed in mesocosm in this wetland, allowed the analysis of the interactions among the mesozooplankton and other planktonic components (Sinistro, 2007). The obtained results support the concept of Hulot et al. (2001) which divides the herbivore zooplankton in two categories (small and big), which in turn predate in two phytoplankton categories (unprotected and protected). The bigger sized algae (generally above 35 μm) and the algae with mucilage may be considered as non protected species. The small sized herbivores (principally rotifers and *nauplii* copepods) feed exclusively of non protected phytoplankton, whereas the big sized herbivores (cladocerans and copepods adult and preadult) may also predate on protected algae. Moreover, our experimental studies concord with Sommer *et al.* (2003) proposal, in the sense that zooplankton is more efficient in reducing the phytoplankton biomass when it has a mayor functional diversity (copepods, cladocerans and rotifers). In particular, our experiments showed that the control of big sized algae only occurred when the zooplankton in the enclosures included adult copepods and cladocerans. This occurs because the zooplankton community is confirmed by a wider range of sizes and organisms with different feeding mechanisms (filtrators, capturers) allowing to control different phytoplankton sized preys. In our studies, we observed that the zooplankton posed a strong predation pressure over the nanoplanktonic fraction. Conversely, it may be concluded that the zooplankton would not selectively predate on the micro- phytoplanktonic fraction (algae > 30 μm), because the algae of this size increased during the course of the experiments, even under the presence of zooplankton, albeit they decreased when the zooplankton community was composed by organisms whose sizes and feeding habits allowed them to predate on this fraction.

On the other hand, the results of our experiments support the concept of trophic cascades, where the alteration of one of the components impacts in turn in the lower levels. At the microbial chain level, an increase of the HNF and ciliates was observed due to the absence of their own predators (mesozooplancton), and to a decline in the picoplankton. In the enclosures lacking zooplankton, the abundance of ciliates followed the same increase pattern than HNF, but with a delay in the timing of the response, due to differences in the generational times between trophic levels. Moreover, the picoplankton total abundance in the enclosures containing zooplankton, reaches its peak only after a marked decrease on the small sized ciliates. As previously indicated, even if the HNF are the main picoplankton predators, the small sized ciliates may also pose a grazing pressure over this fraction (Callieri & Stockner, 2002; Pernthaler, 2005; Pernthaler, 1996). Nevertheless, the impact of zooplankton over the microbial communities not always is straightforward due to complex interactions between the components, and there is some certain controversy about the effects of the trophic cascade towards the lower levels (Pace & Funke, 1991; Schnetzer & Caron, 2005).

Other of the experiments with mesocosms performed in this wetland focused in analyzing the combined effect of the nutrients supply from sediments (bottom-up) and the predation (top-down) produced by planktivorous fish over zooplankton (Sinistro, 2007).

It was observed that in all the treatments that included either planktivorous fish or wetland's sediments or both, the algal growth had a positive response when compared to the control, particularly the algae < 30 μm. No significant differences in total phytoplankton densities were found among treatments by the end of the experiment, nevertheless there were differences for the nano-phytoplanktonic fractions. This latter fraction was the predominant in the system at the onset of the experiment and it is also the fraction more vulnerable to zooplankton predation. Thus, it is probable that this is the reason why maximal algal densities were encountered in the combined treatment (sediments plus planktivorous fish), where in addition to count with an extra supply of sediments, the grazing pressure was lessened due to the cascading effects of fish on zooplankton.

In the treatment where only sediments were added, the lack of fish allowed zooplankton (mainly cyclopois copepods, and cladocerans) to control phytoplankton growth, and thus a top-down effect was evident. This, in turn, impacted on the phytoplankton size structure, with an increase in the density of micro-phytoplankton, mainly big filamentous and colonial cyanobacteria (*Planktolyngbya limnetica*, *Anabaena* sp., *Microcystis* sp. y *Aphanothece* sp.), because theses species received less predating pressure by zooplankton when compared with the nano-planktonic fractions.

Analyzing the treatment were fish were added but no sediment supplement was provided, phyhoplankton was controlled more by the "bottom-up" effect that by the "top-down" effect. By the end of the experiment, this treatment, presented a low density of cyclopoid copepods, cladocerans and rotifers, together with a high density of ciliates, when compared to all other treatments. In this treatment, the density of the micro-phytoplanktonic fraction was lower, probably due to a competition effect for nutrients between both algal size fractions, where probably the smallest algae would be favored.

Comparing the treatments where both sediments and planktivorous fish were added (SP), versus the treatments solely with sediments (S), in the latter a higher proportion of algae of size >30 μm was observed. Thus, in the absence of fish, zooplankton controlled more efficiently the nanoplanktonic algae. Even if it was expected that the SP treatments presented the highest values of algal densities, the higher nutrient availability and the lessened grazing pressure, it is interesting to analyze the implications of the top-down controls and bottom-up for this kind of systems. Even if in eutrophic environments it may be assumed that the phytoplankton is regulated more by a top-down effect, because nutrients would not be limiting, this not always occurs. Under certain circumstances, an intense algal growth may turn nutrients limiting for phytoplankton and this may be the main factor controlling algae. This situation may occur throughout the year in the lake, with nitrogen being the nutrient generally limiting for algal growth. Such nitrogen limitation in shallow vegetated lakes has been described by diverse authors (Jansson, 1998; Saunders *et al.*, 2000; Scheffer, 1998). On the other hand, if the nitrogen limitation favors the development of nitrogen fixing algae, among which big sized species are found, this is another important factor to take into account.

In this sense, the conditions of top down regulations play a minor role because these algae are generally not controlled by zooplankton.

As regards the role of fish, many studies have pointed out their impact over the plankton community (Brooks & Dodson, 1965; Holopainen et al., 1992; Hrbácek et al., 1961; Mittelbach et al., 1995; Reinertsen et al., 1990). The big sized zooplankton species are vulnerable to visual predation by planktivorous fish, and thus, lakes with abundant planktivorous fish usually are dominated by small sized zooplankton species, including nano and microzooplankton (Järvinen, 2002). In our experiments we observed a marked effect of the planktivorous fish over the zooplankton; in the treatments where planktivorous fish were added, the mesozooplankton fraction (cyclopoid copepods and cladocerans) was almost cero. On the contrary, rotifers did not showed significant changes in their abundances among treatments, probably because they are beyond the threshold of vision of the planktivorous fish introduced. The effect of the fish predation over the mesozooplankton produces a cascading effect liberating the phytoplankton (mainly algae < 30 µm) from their predating pressure.

In the trophic cascades, the body size is a key factor in the interaction predator-prey (Scheffer, 1998). In our experiment, the planktivorous fish introduced were of a fixed size (1.5 to 2 cm), and thus they selectively predated over the largest zooplankton preys. Thus, the cascade effect was in part conditioned by the experimental conditions. In nature, the same response should be observed when the cohort of fish was of similar sizes. Nevertheless, it must be highlighted that throughout the year in the natural environment, it is possible to observe different cascading responses triggered mainly by the planktivorous fish size structure. It is important to mention that one individual may, along its life cycle change its diet, not only due to an increase in its body size, but also depending on the season and the sex. For example, a *nauplii* cyclopoid copepod feed on bacteria and algae, but when it turns an adult it may feed on other zooplankton, even if its main diet continues to be based on algae, bacteria and ciliates.

The scheme (Figure 3) represents graphically two possible scenarios- fish absence or presence - that resume the cascading effects observed in the manipulative experiment for the planktivorous fish size fractions utilized. The height of the black band represents the abundance of each of the components and its width represents the variation in the size fraction or the different species within the component. On the other hand, the gray areas represents the grazing pressure posed by one given predator size over the different plankton communities, producing a decrease on the abundance of the component (high of the band).

As regards nutrients, it is worth mentioning that the zooplankton grazing effect increases the available nutrients for the bacterial and phytoplanktonic growth, and in this way compensate its decreases due to predation. In this sense, the interactions predator- prey, the superior trophic levels may affect the availability of nutrients for the primary producers, through the stoichiometry of the recycling by consumers, triggering a "bottom-up" effect, as a response of the predation over preys. In the experiment with addition of zooplankton, an increase in the nutrient concentrations was observed, mainly of dissolved inorganic nitrogen, in those treatments including zooplankton.

Ciliates densities were highest in the treatments with fish, probably due to a cascading effect (fish- mesozooplankton – ciliates- HNF). Ciliates not only increased in numbers, but also bigger body sizes occurred. For this same reason, the HNF showed a rather unchanged dynamics, with a tendency to decrease throughout the experiment in the treatments with planktivorous fish.

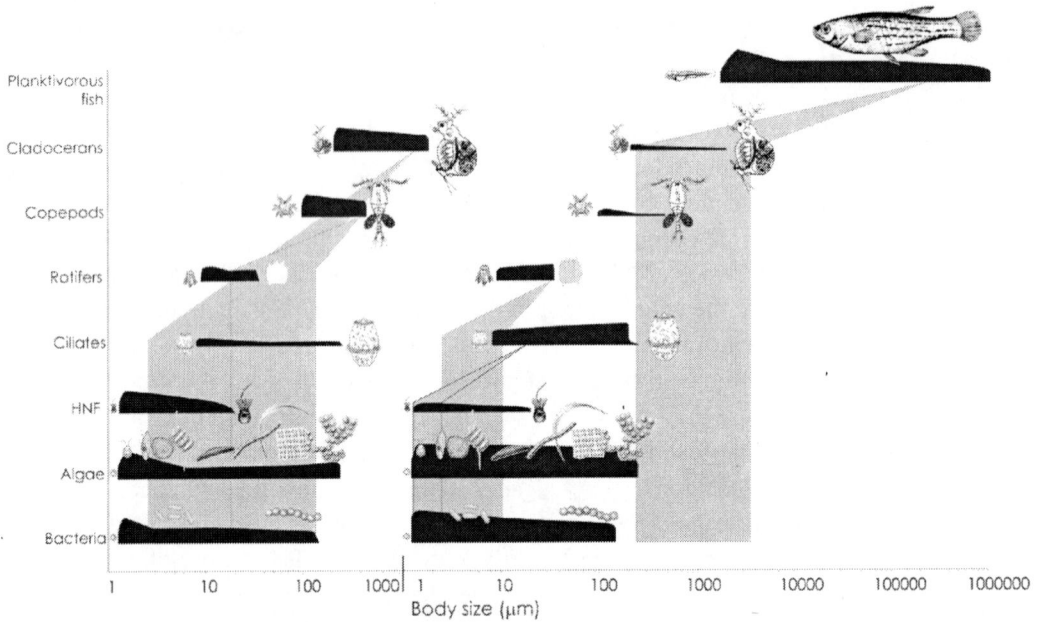

Figure 3. Scheme of the trophic cascades, under two possible situations (with and without planktivorous fish), where the body size plays an important role in the predator- prey interactions. The effect of a given predator on its potential preys is highlighted in gray color. The high of the black band indicates the abundance of the component.

Furthermore, in the treatments without planktivorous fish, as the mesozooplankton was present, the interactions between ciliates and HNF resulted less evident. As before mentioned, many authors explain that these trophic cascade interactions may become truncated at the protozoan level (Pace & Funke, 1991). Because the mesozooplankton is present, the observed decreases may be explained by direct effects (predation, interference) or by indirect effects (competence for resources) produced by copepods, cladocerans and even by rotifers (Arndt, 1993; Burns & Gilbert, 1993; Jack & Gilbert, 1994; Perrow et al., 1999).

REFERENCES

Arndt, H. (1993). Rotifers as predators on components of the microbial web (bacteria, heterotrophic flagellates, ciliates) - a review. *Hydrobiologia 255/256*, 231-246.

Brooks, J. L. & Dodson, S. I. (1965). Predation, body size, and composition of plankton. *Science 150*, 28-35.

Burns, C. W. & Gilbert, J. J. (1993). Predation on ciliates by freshwater calanoid copepods: rates of predation and relative vulnerabilities of prey. *Freshwater Biology 30*, 377-393.

Callieri, C. & Stockner, J. G. (2002). Freshwater autotrophic picoplankton: a review. *Journal of Limnology 61*, 1-14.

de Tezanos Pinto, P., Allende, L. & O'farrell, I. (2007). Influence of free-floating plants on the structure of a natural phytoplankton assemblage: an experimental approach. *Journal of Plankton Research 29*, 47-56.

Domaizon, I., Viboud, S. & Fontvieille, D. (2003). Taxon-specific and seasonal variations in flagellates grazing on heterotrophic bacteria in the oligotrophic Lake Annecy-importance of mixotrophy. *FEMS Microb. Ecol. 46*, 317-329.

Flöder, S., Urabe, J. & Kawataba, Z. (2002). The influence of fluctuating light intensities on species composition and diversity of natural phytoplankton communities. *Oecologia 133*, 395-401.

Franco, D. A. & Heath, R. T. Abiotic uptake and photodependent release of phosphate from high-molecular-weight humic-iron complexes in bog lakes. In: R. F. Christman & E. Gjessing, *Aquatic and terrestrial humic materials*. Ann Arbor: Ann Arbor Sci. Publ.; 1983: 467-480.

Hahn, M. W. & Hofle, M. G. (2001). Grazing of protozoa and its effect on populations of aquatic bacteria. *FEMS Microb. Ecol. 35*, 113-121.

Holopainen, I. J., Tonn, W. M. & Paszkowski, C. A. (1992). Effects of fish density on planktonic communities and water quality in a manipulated forest pond. *Hydrobiologia 243/244*, 311-321.

Hrbácek, J., Draková, B., Korínek, V. & Procházkóva, L. (1961). Demonstration of the effects of fish stock on the species composition of zooplankton and the intensity of metabolism of the whole plankton association. *Verh. Interant. Verein. Limnol. 14*, 192-195.

Huisman, J., Jonker, R. R., Zonneveld, C. & Weissing, F. J. (1999). Competition for light between phytoplankton species: experimental tests of mechanistic theory. *Ecology 80*, 211-222.

Hulot, F. D., Morin, P. J. & Loreau, M. (2001). Interactions between algae and the microbial loop in experimental microcosms. *Oikos 95*, 231-238.

Izaguirre, I., O'Farrell, I. & Tell, G. (2001a). Variation in phytoplankton composition and limnological features in a water-water ecotone of the lower Paraná Basin (Argentina). *Freshwater Biology 46*, 63-74.

Izaguirre, I., O'Farrell, I., Unrein, F., Sinistro, R., dos Santos Afonso, M. & Tell, G. (2004). Algal assemblages across a wetland, from a shallow lake to relictual oxbow lakes (Lower Paraná River, South America). *Hydrobiologia 511*, 25-36.

Jack, J. D. & Gilbert, J. J. (1994). Effects of Daphnia on microzooplankton communities. *Journal of Plankton Research 16*, 1499-1512.

Jansson, M., Blomqvist, P., Jonsson, A. & Bergström, A.-K. (1996). Nutrient limitation of bacterioplankton, autotrophic and mixotrophic phytoplankton, and heterotrophic nanoflagellates in Lake Örträsket. *Limnology and Oceanography 41*, 1552-1559.

Jansson, M. Nutrient limitation and bacteria – phytoplankton interactions in humic lakes. In: D. O. Hessen & L. J. Tranvik, *Aquatic humic substances: ecology and biogeochemistry*. Berlin Heidelberg: Springer-Verlag; 1998.

Jansson, M., Bergström, A.-K., Blomqvist, P., Isaksson, A. & Jonsson, A. (1999). Impact of allochthonous organic carbon on microbial food web carbon dynamics and structure in Lake Örträsket. *Archiv für Hydrobiologie 144*, 409-428.

Järvinen, M. (2002). *Control of plankton and nutrient limitation in small boreal brown-water lakes: evidence from small- and large-scale manipulation experiments.* Department of Ecology and Systematics. Helsinki, University of Helsinki: 45.

Jones, R. I. (1994). Mixotrophy in planktonic protists as a spectrum of nutricional strategies. *Marine Microbial Food Webs 8*, 87-96.

Jones, R. I. Phytoplankton, primary production and nutrient cycling. In: D. O. Hessen & L. J. Tranvik, *Aquatic humic substances: ecology and biogeochemistry.* Berlin Heidelberg: Springer-Verlag; 1998: 145-175.

Jones, R. I. (2000). Mixotrophy in planktonic protists: an overview. *Freshwater Biology 45,* 219-226.

Junk, W. J., Bayley, P. B. & Sparks, R. E. (1989). The flood pulse concept in River-floodplain systems. Proceeding of the International Large River Symposium, Canadian Spec. Publ. Fisheries and Aquatic Science.

Langenheder, S. & Jürgens, K. (2001). Regulation of bacterial biomass and community structure by metazoan and protozoan predation. *Limnology and Oceanography 46*, 121-134.

Malvares, A. I. (1999). El delta del río Paraná como mosaico de humedales. Tópicos sobre humedales subtropicales y templados de Sudamérica, Montevideo, UNESCO.

Mittelbach, G. G., Turner, A. M., Hall, D. J., Rettig, J. E. & Osenberg, C. W. (1995). Perturbation and resilience: a long-term, whole-lake study of predator extinction and reintroduction. *Ecology 76*, 2347-2360.

Neiff, J. J., Iriondo, M. H. & Carignan, R. (1994). Large Traopical South American Wetlands: An Overview. Procedings of the International Workshop on The ecology and management of aquatic-terrestrial ecotones.

Neiff, J. J. & Malvares, A. I. Grandes humedales fluviales. In: A. I. Malvares, *Documentos del curso-taller "Bases ecológicas para la clasificación e inventario de humedales en Argentina".* Buenos Aires: Talleres Gráficos Leograf S.R.L.; 2004: 119.

O'Farrell, I., Sinistro, R., Izaguirre, I. & Unrein, F. (2003). Do steady state assemblages occur in shallow lentic environments from wetlands? *Hydrobiologia 502*, 197-209.

O´Farrell, I., de Tezanos Pinto, P. & Izaguirre, I. (2007). Phytoplankton morphological response to the underwater light conditions in a vegetated wetland. *Hydrobiologia 578*, 65-77.

Pace, M. L. & Funke, E. (1991). Regulation of planktonic microbial communities by nutrients and herbivores. *Ecology 72*, 904-914.

Panuska, J. C. & Robertson, D. M. (1999). Estimating phosphorus concentrations following alum treatment using apparent settling velocity. *Lake and Reservoir Management 15*, 28-38.

Pernthaler, J., Sattler, B., Šimek, K., Schwarzenberger, A. & Psenner, R. (1996). Topdown effects on the size-biomass distribution of a freshwater bacterioplankton community. *Aquatic Microbial Ecology 10*, 255-263.

Pernthaler, J. (2005). Predation on prokaryotes in the water column and its ecological implications. Nature Reviews Microbiology, 1–10. *Nature Reviews Microbiology* 1-10.

Perrow, M. R., Jowitt, A. J. D., Stansfield, J. H. & Phillips, G. L. (1999). The practical importance of the interactions between fish, zooplankton and macrophytes in the shallow lake restoration. *Hydrobiologia 395/396*, 199-210.

Porter, K. G., Sherr, E. B., Sherr, F., Pace, M. L. & Sanders, R. W. (1985). Protozoa in planktonic food webs. *Journal of Protozoology 32*, 409-415.

Raven, J. A. (1997). Comment: Phagotrophy in phototrophs. *Limnology and Oceanography 42*, 198-205.

Reinertsen, H., Jensen, A., Koksvik, J. I., Langeland, A. & Olsen, Y. (1990). Effects of fish removal on the limnetic ecosystem of a eutrophic lake. *Canadian Journal of Fisheries and Aquatic Sciences 47*, 166-173.

Rodríguez, P. & Pizarro, H. (2007). Phytoplankton productivity in a highly colored shallow lake of a South American Floodplain. *Wetlands 27*, 1153-1160.

Sanders, R. W., Porter, K. G., Bennett, S. J. & DeBiase, A. E. (1989). Seasonal patterns of bacterivory by flagellates, ciliates, rotifers, and cladocerans in a freshwater planktonic community. *Limnology and Oceanography 34*, 673-687.

Saunders, P. A., Shaw, W. H. & Bukaveckas, P. A. (2000). Differences in nutrient limitation and grazer suppression of phytoplankton in seepage and drainage lakes of the Adirondack region, NY, U.S.A. *Freshwater Biology 43*, 391-407.

Scheffer, M. (1998). *Ecology of shallow lakes*. London: Chapman & Hall.

Schnetzer, A. & Caron, D. A. (2005). Copepod grazing impact on the trophic structure of the microbial assemblage of the San Pedro Channel, California. *Journal of Plankton Research 27*, 959-971.

Sherr, E. B. & Sherr, B. F. (1987). High rates of consumption of bacteria by pelagic ciliates. *Nature 325*, 710-711.

Sinistro, R., Izaguirre, I. & Asikian, V. (2006). Experimental study on the microbial plankton community in a South American wetland (Lower Paraná River Basin) and the effect of the light deficiency due to the floating macrophytes. *Journal of Plankton Research 28*, 753-768.

Sinistro, R. (2007). Interacciones tróficas planctónicas en un humedal de la cuenca del Paraná Inferior (Argentina): Análisis de las comunidades naturales y estudios experimentales. *Departamento de Ecología, Genética y Evolución*. Buenos Aires, Universidad de Buenos Aires. Doctor: 128.

Sinistro, R., Sánchez, M. L., Marinone, C. & Izaguirre, I. (2007). Experimental study of the zooplankton impact on the trophic structure of phytoplankton and the microbial assemblages in a temperate wetland (Argentina). *Limnologica 37*, 88-99.

Sommer, U., Sommer, F., Santer, B., Zöllner, E., Jürgens, K., Jamieson, C., Boersma, M. & Gocke, K. (2003). *Daphnia* versus copepod impact on summer phytoplankton: functional compensation at both trophic levels. *Oecologia 135*, 639-647.

Stoecker, D. K. (1998). Conceptual models of mixotrophy in planktonic protist and some ecological and evolutionary implications. *European Journal of Protistology 34*, 281-290.

Tranvik, L. J., Porter, K. G. & Sieburth, J. M. (1989). Occurrence of bacterivory in Cryptomonas, a common freshwater phytoplankter. *Oecologia 78*, 473-476.

Unrein, F. (2001). Efecto de los nutrientes y el pH sobre el crecimiento y la estructura del fitoplancton en ambientes de la llanura aluvial del Paraná Inferior. *Departamento de Ecología, Genética y Evolución*. Buenos Aires, Universidad de Buenos Aires: 162.

Williamson, C. E., Morris, D. P. & Pace, M. L. (1999). Dissolved organic carbon and nutrients as regulators of lake ecosystems: resurrection of a more integrated paradigm. *Limnology and Oceanography 44*, 795-803.

Chapter 7

THE APPLICATION OF CONSTRUCTED WETLANDS FOR MUNICIPAL WASTEWATER TREATMENT IN EAST CHINA

Liu Hanhu[1,2,], Han Baoping[1,2], Feng Qian[3], Bai Xiangyu[1,2] and Niu Jinhai[3]*

[1]School of Environment and Survey, China University of Mining and Technology,
Xuzhou, Jiangsu Province, P.R.China
[2] Jiangsu Key Laboratory of Resources and Environmental Information Engineering,
Xuzhou, Jiangsu Province, P.R.China
[3]Zaozhuang Environment Protection Agency, Zaozhuang, Shandong Province, P.R.China

ABSTRACT

In order to alleviate the water shortage problem in North China, China government has been building the South-to-North Water Diversion Project since 2000. The project included three systems: clean water passages, water using security system and water quality improvement engineering system. No municipal wastewater along the line would allow discharging to the passages directly. Constructed wetlands were adopted for municipal wastewater treatment along the line.

Sihong County is located in the northwest of Jiangsu province, P.R.China. In order to meet the demands of the South-to-North Water Diversion Project, under the national finance support, Sihong County Sewage Water Treatment Plant was built in 1999. Subsurface flow constructed wetland system (SSFCW) was adopted in this plant. It was used for treating municipal wastewater. The design capacity is 50,000 cubic meter per day.

According to the monitoring data, the kinetic equation of COD_{Cr} removal with hydraulic detention time in SSFCW fitted well with the first order reaction. The total removal rates of COD_{Cr} in both spring and autumn were 74.1% and 50.9%, respectively. The change of seasons had a great effect on the purifying effect. In spring and autumn,

* E-mail:hanhucumt@sina.com

the removal rates for the total nitrogen were 64.6% and 49.3%, 70.7% and 50.5% for ammonia, respectively.

Tengzhou City is also along the South-to-North Water Diversion Project. It is located in the south of Shandong province, P.R.China. Surface flow constructed wetland system (SFCW) was adopted for treating micro-polluted river water. In 2006, a pilot-scale SFCW was built on coal mine subsidence area.

The results showed that hydrophytes played an important role, and the SFCW had good purification to micro-polluted river water. Comparing the reed system and the cattail system, the former had good removal effect on BOD_5, COD_{Cr}, ammonia and phosphorous. The effluent of two systems can reach the third level of China Surface Water Quality Standard (GB3838-2002).

As an ecological wastewater treatment system, constructed wetland system adapts to municipal wastewater treatment plants in small towns. This technology is in low cost at construction and operation, also the effluent quality is well. And the efficiency of removal of nitrogen and phosphorus is higher than conventional activated sludge. So it has a great value of popularization.

Keywords: South-to-North Water Diversion Project, SSFCW, SFCW, East China, Municipal wastewater

1. INTRODUCTION

Although China's total amounts of water resources is rich, but each person'possesion is poor. China's total water resources is about 2.8 trillion cubic meter, ranking sixth in the world, but per capita is only about 2,200 cubic meter, is the quarter of the world's average level, and is listed as 13 water-deficient countries in the world. China has a vast territory; the differnce of space-time distribution of rainfall affects the spatial distribution of water resources. In North China, water resource is scarcity and South China is rich. The runoff of Yangtze River accounts for 81% of the country's seven major rivers runoff, its per capita and per square of water resource is also higher than the national average, belongs to abundant areas. And the Yellow River-Huaihe-Haihe rivers account for only 7.2% and its per capita water resource only 15% of the national average.So, in North China, the water resource is short, rivers dried up and groundwater was exploited seriously, groundwater table has dropping increasingly.

In order to alleviate the water shortage in North China, to achieve the harmonious development of economic with society, China government has been building the South-to-North Water Diversion Project since 2000. The project included East, Middle and West lines. In East Line, Yangtze River water is pumped from Yangzhou, then through Grand Canal, Hongze Lake and Weishan Lake, the water flows to Tianjin and Beijing. The project included three systems□clean water passages, water using security system and water quality improvement engineering system. No municipal wastewater along the line would allow discharging to the passages directly. The water quality should up to the third national criteria, which calls for the municipal wastewater in the towns along the East Line must meet the discharging criteria.

Figure 1. The geography location of Sihong County and Tengzhou City.

Constructed wetland technology has advantages of high efficiency, low investment and low operation expense. So it is suitable for municipal wastewater treatment in small towns or further treatment of the micro-polluted water [1].

In order to ensure the water quality safely along the East Line, Sihong County Sewage Water Treatment Plant adopted subsurface flow constructed wetland system (SSFCW), and Tengzhou City in Shandong Province selected surface flow constructed wetland system (SFCW) to purify the micro-polluted water in ChengGuo River.

Figure 1 shows the geography location of Sihong County and Tengzhou City.

2. APPLICATION OF SUBSURFACE FLOW CONSTRUCTED WETLAND SYSTEM ON WASTEWATER TREATMENT OF SIHONG COUNTY

2.1. The Geography Location of Sihong County

(1) Geography Condition

Sihong County is located in the northwest of Jiangsu Province. Its east is near Hongze Lake which is one of the biggest five freshwater lakes in China, and its west links up Wuhe County, Si County and Jiashan County of Anhui Province.Its geography coordinate is at latitude of 33°08' to 33°47'N, and longitude 117°56' to 118°46'E. Its terrain is plain and monticule mostly, and has a few of hills. The hypsography is that southwest and west is high and southeast and south is low. Its highest altitude is 62.8 meter, and the lowest altitude is 12.1 meter.

(2) The Climate Condition

Sihong County is in the East Asia monsoon area, and the transition area of north-Asia tropical zone and north-warm temperate zone. So, the monsoon is evidence, the four seasons are demarcated clearly, the climate is mildness and the sunshine is abundant. The average annual temperature is 14.3°C, and lowest temperature is -22.9 °C, the highest temperature is 41°C. The average precipitation is 893.9mm. The average time of sunlight is 2356.4 hours in one year, no frost period is 213 days, snowfall period is 9.2 days and the averge wind speed is 3.7m/s.

2.2. The Technology of Sihong County Wastewater Treatment Plant

Sihong County Wastewater Treatment Plant is located in the south suburb of Sihong County, the area of the plant is 59503m², and the design capacity is 50,000 cubic meter per day. It was built in 1999, and was completed in July, 2002.

Sihong County Wastewater Treatment Plant adopted SSFCW technology, and its processing flow was as Figure 2. SSFCW technology made use of the co-effect of substrate, hydrophyte (reed) and microorganism to form an ecosystem.

SSFCW included 8 units. Each unit was in parallelogram shape; its length was 73.5 meters and width 49 meters. The height of substrate was 0.8 meter. The substrate size is about 10□30mm. After investigation, reed was selected as hydrophyte.

2.3. Experiments on Adsorbing Capability of Substrate

Substrate is an important part of constructed wetland system. It not only can provide attaching surfaces for microorganism, but also offer nutritions for hydrophytes. Generally speaking, substrate removed contaminations in the wastewater through adsorption, absorbability, filtration, ion exchange and so on.

Sihong Wastewater Treatment Plant selected basalt for substrate. The price of basalt was about 500-600 yuan (RMB) each ton, so it was cheaper than others. The physical-chemistry parameters of substrate affected the adsorbing capability.

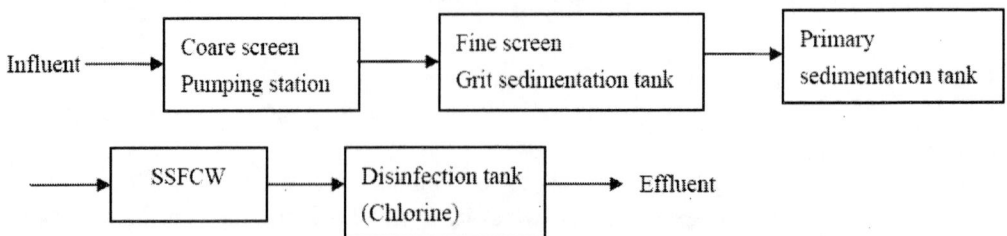

Figure 2. The flow diagram of Sihong Wastewater Treatment Plant.

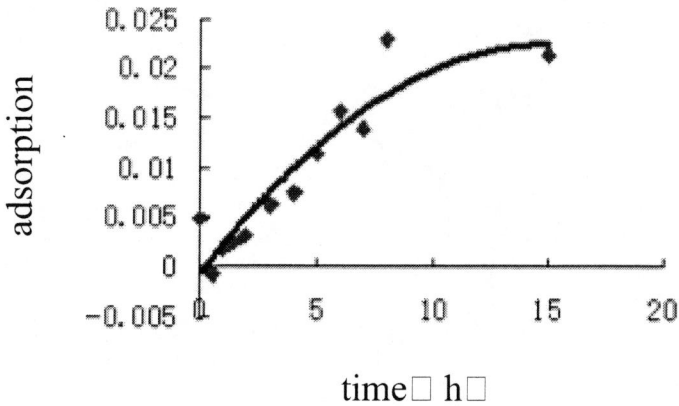

Figure 3. The adsorption balance time of basalt on nitrogen.

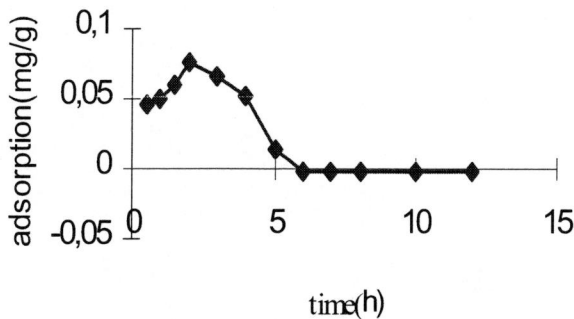

Figure 4. The adsorption balance time of basalt on phosphorus.

The adsorption of nitrogen and phosphorus was the main function of substrate in constructed wetland system [2]. So we did a series of experiments to test the adsorbing capabilities of basalt.

(1) The Measurement of Adsorption Balance Time

Figure 3 and Figure 4 showed the adsorption balance time of basalt on nitrogen and phosphorus.

From two diagrams, we could see that the adsorption balance time of basalt on nitrogen and phosphorus was about 10-15 hours and 6 hours respectively. It was noticeable that there was a maximum value on phosphorus, and then there was a releasing process. The maximum value emerged at about half of the adsorption balance time. The reasons needed more study.

(2) The Measurement of Adsorption Isotherm

According to the fore-and-aft liquor concentration changes, the adsorbing quantity of basalt under different concentrations could be caculated. The liquor balance concentration was as x-axis and adsorbing quantity as y-axis, the adsorption isotherm could be drawn. Figure 5 and Figure 6 showed the adsorption isotherm of basalt on nitrogen and phosphorus respectively.

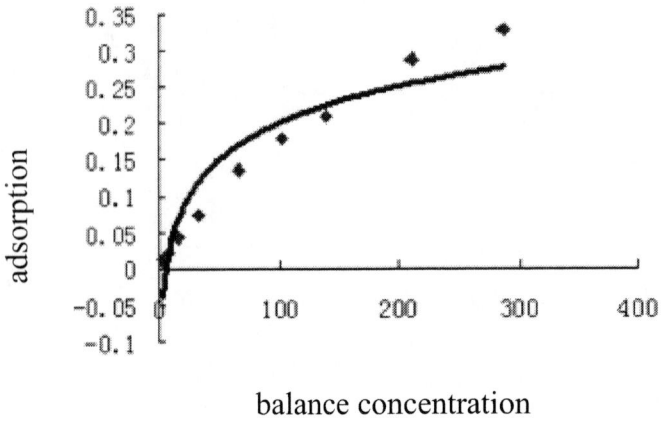

Figure 5. The adsorption isotherm of basalt on nitrogen.

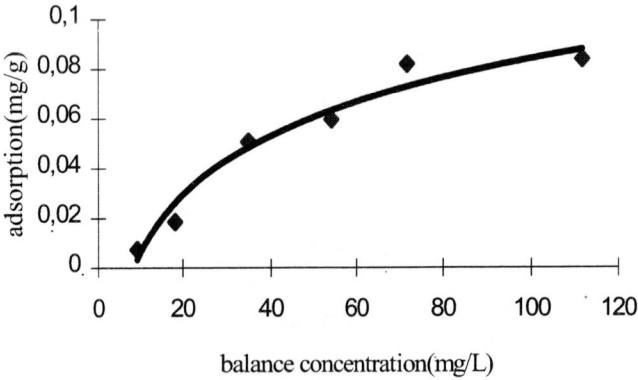

Figure 6. The adsorption isotherm of basalt on phosphorus.

According to the Langmuir and Freundlich adsorption isotherm model, through regression of the experiment data, the regression modulus of equation could be caculated. Figure 7 and Figure 8 showed the regression results of basalt on nitrogen and phosphorus.

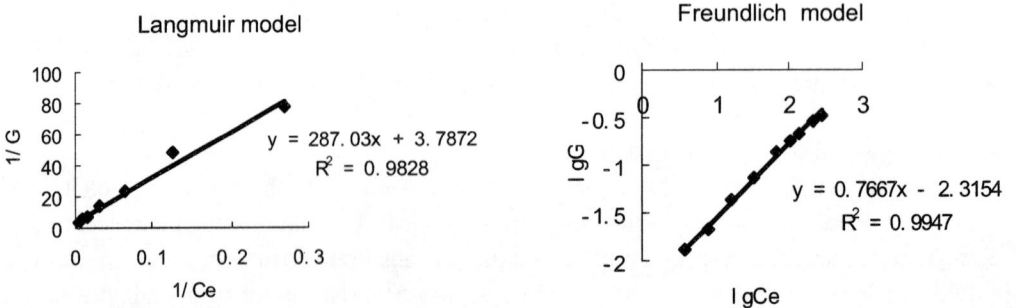

Figure 7. The regression curve of basalt on nitrogen.

Langmuir model

$y = 1289.9x - 8.2217$

$R^2 = 0.9772$

Freundlich model

$y = 1.0236x - 3.0296$

$R^2 = 0.9366$

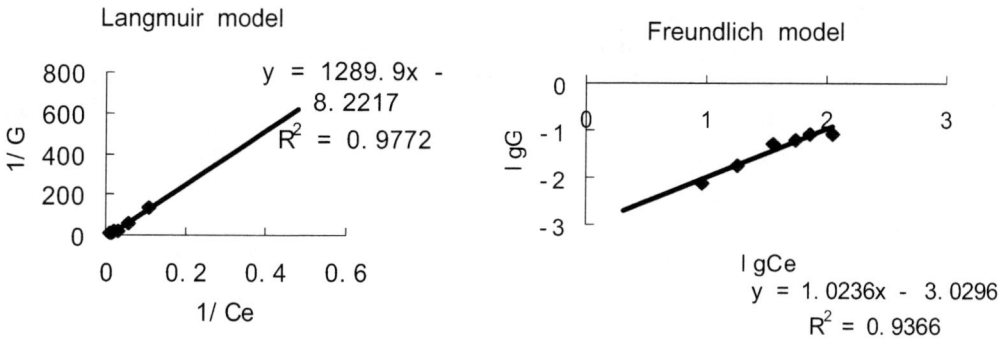

Figure 8. The regression curve of basalt on phosphorus.

According to the regression results and the comparison of the regression modulus R^2, the adsorption isotherm of basalt on nitrogen subjected to the Freundlich model, the regression equation was $lgG=0.7667lgCe-2.3154$; the adsorption isotherm of basalt on phosphorus subjected to the Langmuir model, the regression equation is $1/G=1289.9/Ce-8.2217$.

2.4. Studies on the Microorganism Species in Constructed Wetland System

(1) Location of Sampling Points

Constructed wetland system had complicated microorganism species. Microorganism was very important to remove organic pollutants and nitrogen. Quantificational estimating the quantity and activity of microorganism had important scientific significance and practical worth to construct constructed wetland ecosystem, enhance the wastewater treatment efficiency; and study the wetland mechanism.

Sihong SSFCW had 8 parallel beds. Selecting anyone for study site (each is parallelogram, length is 73.5 meters, width is 49 meters and depth is 0.8 meter),seven sampling points were set up in the bed and the distance was 0 meter (influent, 1#), 9 meter (2#), 27 meter(3#)□37 meter (middle oberservation well, 4#), 46 meter (5#), 64 meter (6#) and 73 meter (7#) respectively. Figure 9 showed the location of the sampling points.

Through determination of ammonifier, nitrosomonas and denitrifying bacteria on reed roots and substrate at sampling points 2#, 3#, 5# and 6#, we studied the distribution of microorganism and the relationship of microorganism with nitrogen removal.

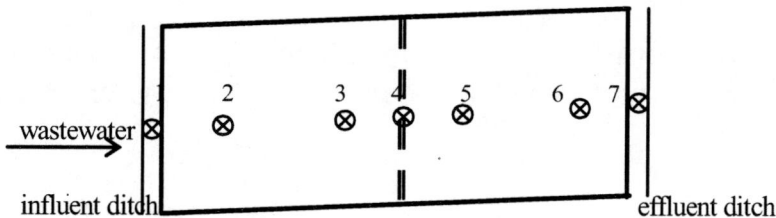

Figure 9. The location of sampling points.

Determination of microorganism: the determination of ammonifier and nitrosomonas was referred to "Analysis Method Manual of Soil Microorganism" [3]. The determination of denitrifying bacteria was referred to "Analysis Method of Soil Agriculture and Chemistry" [4]. The number of ammonifier was counted by plate counting method. Nitrosomonas and denitrifying bacteria were counted by most probably number (MPN) method.

(2) Experiment Results and Analysis

The amount of three kinds of microorganisms for nitrogen removal in different sites and seasons in SSFCW is shown in Table 1.

Table 1. Amount of microorganism in different sites and seasons in SSFCW (Year 2004)

Sampling sites	Kind of bacteria	Spring (5.22) Substrate (Entries/g)	Surface of root (Entries/g(wet root))	Autumn (10.24) Substrate (Entries/g)	Surface of root [Entries/ g (wet root)]
2	Ammonifier	3.5×10^7	1.1×10^9	1.76×10^9	1.1×10^{12}
	Nitrosomonas	7.0×10	3.5×10^2	1.5×10^3	1.5×10^4
	Denitrifying bacteria	3.0×10^4	3.0×10^4	7.5×10^4	2.0×10^5
3	Ammonifier	3.0×10^7	1.3×10^{11}	1.3×10^9	3.0×10^{11}
	Nitrosomonas	1.5×10^3	1.1×10^4	1.4×10^2	7.5×10^3
	Denitrifying bacteria	3.5×10^3	4.5×10^6	3.0×10^4	4.0×10^4
5	Ammonifier	1.0×10^7	3.9×10^{10}	3.0×10^{10}	3.0×10^{11}
	Nitrosomonas	3.0×10^3	3.0×10^3	1.6×10^2	4.0×10^3
	Denitrifying bacteria	4.5×10^5	9.5×10^5	3.0×10^3	3.0×10^4
6	Ammonifier	2.9×10^7	3.5×10^{11}	3.2×10^9	3.0×10^{11}
	Nitrosomonas	3.0×10^2	1.15×10^3	4.0×10^3	3.5×10^2
	Denitrifying bacteria	1.4×10^7	3.0×10^4	2.0×10^4	3.0×10^5

Figure 10 and Figure 11 showed the amount of microorganisms (logarithm) vs distance.

From Figure 10 and Figure 11, it can be concluded that both in spring and autumn the amount of ammonifier in SSFCW was larger than that of denitrifying bacteria on the whole, and the amount of denitrifying bacteria was larger than that of nitrosomonas. The amount of bacteria on the surface of reed roots was larger than that on the surface of basalt, which can be ascribed to the oxygen transportation structure of reed. It also indicated that reeds play an important role in the constructed wetland, which is similar to the findings of Jiayao Zhang [5].

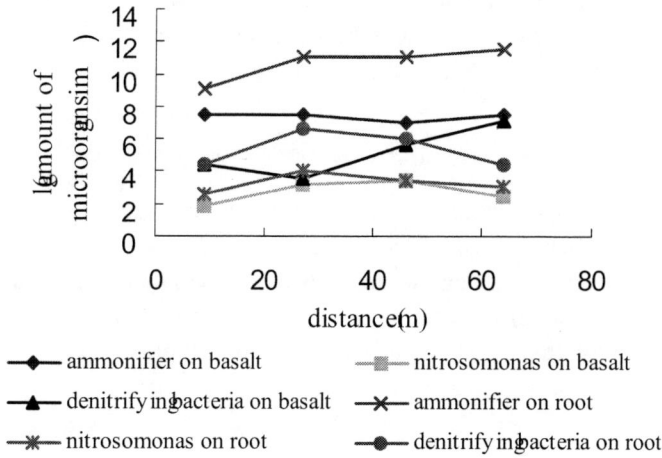

Figure 10. The amount of microorganisms vs distance (spring, year 2004).

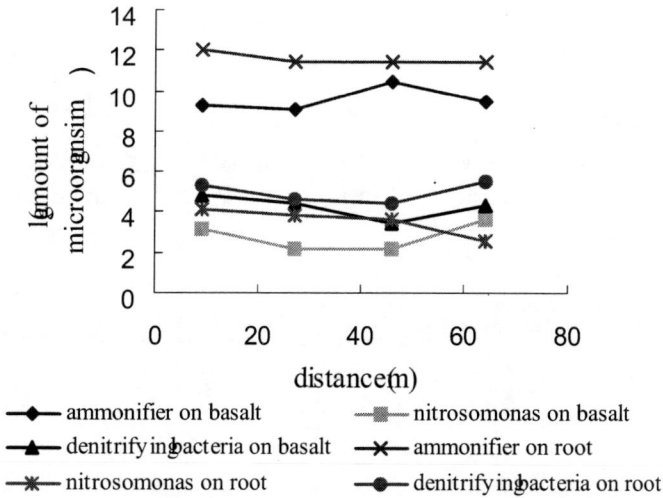

Figure 11. The amount of microorganisms vs distance (autumn, year 2004).

It can be concluded that from the comparison of amount of microorganism in different sites and seasons in SSFCW, on the whole, the amount of ammonifier in autumn was larger than that in spring. And the amount of ammonifier on the surface of basalt in autumn was 10-100 times higher than that in spring, and the amount on the surface of reed roots in autumn was 10 times than that in spring. In every season there was no obvious change in the amount of ammonifier in different sites. But the change range of amount between different sites was not beyond 100 times yet. No obvious change was observed for the amount of nitrosomonas between spring and autumn. In spring, no matter on the surface of basalt or the root, the amount of nitrosomonas in the central part was higher than those both in inlet part and outlet part. But the amount change in autumn was not so obvious. Similarly, there was no obvious change of the amount of denitrifying bacteria between spring and autumn. But the amount of denitrifying bacteria in central part was lower than those in inlet and outlet parts in autumn.

(3) Conclusions

1. No matter in spring or autumn, the amount of microorganisms in SSFCW was as follows: ammonifier > denitrifying bacteria > nitrosomonas. For the same kind of microorganism, the amount on the root was larger than that on the substrate.
2. The amount of ammonifier in autumn was 10-100 times higher than that in spring. There was no obvious change in the amount of nitrosomonas and denitrifying bacteria between spring and autumn.

2.5. Studies on Removal of pollutants in Constructed Wetland System

(1) Location of Sampling Points
Sihong SSFCW had 8 parallel beds.Selecting the same bed which was used for microorganism study; seven sampling points were set up. The water samples were got from depth of 30 centimeter below the substrate. The detail location of sampling points was as Figure 9.

(2) Sampling Time and Frequency
The change of seasons has great effect on the microorganism and water quality purification. So, the May (sping) and October (autumn), 2004 were selected for sampling time.

(3) Monitoring Item and Method
The monitoring items included dissolved oxygen (DO), chemical oxygen demand (COD), total nitrogen (TN), Kjeldahl nitrogen (KN), ammonia nitrogen (NH_4-N) and total phosphorus (TP). The method referred to "Water and Wastewater Monitoring and Analysis Methods (fourth edition)"[6].

(4) Experiment Results and Analysis
In May, 2004, the reed was growing quickly, and the height of the reed in SSFCW was about 1.5-2.0 meter. The sampling date was on May 22, 2004, and the air temperature was 28.7 °C. Table 2 showed the water temperature, dissolved oxygen and pH on different points.

In October, 2004, the height of the reed had been 2-3 meter, and about 40-60 percentages of reeds had faded. The sampling date was on October 22, 2004, the water temperature was 14.2-12.1 °C. The concentration of dissolved oxygen was 1.0-0.4 mg/L, and the pH was 6.4-6.5. Comparing with May, 2004, there were no great changes in DO and pH, but the water temperature had decreased.

Table 2. The water temperature, dissolved oxygen and pH at different sampling points

sampling points	water temperature (°C)	dissolved oxygen (mg/L)	pH
1#	21.6	1.3	6.4
2#	21.2	0.3	6.4
3#	19.7	0.4	6.2
4 #	18.9	0.3	5.8
5#	18.0	0.2	6.0
6#	17.9	0.3	6.2
7#	17.0	0.4	6.2

1. The Removal of Organic Pollutants

Table 3, Figure 12 and 13 showed the COD_{Cr} variety along the bed in spring and autumn.

Table 3 showed that the removal of COD_{Cr} was 74.1% in May. The removal of organic was 55.7% in the former one-third part in the constructed wetland. Also Table 3 showed, in October, the total removal of COD_{Cr} was 50.9%, at the same time that was 40.3% in former one-third part. The removal in former one-third part accounted for 80% of the total COD_{Cr} removal in two seasons. The change of pollutant concentration decreased with the increasing distance, degradation rate gradually reduced correspondingly.

Table 3. COD_{Cr} variety at different sampling points in spring and autumn (2004)

Date	May 22			October 24		
Sampling points	COD_{Cr} (mg/L)	lnCOD	Removal rate (%)	COD_{Cr} (mg/L)	lnCOD	Removal rate (%)
1#	69.54	4.24	—	104.24	4.65	—
2#	63.66	4.15	8.46	76.50	4.34	26.61
3#	30.82	3.43	55.68	61.97	4.13	40.55
4 #	24.84	3.21	64.28	58.12	4.06	44.24
5#	15.86	2.76	77.19	48.72	3.89	53.26
6#	14.04	2.64	79.81	50.40	3.92	51.65
7#	18.00	2.89	74.11	51.13	3.93	50.95

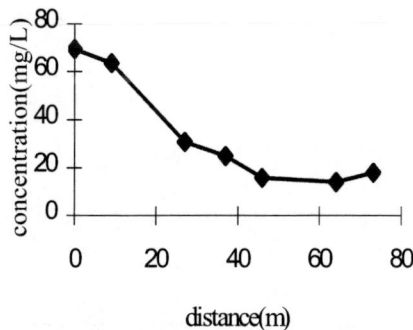

Figure 12. The curve of COD_{cr} vs distance (in spring).

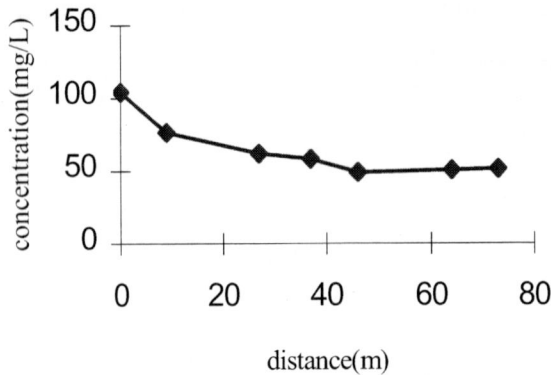

Figure 13. The curve of COD$_{cr}$ vs distance (in autumn).

After COD$_{Cr}$ value was taken logarithm in different sampling points, it was concluded that the relationship between lnCOD and distance L was linearity. The equation of linear regression in spring was lnCOD=-0.0279L+4.2586, and the correlation coefficient was R^2=0.9559. The equation in autumn was lnCOD=-0.0148L+4.5646, R^2=0.9438. Through hypothesis testing it was proved that the two equations were highly linearly correlated. Because in constructed wetland water flow speed is almost even, the relationship between COD and distance indicated that the kinetic equation of COD$_{Cr}$ removal with hydraulic detention time in constructed wetland fitted well with the first order reaction. In spring reaction rate constant k was 0.0279, the reaction kinetics equation was C=70.71e$^{-0.0279t}$. In autumn reaction rate constant k was 0.0148, the reaction kinetics equation was C=96.02e$^{-0.0148t}$. In 2004, the average temperature of May and July in Sihong County was 21.8 °C. The average temperature of October and November in Sihong County was 12.5 °C. So it can be concluded that the reaction rate constant k decreased evidently with the decreasing of temperature.

Table 4. The concentration of nitrogen at different sampling points (May 22, 2004) (unit,mg/L)

Sampling points	TN	KN	NH$_4$-N	Organic nitrogen	NO$_3$-N	Removal of TN
1#	36.29	31.79	27.89	3.90	4.50	—
2#	29.78	24.32	21.02	3.30	5.46	6.51
3#	28.20	17.48	17.52	—	10.72	1.58
4 #	24.11	18.59	17.30	1.29	5.52	4.09
5#	19.21	13.56	11.90	1.66	5.65	4.90
6#	15.46	11.72	9.75	1.97	3.74	3.75
7#	12.84	8.79	8.17	0.62	4.05	2.62
Removal rate (%)	64.6	72.4	70.7	84.3	—	—

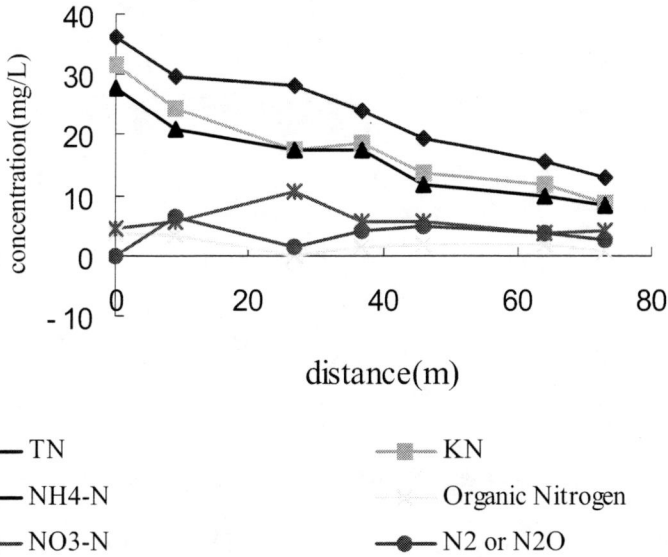

Figure 14. The concentration of nitrogen vs distance (May 22, 2004).

2. The Removal of Nitrogen and its Conformation

The relevant indicators of nitrogen (TN, KN and NH_4-N) were measured in late May and late October. Table 4 and Figure 14 showed the removal of nitrogen and its conformation in late May.

.Table 4 showed that the removal of the organic nitrogen was mainly in the former one-third part in the constructed wetland.The removal rate of TN was 64.6% and ammonia 70.7% respectively in spring.

Table 5 and Figure 15 showed the removal of nitrogen and its conformation in late October.

Table 5. The concentration of nitrogen at different sampling points (October 24, 2004) (unit,mg/L)

Sampling points	TN	KN	NH_4-N	Organic nitrogen	NO_3-N	Removal of TN
1#	49.84	41.99	37.87	4.12	7.85	—
2#	41.23	33.49	32.37	1.12	7.74	8.61
3#	37.68	32.62	28.37	4.25	5.06	3.55
4 #	40.14	26.50	32.60	—	13.64	—
5#	37.74	28.50	29.25	—	9.24	2.40
6#	31.62	25.62	21.00	4.62	6.00	6.12
7#	25.41	27.12	18.75	8.37	—	6.21
Removal rate (%)	49.3	35.42	50.5	—	—	—

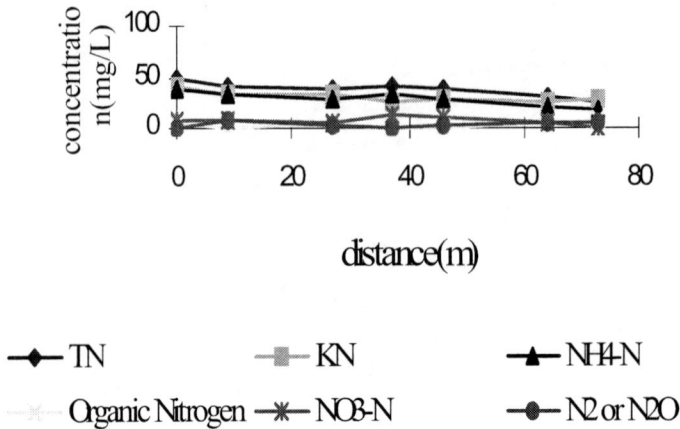

Figure 15. The concentration of nitrogen vs distance (October 24, 2004).

Table 5 showed that the removal rate of TN was 49.3% and ammonia 50.5% respectively in autumn.

Comparing the two seasons, it was concluded that the seasons had great effect on the removal of nitrogen, in autumn it was 15-20% lower than spring.This phenomenon was resulted from temperature and hydrophyte. Firstly, nitrosomonas and nitrifying bacteria were extremely sensitive to temperature.The decline of temperature can restrain growth rate and activity of two bacteria directly. At the same time the decline of temperature can affect hydrophyte growth. In autumn, the growth of hydrophyte was in stagnation, and the absorption and assimilation for ammonia reduced significantly. Hydrophyte's oxygen transportation to SSFCW decreased. So, the removal rate of nitrogen decreased.

3. The Removal of Total Phosphorus

Some researchers have found that the removal of phosphorus was caused by plants absorption, micro-organisms, and physical-chemical process in constructed wetland. Table 6 and Figure 16 showed the concentration of total phosphorus at different sampling points.

Table 6. The concentration of TP at different sampling points (2004)

Sampling points	spring	autumn
1#	2.49	3.30
2#	2.30	3.79
3#	2.27	3.07
4 #	2.23	3.37
5#	1.60	3.04
6#	1.55	3.25
7#	1.39	2.86
Removal rate (%)	44.1	13.3

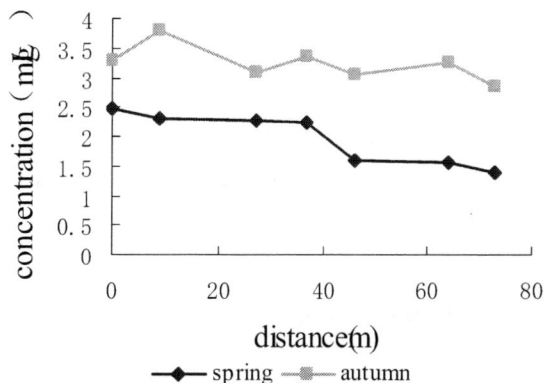

Figure 16. The concentration of total phosphorus vs distance (2004).

From Table 6 and Figure 16, it was concluded that the removal rate of total phosphorus was 44.1% in the spring, 13.3% for the autumn. In the spring the rate was high, because the reed was growing rapidly.In the autumn, the temperature dropped, microbial activity reduced, and hydrophyte stopped growing.

(5) Conclusions

1. The kinetic equation of COD_{Cr} removal with hydraulic detention time in Sihong SSFCW fitted well with the first order reaction. The total removal rates of COD_{Cr} in both spring and autumn were 74.1% and 50.9%, respectively.
2. The change of seasons had a great effect on the purifying effect of the constructed wetland. In spring and autumn, the removal rates for the total nitrogen were 64.6% and 49.3%, 70.7% and 50.5% for ammonia, respectively. The removal of organic nitrogen decreased by 15-23 percent in autumn compared with spring. The removal rate decreased with the decreasing temperature.
3. In spring and autumn, the removal rates for the total phosphorus were 44.1% and 13.3%,respectively.
4. It is critical to control the dissolved oxygen concentration for the process of nitrification and denitrification in the constructed wetland.

3. APPLICATION OF SURFACE FLOW CONSTRUCTED WETLAND SYSTEM ON MICRO-POLLUTED RIVER IN TENGZHOU CITY

3.1. Location of Tengzhou City

(1) Nature Conditions
Tengzhou City is located at south of Shandong province, its geography coordinate is at latitude of 34°50′ to 35°17′N, and longitude 116°48′ to 117°23′. Yimeng Mountains, Zaozhuang city, Weishan Lake and Zoucheng city are round it, respectively. Extent of Thengzhou city is 45km×46km and the total area is 1,485km^2.

(2) Climate Conditions

Tengzhou City is located in the southern temperate semi-humid areas, and has a continental monsoon climate significantly. Tengzhou City has annual average temperature of 14.1 °C, extreme maximum temperature of 39.6°C, and extreme minimum temperature of -19.2 °C. The highest annual rainfall of Tengzhou City is 1,245.8 mm (1964), and the minimum is 388.9 mm (1981), while the average annual rainfall is 773.1 mm.

3.2. Pollution Ratings Evaluation of Chengguo River

Tengzhou City is along the East Line of South-to-North Water Diversion Project and mainly surface rivers of it converge into Chengguo River. Chenghe, Guohe and the confluence of them constitute Chengguo River; its water quality suffered certain pollution of agricultural products and papermaking wastewater discharging, resulted that the water quality can't meet the request of South-to-North Water Diversion Project.

According to the monitoring data from 2005 to 2006, Chengguo River water quality was shown as Table 7.

Overall rating, water quality of Chengguo River is less than the third level of China Surfacewater Quality Standard (GB3838-2002). It belongs to typical micropolluted water.

Table 7. The water quality of Chengguo River (unit, mg/L)

item	minimum	maximum	average	National standard (the third level)
COD_{Cr}	19.00	49.30	40.0	20.0
BOD_5	6.20	16.90	12.80	4.0
NH_4-N	0.39	6.58	4.32	1.0
TN	0.40	7.11	4.97	1.0
TP	0.029	1.30	0.59	0.2

3.3. Pilot-scale Surface Flow Constructed Wetland System Treating for Micropolluted Water

(1) System Layout

According to the request of South-to-North Water Diversion Project, Chengguo River water quality should achieve the third level of China Surfacewater Quality Standard (GB3838-2002) in 2010. Therefore, surface flow constructed wetland system(SFCW) in Tengzhou City should aim to achieve following criterion, that is, $COD_{Cr} \leq 20$ mg / L, $BOD_5 \leq 4$ mg / L, NH_4-N ≤ 1.0 mg / L, TN ≤ 1.0 mg / L, oil ≤ 0.05 mg / L, respectively.

Chengguo River constructed wetland pilot system adopted surface flow constructed wetland system.It was built on coal mining subsidence area.

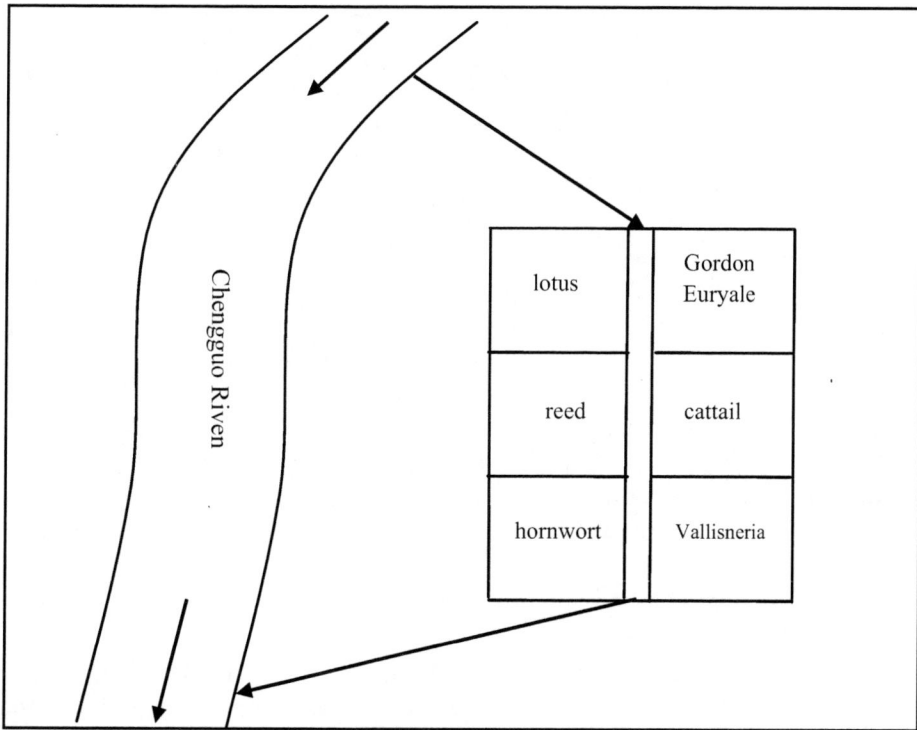

Figure 17. Layout of pilot-scale constructed wetland system in Tengzhou City.

Chengguo river pilot system was divided into two parallel units, the size of each unit was 4 m x 12m, and each unit was divided into three parts, floating plants, straight plants, submerged plants were planted in turn. In Unit I, floating plants, straight plants, submerged plants were lotus root, reed and hornwort, respectively, and in unit Ⅱ were Gordon Euryale, cattail and Vallisneria.

Two units of pilot system were established side-by-side, and separated by pan soil from the middle. Pan Soil was 0.3 m above the ground, and depth of the underground test system was 0.7 m. Surface water was pumped from Chengguo River and entered SFCW through an irrigation channel with 200m long, the hydraulic load was controlled by inflow pump, and water depth in the system was controlled by outflow pump. The irrigation channel also served as primary sedimentation tank.

Figure 17 showed the layout of SFCW in Tengzhou City.

(2) Layout of Sampling Points

The sampling points of two units had symmetrical layout. Each unit had 18 sampling points, and had 6 sampling points along the direction of flow in each of aquatic plants respectively. The line space and space from the border along the direction of flow were both 1.0 m, and that in a line were 2.0 m and 1.0m, respectively. Layout of sampling points was shown as Figure 18.

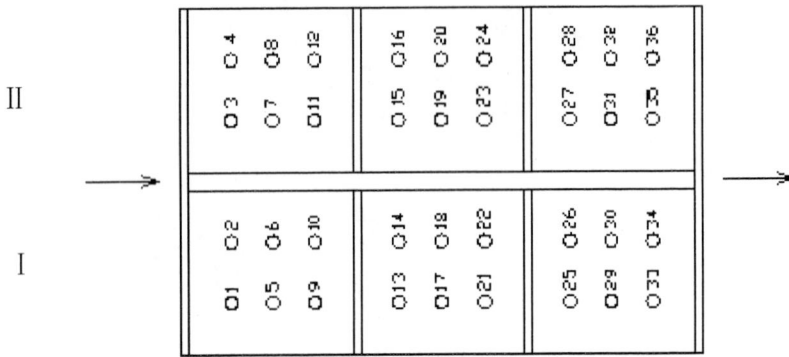

Figure 18. Layout of sampling points in Tengzhou SFCW.

(3) Operation of Pilot-scale System

When the pilot-scale system was stable, a series of researches had been done from June 2006 to July 2007.

The research plan was as follows [7]:

(1) Overall comparison of two units' removal effects: inlet and outlet sampling in each unit from July 2006 to July 2007 every eight days, every point sampling 1000mL for water quality analysis.

(2) Comparison of two units' removal effects along the direction of flow: sampling nine mixed samples along the direction of flow from the 18 sampling points setting in each unit every month (Sampling mixed samples with vertical direction of flow).

(3) Comparison of removal effects at different water depths and HRT: inlet and outlet sampling at various water depths, HRT conditions.

- Made the system operating in water depth h = 0.3m, hydraulic retention time = 5 days (state I) for 25 days, mensurated the removal effect, and analyzed the impact of different hydrophytes on removal efficiency.

- Made the system operating in water depth h = 0.5m, hydraulic retention time = 9 days (state II) for 25 days, mensurated the removal effect.

- Made the system operating in water depth h = 0.5m, hydraulic retention time = 7 days (state III) for 25 days, mensurated the removal effect.

- Made the system operating in water depth h = 0.5m, hydraulic retention time = 7 days (state IV) for 25 days, mensurated the removal effect, and analyzed the impact of different hydrophytes on removal efficiency.

- Made the system operating in water depth h = 0.7m, hydraulic retention time = 5 days (state V) for 25 days, mensurated the removal effect, and analyzed the impact of different hydrophytes on removal efficiency.

- Made the system operating in water depth h = 0.9m, hydraulic retention time = 3 days (state VI) for 25 days, mensurated the removal effect, and analyzed the impact of different hydrophytes on removal efficiency.

(4) Comparison of Two Units for Treating Micropolluted River

From July 2006 to July 2007, the effluent qualities of two units were shown as Table 8.

Table 8. Comparison of effluent water quality of two Units (unit, mg/l)

	DO		NH₄-N		TN		TP		COD_{cr}		BOD₅	
date	unit I	unit II	unit I	unit II	unit I	unit II	unit I	unit II	unit I	unit II	unit I	unit II
2006.07.08	6.5	6.65	0.69	0.9	1.23	1.23	0.21	0.23	12.3	13.93	1.8	1.84
2006.07.16	6.1	6.41	0.71	0.8	0.99	1.04	0.19	0.20	14.9	16.32	2.0	2.11
2006.07.24	6.0	6.61	0.67	0.9	0.87	0.92	0.21	0.23	14.8	16.24	2.3	2.51
2006.08.01	7.8	6.55	0.59	0.9	0.76	0.87	0.10	0.09	18.8	16.78	1.8	1.96
2006.08.09	7.2	6.61	0.58	0.6	0.79	0.90	0.09	0.10	16.7	18.62	3.2	3.43
2006.08.17	6.9	6.06	0.51	0.7	0.72	0.73	0.10	0.10	15.1	14.96	2.9	3.14
2006.08.25	6.7	6.16	0.61	0.7	0.82	1.02	0.06	0.08	15.3	16.61	3.2	3.53
2006.09.02	7.5	5.72	0.65	0.8	0.75	0.80	0.19	0.18	14.8	16.15	3.5	4.59
2006.09.10	7.6	6.04	0.58	0.6	0.94	0.95	0.20	0.34	14.3	15.20	3.6	3.95
2006.09.18	7.3	5.85	0.57	0.5	0.87	0.84	0.18	0.29	13.6	15.99	3.3	3.83
2006.09.26	6.2	5.28	0.56	0.5	0.89	0.84	0.14	0.16	14.8	18.40	3.6	3.59
2006.10.04	5.0	4.64	0.54	0.7	0.99	0.91	0.13	0.13	16.6	18.86	3.2	3.08
2006.10.12	5.2	4.77	0.69	0.7	0.88	0.79	0.13	0.12	18.7	19.32	4.1	3.87
2006.10.20	4.9	4.13	0.71	0.8	0.86	0.86	0.13	0.12	18.3	19.89	4.2	4.60
2006.10.28	4.8	4.28	0.73	0.8	0.91	0.99	0.15	0.14	17.1	19.07	3.8	4.02
2006.11.05	4.1	4.17	0.75	0.7	0.88	0.95	0.20	0.22	17.5	21.09	3.5	3.82
2006.11.13	4.0	3.80	0.90	0.9	0.82	0.79	0.18	0.19	19.9	21.65	3.7	3.87
2006.11.21	3.7	3.31	0.95	0.9	0.97	0.86	0.14	0.15	20.0	24.55	4.1	4.20
2006.11.29	3.5	3.14	1.05	1.1	0.92	0.88	0.17	0.18	21.2	23.35	3.9	3.87
2006.12.07	3.3	3.3	1.01	1.0	1.46	1.08	0.16	0.15	20.6	20.99	3.7	3.72
2006.12.15	3.5	3.6	0.97	0.8	0.73	0.73	0.11	0.11	21.5	21.65	3.6	3.71
2006.12.23	3.6	3.8	0.99	1.0	0.86	0.82	0.14	0.15	20.2	21.01	3.9	4.54
2006.12.31	3.7	3.5	0.87	1.0	0.86	0.82	0.12	0.13	22.3	22.22	4.1	4.72
2007.01.08	3.5	3.4	1.01	1.1	1.06	1.01	0.14	0.14	24.3	24.15	4.0	5.64
2007.01.16	3.4	3.6	0.99	1.2	0.82	0.86	0.20	0.21	21.7	21.79	4.0	4.19
2007.01.24	3.6	3.8	1.10	1.2	0.88	0.88	0.19	0.22	22.7	23.27	3.7	3.85
2007.02.01	4.0	4.1	1.00	1.0	0.88	0.94	0.19	0.18	22.8	23.52	3.8	3.86
2007.02.09	4.5	4.6	1.30	1.4	0.87	0.96	0.16	0.16	23.9	24.60	3.9	3.81
2007.02.17	4.3	4.4	1.20	1.4	1.01	0.97	0.10	0.11	23.2	24.48	3.7	3.77
2007.02.25	4.2	4.1	1.30	1.3	0.92	0.90	0.10	0.10	20.9	21.48	3.9	4.13

Table 8. Continued

	DO		NH₄-N		TN		TP		CODcr		BOD₅	
date	unit I	unit II	unit I	unit II	unit I	unit II	unit I	unit II	unit I	unit II	unit I	unit II
2007.03.05	5.0	4.9	1.10	0.9	0.98	0.97	0.10	0.12	19.8	20.90	3.5	3.64
2007.03.13	5.3	5.4	1.10	1.1	0.84	0.79	0.08	0.07	19.0	19.70	3.2	3.40
2007.03.21	5.5	5.6	0.79	0.7	0.79	0.86	0.10	0.10	19.3	19.89	3.1	3.33
2007.03.31	6.0	6.3	0.78	0.8	0.86	0.92	0.10	0.10	17.8	18.07	3.2	3.13
2007.04.08	5.9	6.2	0.72	1.0	0.86	0.91	0.12	0.12	17.4	18.45	3.1	3.28
2007.04.16	6.5	6.6	0.64	0.7	0.64	0.71	0.11	0.12	15.2	16.33	2.9	3.18
2007.04.24	6.1	6.1	0.60	0.6	0.86	0.92	0.11	0.12	15.3	20.37	2.8	3.02
2007.05.01	6.1	6.2	0.66	0.9	0.92	0.67	0.15	0.18	12.6	13.07	2.9	3.18
2007.05.09	6.4	6.5	0.68	0.6	0.92	0.77	0.17	0.19	16.9	18.86	2.7	2.97
2007.05.17	6.5	6.6	0.65	0.6	0.83	0.69	0.22	0.24	10.8	12.33	2.6	2.87
2007.05.25	6.4	6.2	0.70	0.9	0.94	0.82	0.16	0.14	11.2	12.39	2.2	2.35
2007.06.02	6.3	6.4	0.60	1.1	0.87	0.82	0.17	0.15	9.4	10.52	1.8	2.02
2007.06.10	6.5	6.7	0.76	0.8	0.84	0.84	0.16	0.18	11.4	11.22	2.0	2.14
2007.06.18	6.2	6.3	0.84	0.9	0.81	0.94	0.10	0.12	8.2	8.74	2.6	2.59
2007.06.26	6.3	6.4	0.69	0.7	0.73	0.74	0.10	0.11	9.9	10.53	2.5	2.81
2007.07.03	6.6	6.5	0.70	0.7	0.66	0.74	0.07	0.08	7.7	8.09	2.1	2.51
2007.07.11	6.6	6.7	0.67	0.7	0.71	0.67	0.13	0.16	9.5	11.51	1.9	2.06
2007.07.19	6.5	6.4	0.53	0.6	0.74	0.74	0.11	0.12	9.5	10.25	1.8	2.12
2007.07.27	6.8	6.6	0.58	0.7	0.67	0.70	0.10	0.12	8.6	9.12	2.1	2.39
average	5.5	5.33	0.79	0.86	0.87	0.86	0.14	0.15	16.49	17.68	3.12	3.36
standard	5.0		1.0		1.0		0.2		20.0		4.0	

Through comparative analysis, the effluent water quality of unit I (reed)was better than unit II (cattail),but the difference was not notable.And the effluent water quality of two units could meet the requirements of the third level of China Surfacewater Quality Standard (GB3838-2002).

(5) Purification Effects of Surface Fow Constructed Wetland on Micropolluted River

In view of the purification effects of two units to Chengguo River were similar, we will analyze purification effect on Chengguo River in unit one (lotus, reed and hornwort). Purification effects of unit one is shown as Table 9.

Table 9. Purification effects of unit one (lotus, reed and hornwort) (unit, mg/L)

date	DO			COD$_{cr}$			BOD$_5$			NH$_4$-N			TN			TP		
	inlet	outlet	change rate(%)	inlet	outlet	removal rate (%)	inlet	outlet	removal rate (%)	inlet	outlet	removal rate (%)	inlet	outlet	removal rate (%)	inlet	outlet	removal rate (%)
2006.07.08	8.0	6.5	19.0	43.0	12.3	71.30	7.5	1.8	76.30	4.8	0.69	85.63	4.9	1.23	75.00	0.49	0.21	56.40
2006.07.16	7.7	6.1	21.0	48.0	14.9	68.90	8.1	2.0	75.10	4.2	0.71	83.10	4.5	0.99	78.00	0.44	0.19	55.60
2006.07.24	7.7	6.0	22.0	44.0	14.8	66.40	8.5	2.3	72.60	4.6	0.67	85.43	4.6	0.87	81.00	0.50	0.21	58.30
2006.08.01	9.5	7.8	18.0	47.0	18.8	60.00	5.8	1.8	69.20	3.6	0.59	83.61	3.8	0.76	80.00	0.28	0.10	63.20
2006.08.09	8.5	7.2	14.8	49.0	16.7	66.00	10.1	3.2	68.30	3.4	0.58	83.00	4.0	0.79	80.30	0.26	0.09	65.30
2006.08.17	8.3	6.9	16.1	44.0	15.1	65.70	9.8	2.9	70.40	3.7	0.51	86.10	3.6	0.72	80.00	0.29	0.10	66.40
2006.08.25	8.1	6.7	18.1	45.0	15.3	66.00	10.2	3.2	68.80	3.8	0.61	84.00	3.8	0.82	78.50	0.24	0.06	76.80
2006.09.02	9.3	7.5	19.0	41.0	14.8	63.80	12.1	3.5	71.00	4.3	0.65	85.00	4.4	0.75	83.00	0.75	0.19	75.00
2006.09.10	9.6	7.6	20.9	38.0	14.3	62.40	11.4	3.6	68.50	4.0	0.58	85.60	4.1	0.94	77.00	1.30	0.20	85.00
2006.09.18	9.4	7.3	22.0	39.0	13.6	65.20	9.2	3.3	64.20	3.9	0.57	85.30	4.1	0.87	78.90	1.05	0.18	83.00
2006.09.26	7.8	6.2	20.3	40.0	14.8	63.00	12.4	3.6	70.90	4.0	0.56	86.00	4.3	0.89	79.40	0.80	0.14	82.00
2006.10.04	6.5	5.0	22.6	41.0	16.6	59.40	8.7	3.2	63.10	4.3	0.54	87.44	4.4	0.99	77.60	0.70	0.13	81.00
2006.10.12	6.9	5.2	24.8	42.0	18.7	55.40	10.0	4.1	58.90	4.1	0.69	83.20	4.1	0.88	78.50	0.62	0.13	79.00
2006.10.20	6.9	4.9	29.0	41.0	18.3	55.40	10.7	4.2	60.70	4.2	0.71	83.10	4.2	0.86	79.60	0.56	0.13	76.00
2006.10.28	7.9	4.8	39.3	39.0	17.1	56.20	8.6	3.8	55.60	4.0	0.73	81.75	4.0	0.91	77.20	0.54	0.15	71.60
2006.11.05	7.0	4.1	41.1	38.0	17.5	54.00	6.9	3.5	49.20	3.0	0.75	75.00	3.2	0.88	72.40	0.84	0.20	76.50
2006.11.13	6.7	4.0	40.2	39.0	19.9	49.10	7.4	3.7	50.30	3.3	0.90	72.73	3.3	0.82	75.30	0.75	0.18	76.00
2006.11.21	6.9	3.7	46.2	39.6	20.0	49.50	7.5	4.1	45.30	3.2	0.95	70.31	3.3	0.97	70.70	0.68	0.14	79.00
2006.11.29	6.5	3.5	45.8	38.4	21.2	44.70	7.6	3.9	48.70	3.4	1.05	69.12	3.5	0.92	73.60	0.77	0.17	78.50
2006.12.07	5.9	3.3	44.8	39.6	20.6	48.00	7.3	3.7	49.30	4.1	1.01	75.37	4.3	1.46	66.00	0.56	0.16	71.00
2006.12.15	5.8	3.5	40.0	41.0	21.5	47.60	7.0	3.6	48.60	4.3	0.97	77.44	4.3	0.73	83.00	0.34	0.11	69.00
2006.12.23	6.5	3.6	44.4	38.9	20.2	48.00	8.4	3.9	47.00	3.9	0.99	74.35	3.9	0.86	78.00	0.41	0.14	65.00
2006.12.31	6.8	3.7	45.9	42.0	22.3	47.00	8.9	4.1	46.90	4.0	0.87	78.09	4.1	0.86	79.00	0.35	0.12	66.00
2007.01.08	5.8	3.5	40.0	46.0	24.3	47.10	7.5	4.0	47.00	4.3	1.01	76.51	4.4	1.06	76.00	0.39	0.14	64.40
2007.01.16	5.8	3.4	41.2	41.5	21.7	47.60	7.7	4.0	48.60	4.2	0.99	76.50	4.3	0.82	81.00	0.61	0.20	67.50
2007.01.24	5.9	3.6	38.9	42.3	22.7	46.30	6.9	3.7	46.20	4.3	1.10	74.42	4.5	0.88	80.50	0.56	0.19	65.50
2007.02.01	7.6	4.0	47.5	42.0	22.8	45.80	7.0	3.8	45.90	4.0	1.00	74.81	4.1	0.88	78.60	0.53	0.19	64.80

Table 9. Continued

date	DO			CODcr			BOD5			NH4-N			TN			TP		
	inlet	outlet	change rate(%)	inlet	outlet	removal rate (%)	inlet	outlet	removal rate (%)	inlet	outlet	removal rate (%)	inlet	outlet	removal rate (%)	inlet	outlet	removal rate (%)
2007.02.09	8.8	4.5	48.9	44.0	23.9	45.70	7.2	3.9	46.00	4.1	1.30	68.29	4.3	0.87	79.80	0.54	0.16	71.00
2007.02.17	8.8	4.3	51.2	44.5	23.2	47.90	7.1	3.7	48.00	4.2	1.20	71.43	4.4	1.01	77.00	0.42	0.10	76.50
2007.02.25	7.7	4.2	45.2	41.3	20.9	49.50	7.7	3.9	49.60	4.1	1.30	68.29	4.3	0.92	78.50	0.43	0.10	77.80
2007.03.05	9.3	5.0	46.0	39.8	19.8	50.30	7.1	3.5	51.00	4.1	1.10	72.84	4.2	0.98	76.60	0.47	0.10	78.00
2007.03.13	8.8	5.3	39.6	39.0	19.0	51.20	6.6	3.2	51.30	4.3	1.10	74.42	4.4	0.84	81.00	0.39	0.08	79.80
2007.03.21	8.0	5.5	30.9	39.0	19.3	50.60	6.4	3.1	51.60	3.9	0.79	79.74	4.1	0.79	80.70	0.56	0.10	81.60
2007.03.31	8.6	6.0	30.0	36.5	17.8	51.30	6.7	3.2	52.00	3.8	0.78	79.47	3.9	0.86	78.00	0.46	0.10	77.60
2007.04.08	7.4	5.9	20.3	41.0	17.4	57.60	7.4	3.1	58.00	4.8	0.72	85.00	4.8	0.86	82.00	0.47	0.12	75.00
2007.04.16	8.3	6.5	21.5	36.7	15.2	58.60	7.1	2.9	59.00	4.6	0.64	86.00	4.6	0.64	86.00	0.46	0.11	76.40
2007.04.24	7.8	6.1	21.3	38.8	15.3	60.50	7.2	2.8	61.00	4.3	0.60	86.00	4.5	0.86	81.00	0.43	0.11	74.90
2007.05.01	7.3	6.1	16.4	39.0	12.6	67.60	9.2	2.9	68.60	4.2	0.66	84.30	4.2	0.92	78.00	0.61	0.15	75.50
2007.05.09	7.9	6.4	18.8	41.0	16.9	58.90	6.7	2.7	59.80	4.5	0.68	85.00	4.5	0.92	79.60	0.75	0.17	76.70
2007.05.17	8.1	6.5	20.0	36.8	10.8	70.60	9.0	2.6	71.00	4.7	0.65	86.10	4.6	0.83	82.00	1.02	0.22	78.00
2007.05.25	7.4	6.4	14.1	35.9	11.2	68.90	7.3	2.2	69.80	4.1	0.70	82.93	4.1	0.94	77.00	0.75	0.16	78.50
2007.06.02	7.1	6.3	11.1	36.9	9.4	74.60	7.2	1.8	75.00	4.3	0.60	86.05	4.3	0.87	79.80	0.76	0.17	77.60
2007.06.10	7.5	6.5	13.8	38.7	11.4	70.60	7.0	2.0	71.60	3.9	0.76	80.51	4.4	0.84	81.00	0.86	0.16	81.00
2007.06.18	6.7	6.2	8.1	33.6	8.2	75.60	10.8	2.6	76.00	4.1	0.84	79.60	4.1	0.81	80.30	0.46	0.10	78.80
2007.06.26	7.2	6.3	12.7	39.0	9.9	74.60	10.3	2.5	75.80	3.7	0.69	81.30	3.7	0.73	80.40	0.43	0.10	77.60
2007.07.03	7.8	6.6	15.3	33.0	7.7	76.80	10.3	2.1	79.60	3.6	0.70	80.60	3.9	0.66	83.00	0.31	0.07	78.50
2007.07.11	7.1	6.6	7.6	39.0	9.5	75.60	7.8	1.9	75.60	3.4	0.67	80.20	3.8	0.71	81.20	0.60	0.13	77.60
2007.07.19	7.0	6.5	7.7	41.0	9.5	76.80	8.0	1.8	77.60	3.8	0.53	86.00	3.9	0.74	81.00	0.53	0.11	79.00
2007.07.27	7.8	6.8	13.2	38.0	8.6	77.30	9.7	2.1	78.30	3.6	0.58	84.00	3.6	0.67	81.50	0.47	0.10	79.20
average	7.59	5.5	27.7	40.44	16.49	59.00	8.31	3.12	61.10	4.02	0.79	80.20	4.13	0.87	79.00	0.57	0.14	74.00

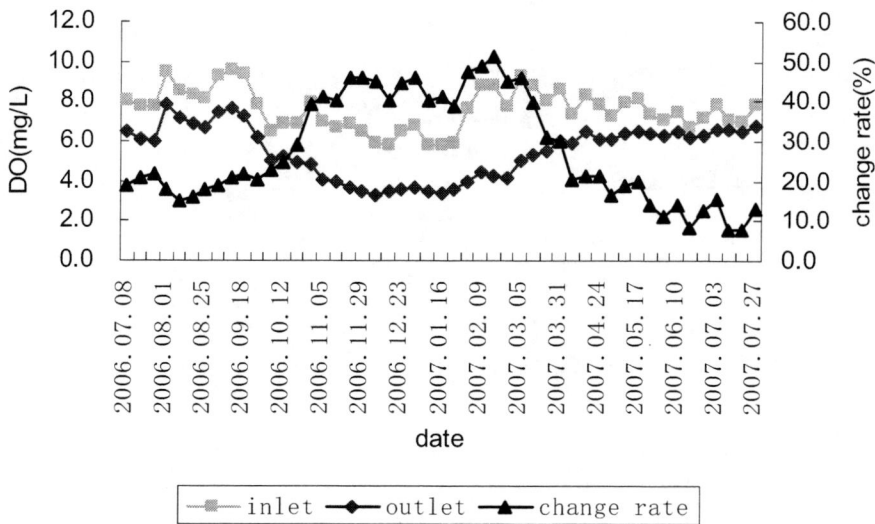

Figure 19. DO concentration of influent /effluent and the change rate of DO.

DO

The influent/effluent dissolved oxygen (DO) and its changes were shown in Figure 19. The influent DO always held on 6~8mg/L, while the effluent DO value changed greatly and was more than 3mg/L. The minimum DO change rate of 7.58% appeared in early July 2007, the maximum DO change rate of 51.2% appeared in mid-February 2007, and DO changed in the average rate of 27.7%: At the beginning of the test, DO change rate remained at around 19%, in October to about 28.9%, and DO change rate gradually increased since then until to February 2007 ,when had the highest DO change rate, and then began to decline.

With the hydraulic retention time (HRT) increased, the effluent DO value gradually increased, but the DO change rate gradually declined. When the hydraulic retention time (HRT) changed from 5 days in July to 9 days in August, 2006, change rate of DO decreased. As HRT decreased since August, DO change rate gradually increased until the maximum value appeared in February 2007.

pH

The influent pH in pilot-scale system varied from 6.8 to 8.1, with a mean of 7.6, which was weakly alkaline. Effluent pH average value was 7.48, which was closer to neutral than influent. It is obvious that the SFCW had a certain role in adjusting pH.

COD$_{Cr}$

The influent/effluent COD$_{Cr}$ and its changes in pilot-scale system were shown in Figure 20.

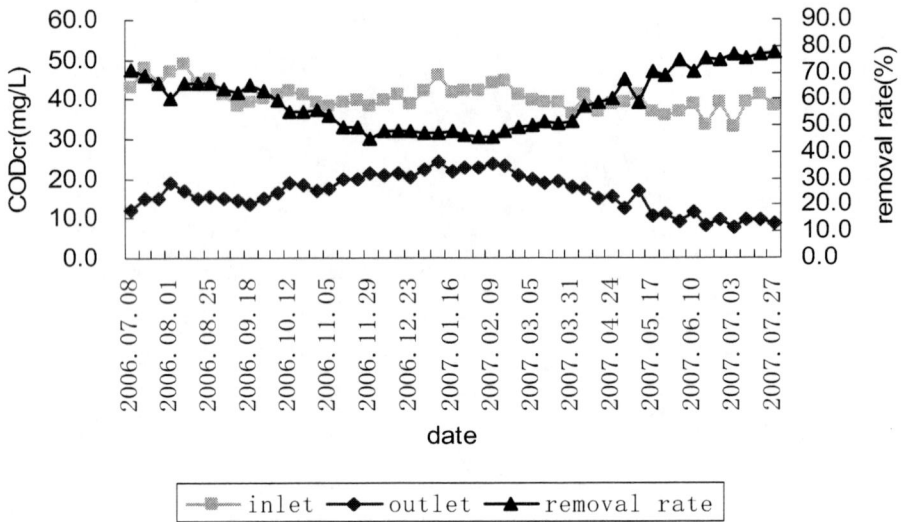

Figure 20. COD$_{Cr}$ concentration of influent/effluent and removal rate of COD$_{Cr}$.

As can be seen in Figure 20, influent COD$_{Cr}$ concentration maintained at $33 \sim 49$ mg/L, and in October 2006, June and July 2007, influent COD$_{Cr}$ concentration decreased slightly.The reason was related with high volume of water runoff. Removal rate of COD$_{Cr}$ had a direct relationship with the growth of hydrophytes. In July and August 2006, hydrophytes were growing rapidly, so the removal rate was high. In October 2006, although hydrophytes stopped growing and were in a stabilization phase, their activity and metabolism were still strong, so the removal rate declined slightly. Since late October 2006 to April 2007, hrdrophytes entered into the decline phase, some plants began to fade, so removal rate declined rapidly.

BOD$_5$

The influent/effluent BOD$_5$ and its changes in pilot-scale system were shown in Figure 21.

As can be seen from Figure 21, BOD$_5$ removal rate was slightly downward during July 2006 to October 2006. The removal rate decreased significantly since October 2006, and that was consistent with the growth of reed.Reeds growed well in July, August and September, and had stronger activity and metabolism, so they could remove BOD$_5$ effectively. After October, some reeds had faded, their activity reduced greatly, and the removal rate had declined markedly.

NH$_4$-N

The influent/effluent NH4-N and its changes in pilot-scale system were shown in Figure 22.

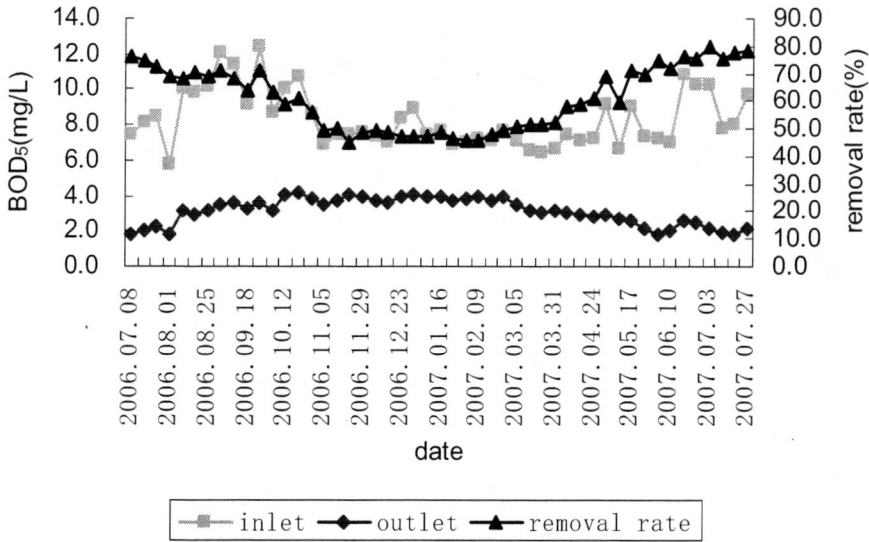

Figure 21. BOD$_5$ concentration of influent / effluent and removal rate of BOD$_5$.

As can be seen from Figure 22, NH$_4$-N concentration increased since October 2006, and didn't stop rising until March 2007. Such a situation was due to a stable source of NH$_4$-N pollution, which came mainly from a number of small manufactories near the river. According to hydrological data, runoff of Chengguo River in summer and autumn was always larger than in winter and spring, so there was less dilution effect in winter and spring.

NH$_4$-N removal rate were maintained at a high level, the maximum rate was 87.44%, the minimum was 68.29%, and the average removal rate reached 80.2%.

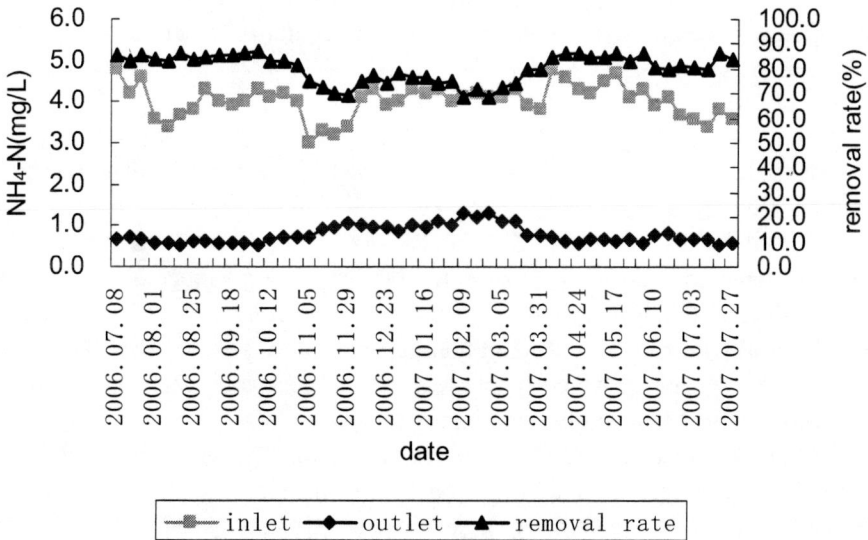

Figure 22. NH$_4$-N concentration of influent / effluent and the removal rate of NH$_4$-N.

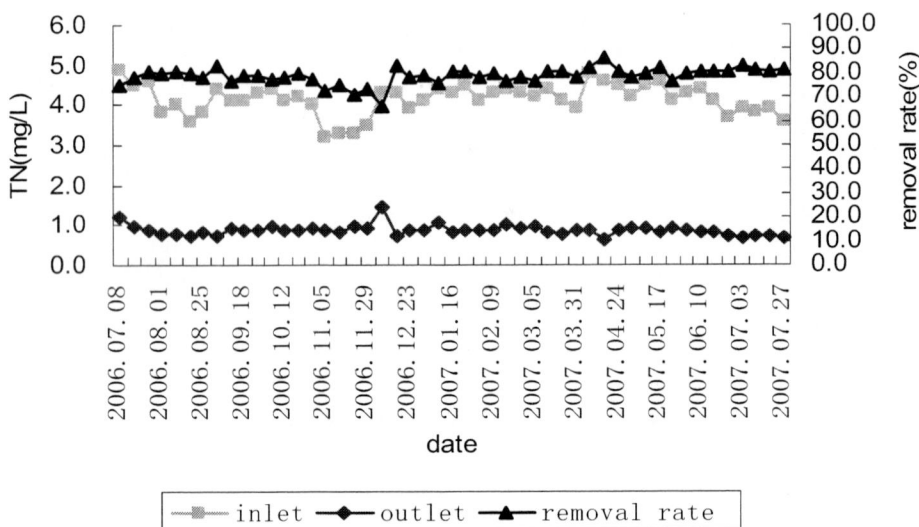

Figure 23. TN concentration of influent / effluent and removal rate of TN.

TN

The influent/effluent TN and its changes in pilot-scale system were shown in Figure 23.

As can be seen from Figure 23, TN concentration increased since November 2006, and didn't stop rising until March 2007. This trend was similar to NH_4-N. Effluent TN was maintained at a relatively low level during the whole test period. On the whole from July 2006 to October 2006, TN concentration of effluent maintained at a relatively low level, had an upward trend since October 2006, and began to decline after April 2007.

Since the removal of nitrogen depends primarily on nitrification and denitrification, and nitrification only alters nitrogen form [8], denitrification is the only way to remove nitrogen, therefore, effluent TN concentration and removal rate should be consistent with nitrate, but differ with ammonia.

TP

The influent/effluent TP and its changes in pilot-scale system were shown in Figure 24.

As can be seen from Figure 24, TP concentration of influent was about 0.24 to 1.30 mg/L, and the peak value occurred in September 2006. With the TP concentration changed, the effluent TP concentration varied consistently, but even in September 2006, when the peak value of the influent TP concentration appeared, the effluent TP concentration still reached 0.2 mg/L. The peak value of TP concentration appeared in September may be due to crops harvest in summer or autumn, which would release phosphorus by decomposed agricultural wastes. Therefore agricultural wastes were important sources of TP.

The maximum removal rate of TP was 85%, the minimum was 55.6%, and the average removal rate was 74%.Removal of TP mainly depended on the substrate adsorption, complex and sedimentation, also plants and micro-organisms played a certain role.

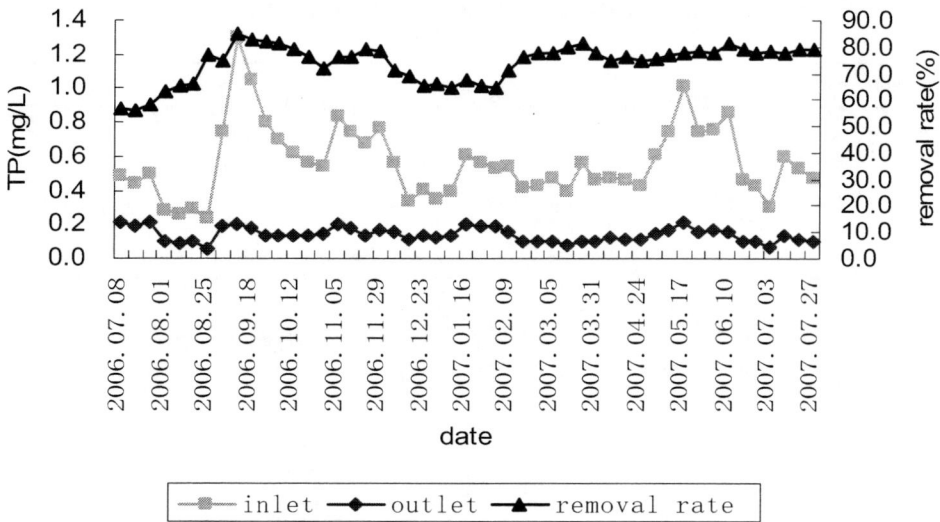

Figure 24. TP concentration of influent/ effluent and removal rate of TP.

(6) Purification Effects under Different Hydraulic Retention Time

Setting HRT as 3, 5, 7, 9 days, respectively and comparing pollutant removal rate under different hydraulic retention time, the results were shown in Table 10.

As can be seen from Table 10, with the HRT increasing, removal rates of COD_{Cr} and BOD_5 had upward trend, NH_4-N, TN, and TP removal rate varied little. Considering about economical factors, the actual hydraulic retention time was selected as five to seven days.

(7) Purification Effects under Different Water Depth

To ascertain the most suitable water depth of surface flow constructed wetland system, removal rates under different water depth of 0.3, 0.5, 0.7, 0.9 meters was compared. The results were shown in Table 11.

As can be seen from Table 11, water depth had great impact on BOD_5 and COD_{Cr} removal rate. With the increasing of water depth, BOD_5 and COD_{Cr} removal rate declined obviously. NH_4-N removal rate increased first and then reduced with the increase of water depth, but TN removal rate had increased trend, which might be due to water depth would affect the growth of submerged plants and floating plants, and affected the ability of oxygen transmission. So water depth affects the growth of hydrophytes and influences the removal of BOD_5 and COD_{Cr}.

Table 10. Removal rates of pollutants under different hydraulic retention time

HRT(d)	COD_{Cr}	BOD_5	NH_4-N	TN	TP
3	48.85%	48.95%	92.00%	73.00%	84.95%
5	62.30%	63.65%	93.30%	71.95%	86.15%
7	64.20%	68.65%	90.90%	65.05%	80.15%
9	62.85%	69.80%	84.40%	67.70%	64.80%

Table 11. Removal rates of pollutants under different water depth

Water depth(m)	COD_{Cr}	BOD_5	NH_4-N	TN	TP
0.3	68.85%	74.45%	91.05%	66.80%	57.35%
0.5	62.30%	63.65%	93.30%	71.95%	86.15%
0.7	55.80%	57.25%	94.35%	76.95%	88.75%
0.9	48.85%	48.95%	92.00%	73.00%	85.25%

For this pilot-scale SFCW, considering growth requirements of different hydrophytes, 0.5 m water depth was reasonable.

3.4. Conclusions

(1) Complex system of lotus, reed, and hornwort had better purification ability to Chengguo River. Water quality, that was NH_4-N$\leq$1mg/L, TN concentration about of 1 mg/L, TP$\leq$0.2 mg/L, $BOD_5\leq$4 mg/L, CODCr$\leq$20 mg/L,which treated by surface flow constructed wetland treatment system can reach the third level of China Surface Water Quality Standard (GB3838-2002),. Moreover, the system had a certain role in adjusting pH.

(2) With the hydraulic retention time increasing from three days to nine days, COD_{Cr} and BOD_5 removal rate had upward trend, NH_4-N, TN, and TP removal rate varied little. Considering about economical factors, the actual hydraulic retention time was reasonable for five to seven days.

(3) Water depth could affect the growth of hydrophytes, thereby affected the removal of the pollutants. As the water depth increased in pilot-scale system, NH_4-N removal rate increased first and then reduced, TN had upward trend, BOD_5 and COD_{Cr} removal rate had significantly decreased. Considering growth requirements of different hydrophytes, 0.5 m water depth was reasonable.

(4) During the test period, the temperature had significant effects on the removal rate of pollutants in surface flow constructed wetland.

REFERENCES

[1]   Liu H.H., Bai X.Y. & Xia N. (2006).*Constructed wetland technology for treating municipal wasterwater* (1st version).Xuzhou,China:China University of Mining and Technology Press.

[2]   Xu L.L., Zhou Q. (2002).Study on purification capability of different substrates in constructed wetland.*Journal of Shanghai Environment Science, 21,603*-605.

[3]   Xu G. H., Zheng H. Y. (1999). *Analysis Method Manual of Soil Microorganism.* Beijing, China: China agriculture Press.

[4]   Nanjing Soil Institute. (1999). *Analysis Method of Soil Agriculture and Chemistry.* Beijing, China: China agriculture Press.

[5]   Zhang J. Y., Xia S. L., Qiu K. M. (1999).Nitrogen removal by a subsurface flow constructed wetland wastewater treatment system and nitrogen transformation bacteria.*Journal of Environmental Science, 19*,323□327.

[6]   National Environmental Protection Agency (2002). *Water and Wastewater Monitoring and Analysis Methods (IV version)*. Beijing, China: China Environmental Science Press.

[7]   Wang S. H., Wang W., Yu Y. (2003).Study on the operating characteristics of subsurface flow constructed wetland. *China Water & Wastewater, 19*,9□11.

[8]   Xia N.,Liu H. H.,Guo R.M.,et al.(2006) .Reserch on nitrogen removal and microorganism in a subsurface flow constructed wetland system in Sihong County.*Journal of China University of Mining & Technology(English Edition), 16*,505-508.

In: International Wetlands: Ecology, Conservation…   ISBN 978-1-60456-999-5
Editor: Jose R. Herrera   © 2009 Nova Science Publishers, Inc.

Chapter 8

COMMUNITY COMPOSITION, SPECIES DIVERSITY AND POPULATION BIOMASS OF THE GAOQIAO MANGROVE FOREST IN SOUTHERN CHINA

Jinping Zhang[1,2,3], Hai Ren[1,], Weijun Shen[1],*
Shuguang Jian[1] and Fanghong Xu[4]

[1]South China Botanical Garden, the Chinese Academy of Sciences,
Guangzhou 510650, P.R China
[2]Graduate University of Chinese Academy of Sciences, Beijing 100039, P.R China
[3]Guangdong Public Laboratory of Environmental Science Technology,
Guangzhou 510650, P.R China
[4]Administration of Zhanjiang Mangrove National Nature Reserve,
Zhanjiang 524088, P.R China

ABSTRACT

The distribution of China's mangrove forests has experienced a considerable loss during the past several decades. Systematic research on current mangrove ecosystems is urgently needed to protect and restore diminishing mangrove forests. The Gaoqiao mangrove forest is the core area of the Zhanjiang Mangrove National Nature Reserve, the largest state mangrove nature reserve in China. We investigated species diversity, population biomass, succession, composition and soil attributes of the major community types of the Gaoqiao mangrove forest. Comparing with the previous controversial studies, we pointed out nine mangrove species within eight genera and four families occurred in Gaoqiao, and there were six main mangrove community types. Community composition indicated that the well-grown *Rhizophora stylosa* + *Bruguiera gymnorrhiza* community perhaps represented the climax community in Gaoqiao. As to species diversity, communities with two canopy layers had relatively richer species diversity, lower ecological dominance and higher evenness. In terms of biomass, big difference existed between different population and different organ. For example, ratios of root

* Corresponding author: Tel +8620-37252916; Fax: +8620-37252916; Email: renhai@scib.ac.cn

biomass to total biomass of pioneer species were relatively larger as they possessed stronger roots to live in large tidal fluxes. Six communities covered early, middle and late succession stages, respectively were distributed at different intertidal sites. Soils under mangrove communities at different succession stages showed that with succession soil acidification, organic matter content and salinity increased, but the total N, P content didn't change along the succession series. Local mangrove succession mode was presented, based on different species' abilities to suit the changing soil attributes and tidal influence at different intertidal sites. This study provides a valuable prerequisite for mangrove conservation and management.

Keywords: Mangrove; Community composition; Diversity; Biomass; Succession; Southern China

1. INTRODUCTION

Mangrove forests provide crustacean and mollusk nursing habitat, serve to maintain biodiversity, support complex food webs, stabilize coastal areas and contribute to local ecotourism development (Kathiresan and Bingham, 2001; Ashton and Macintosh, 2002; Gunawardena and Rowan, 2005; Méndez Linares et al., 2006). Despite these values, mangrove forests are disappearing at an alarming rate as a result of human encroachment such as pollution, aquaculture, wood extraction, tin mining, costal industrialization and urbanization. Among them shrimp farming serves as the major reason behind mangrove loss and degradation (Ashton and Macintosh, 2002; Macintosh et al., 2002; Adeel and Pomeroy, 2002; Benfield et al., 2005). Recently, mangrove forest protection, restoration and management for sustainable development have attracted widespread attention.

Mangroves occur in 112 countries and territories, and cover approximately 8% of the world's coastline in terms of length (Kathiresan and Bingham, 2001; Adeel and Pomeroy, 2002). The United Nation noted that about 41% mangrove area is distributed in Indonesia, Australia, Brazil and Nigeria (FAO, 2003). Indonesia covers the largest area of mangroves in the world, accounting for 22% of the total. Mangrove structure in Indonesia is simple, and deforestation had greatly disturbed the natural regeneration. Thirty-nine species have been reported in Australia, which makes it one of the richest countries in the world in terms of mangrove flora. Differences in vegetation composition and structure between hurricane damaged and clear-cut mangrove stands 25 years after clearing in Australia were also been reported (Ferwerda et al., 2007). The most common species in Brazil are *Laguncularia racemosa, Rhizophora mangle* and *Avicennia nitida*. Fernandes (1999) studied the phonological characteristics of flowering and fruiting in these species. Common species in Nigeria include *Rhizophora racemosa, R. harrisonii* and *R. mangle*. The relationship between environmental conditions and species distribution in a Nigeria mangrove swamp was discussed by Ukpong, 1997. Floral composition, distribution, structure, classification, conservation, rehabilitation, biodiversity and biomass of mangrove had also been studied in Sri Lanka (Jayatissa et al., 2002), Mexico (Cruz-Escalona et al., 2006; Méndez Linares et al., 2006), the Philippines (Primavera, 2000), the Dominican Republic (Sherman et al., 2003), Pakistan (Rasool et al., 2002), Thailand (Macintosh et al., 2002), India (Kathiresan and Bingham, 2001), Malaysia (Ashton and Macintosh, 2002), Tanzania (Shunula, 2002), U.S.A

(Ward et al., 2006), Panama (Benfield et al., 2005), and Belize (Murray et al., 2003). The socio-economic role of mangrove ecosystems has been elucidated at the local, regional and global scales. Compared with well-studied mangrove community in other countries, China only has a preliminary assessment on the number and distribution of mangrove species present in China.

China's mangrove forests are largely distributed along the southeastern coastal provinces including Hainan, Guangxi, Guangdong, Fujian, Zhejiang, Taiwan and two Special Administrative Regions (Hong Kong and Macao). The total area occupied by mangrove ecosystems has decreased from 50,000 ha in the 1950s' to 22,000 ha in 2001 (Wang and Wang, 2007). Deforestation and reclamation have been the major causes for the decline of mangrove forests in China (Wang et al., 2002). As a result of mangrove forest degradation, floral and faunal resources in the original mangrove forests have seriously declined. Since mangroves stabilize coastal areas, loss of mangroves has resulted in great economic loss due to frequent storm and tide impacts. The Gaoqiao mangrove forest, located in Yingluo Port, is the core area of the Zhanjiang Mangrove National Nature Reserve, the largest state mangrove nature reserve in Guangdong, China. Some studies about the ecological aspects of mangrove forest in Gaoqiao have been carried out. Yingluo Port, where Gaoqiao was located, can be divided into two adjacent parts of Guangdong and Guangxi. Gaoqiao is the Guangdong part. Miao (1999) concluded that there were 5 types of communities in Gaoqiao. Liang (1996) showed that there were 11 types of communities in the Guangxi part. It's impossible to explain the differences between the numbers of mangrove community types under the similar environmental condition. A total of approximately 270 ha of *Rhizophora stylosa* and *Bruguiera gymnorrhiza* in the Gaoqiao mangrove forest were kept in pristine condition. Its average tree height was 6 m, and the average canopy coverage was above 95%, aged more than 80 years (Zhang et al., 2002). As mentioned above, relatively few studies are available for mangrove community characteristics in China, controversial number of mangrove community types in the previous studies, and Gaoqiao as a rarely undisturbed pristine mangrove forest, all these make it necessary to carry on a baseline ecological study of the Gaoqiao mangrove forest. This paper systematically studied species diversity, population biomass, succession, composition and soil attributes of the major community types of the Gaoqiao mangrove forest to provide a valuable prerequisite for mangrove conservation and management.

2. MATERIALS AND METHODS

2.1. Study Site

The Gaoqiao mangrove forest is situated in Yingluo Port, Beibu Gulf (central coordinates 109°41'E, 21°30'N). The total area is 2,470 ha, and the average belt-like forest width is approximately 500 m. The Gaoqiao mangrove forest supports rich freshwater resource. Four freshwater rivers, including Gaoqiao, Ximi, Jiangbei and Maizhuo Rivers meet in this mangrove forest. The Gaoqiao mangrove forest is located in the transition between the northern tropics and southern subtropics. The mean annual temperature is 23.4°C, with average temperature in the coldest month, January, ranging between 14.2°C and 14.5°C. The annual precipitation ranges between 1500 mm and 1700 mm, with the rainfall typically occurring from May to September. The average annual relative humidity is 80%. The tidal regime is diurnal, which is rarely found in other regions. The average tidal amplitude is 2.52 m, with a maximum of 6.25 m (He et al., 2007).

2.2. Data Collection and Analyses

Based on the previous studies of Miao (1999) and Liang (1996), a full investigation of the community types was undertaken initially. Finally nine representative sites were selected. For each site, we established a 10 m ×10 m quadrant, which was further divided into four adjacent 5 m × 5 m sub-quadrants. Due to relatively poor mangrove species diversity, the sub-quadrant size was large enough for this study (Miao, 1999).

Soil samples from surface to 20 cm depth were collected in all communities in the field work. Soil attributes such as pH, organic matter content, total N content, total P content and salinity were measured in the laboratory. We recorded the number of species, plant height, DBH (trees only) and crown size (trees only) of each individual mangrove in the quadrants. The coverage of each community was simultaneously measured.

Shannon-wiener index, evenness index, and ecological dominance were used to determine the species diversity among the communities.

Population biomass was determined by a stratified harvesting method. Based on the mean DBH and mean height, standard sample trees were identified. Ten representative individuals for each species were harvested at ground level. The above-ground components were further separated into stems, branches and leaves. Stem sections were collected every 30 cm. To estimate the underground biomass, all roots and mud were excavated within a 1.2m radius from the center of the trees. Root sections were collected every 30 cm. All the samples were washed to remove mud and other debris, weighed and then air-dried. Representative samples of each component were taken to the laboratory to determine the ratios for fresh to oven-dry weight at 105°C. Using these ratios, the oven-dry weight of the different components were calculated. Through stem analysis we generated the volume of a standard sample tree, from which the mean form factor was subsequently deduced. By using the ratio of the standard sample tree volume to all sample volume and the above data, biomass of different components were determined.

3. RESULTS

3.1. Community Composition and Species Diversity

Our investigation shows the Gaoqiao mangrove naturally supports nine species representing eight genera and four families. The species include *Aegiceras corniculatum* (Myrcinaceae), *Kandelia candel* (Rhizophoraceae), *Bruguiera gymnorrhiza* (Rhizophoraceae), *Rhizophora stylosa* (Rhizophoraceae), *Acanthus ilicifolius* (Acanthaceae), *Acanthus ebracteatus* (Acanthaceae), *Avicennia marina* (Rhizophoraceae), *Excoecaria agallocha* (Euphorbiaceae) and *Acrostichum aureum* (Pteridaceae). There are also two mangrove associate species, *Clerodendrum inerme* and *Derris trifoliata. Sonneratia apetala*, an exotic species, was introduced from Bengal in 1993. Among the above nine naturally distributed true mangrove species, *A. corniculatum, K. candel, B. gymnorrhiza, A. ilicifolius, A. ebracteatus, A. marina, E. agallocha* and *A. aureum*also occur in the Indo-West Pacific region; *A. corniculatum, B. gymnorrhiza, R. stylosa, A. ilicifolius* and *E. agallocha* occur in the Australia-Pacific region. The widespread thermophilic species include *B. gymnorrhiza, R. stylosa, E. agallocha and A. aureum*; the widespread anti-hypothermia species include *K. candel, A. marina* and *A. corniculatum.*

Six main community types identified in Gaoqiao (Table 1) include: *Aegiceras corniculatum* community, *Avicennia marina + A. corniculatum* community, *Kandelia candel + A. corniculatum* community, *Bruguiera gymnorrhiza + A. corniculatum* community, *Rhizophora stylosa + B. gymnorrhiza* community and *Excoecaria agallocha + B. gymnorrhiza + Clerodendrum inerme* community.

From the outer middle, and inner beach to land margin, six main mangrove communities covered early, middle and late stages of succession. *Aegiceras corniculatum* community, *Avicennia marina + A. corniculatum* community belonged to early successional stage; *Kandelia candel + A. corniculatum* community, *Bruguiera gymnorrhiza + A. corniculatum* community were in the mid-succession stage; *Rhizophora stylosa + B. gymnorrhiza* community stood both in the middle and late succession stages; *Excoecaria agallocha + B. gymnorrhiza + Clerodendrum inerme* community was found in the late succession stage.

The most common species of six communities was *Aegiceras corniculatum*. Four out of six communities contained this species. The total number of individuals in community of *Bruguiera gymnorrhiza + A. corniculatum* was 405, which represented the largest among the six communities. The average number of individuals of in the six communities was 195. Number of individuals in the community of *Rhizophora stylosa + B. gymnorrhiza* and the community of *Excoecaria agallocha + B. gymnorrhiza + Clerodendrum inerme* were greatly less than the other communities. From the height data, we concluded that the community of *R.stylosa + B. gymnorrhiza* was much taller than the other communities. *R.stylosa* in *B. gymnorrhiza + A. corniculatum* community had the maximum average DBH 10.5 cm and crown size 3.00 m ×3.00 m.

Soil attributes of the communities were listed in Table 1. Soil pH was acidic, and varied between 4.30 and 6.50. The average soil organic matter content, total N, P content and salinity in six communities was 3.79%, 0.129%, 0.029% and 2.045%, respectively. *Rhizophora stylosa + Bruguiera gymnorrhiza* community had the lowest, and *Avicennia marina + Aegiceras corniculatum* community had the highest soil pH value, respectively. *A. marina+A. corniculatum* community and *Excoecaria agallocha + B. gymnorrhiza + Clerodendrum inerme* community had much lower soil organic matter content compared to the other four communities. Soil under *E. agalloch+B. gymnorrhiza+C. inerme* community had the lowest total N, P content and salinity.

Table 1. Community composition and species diversity

Community type	Common species	Number of individuals	Average height (m)	Max. height (m)	Min. height (m)	Average DBH (cm)	Corn size (m×m)	Coverage (%)	pH	Organic matter (%)	Total N (%)	Total P (%)	Salinity (%)	Shannon-Wiener index	Evenness index	Ecological dominance
A.corniculatum	A.corniculatum	190	1.00	1.60	0.50	1.5	0.45×0.45	96	6.00	3.90	0.147	0.034	1.958	0.2	0.01	0.9
A.marina+ A.corniculatum	A.marina	132	0.90	1.60	0.48	2.7	0.90×0.90	60	6.50	2.14	0.113	0.035	1.866	0.3	0.02	0.9
	A.corniculatum	20	1.00	1.30	0.80	2.0	0.45×0.45									
K.candel+ A.corniculatum	K.candel	144	2.70	3.10	2.10	3.7	0.70×0.75	85	6.03	5.87	0.140	0.024	2.259	1.0	0.06	0.6
	A.corniculatum	101	2.25	2.90	1.50	2.3	0.45×0.45									
B.gymnorrhiza+ A.corniculatum	B.gymnorrhiza	140	2.20	2.45	1.90	5.6	2.00×1.80	90	5.78	4.56	0.179	0.046	2.596	0.9	0.06	0.7
	A.corniculatum	260	1.50	1.80	1.10	1.5	0.45×0.45									
	R.stylosa	5	3.00	4.40	2.28	10.5	3.00×3.00									
R.stylosa+ B.gymnorrhiza	R.stylosa	80	4.30	4.60	3.70	5.0	1.10×1.20	90	4.30	5.06	0.126	0.022	2.767	0.8	0.11	0.6
	B.gymnorrhiza	15	4.10	4.50	3.70	6.1	1.30×1.30									
E.agallocha+ B.gymnorrhiza+ C.inerme	E.agallocha	16	2.20	4.90	0.40	4.4	1.50×1.30	90	6.15	1.20	0.071	0.014	0.825	1.4	0.07	0.5
	B.gymnorrhiza	25	0.90	1.20	0.50	4.1	1.00×0.80									
	C.inerme	41	1.06	1.10	1.00	1.2	1.20×1.00									

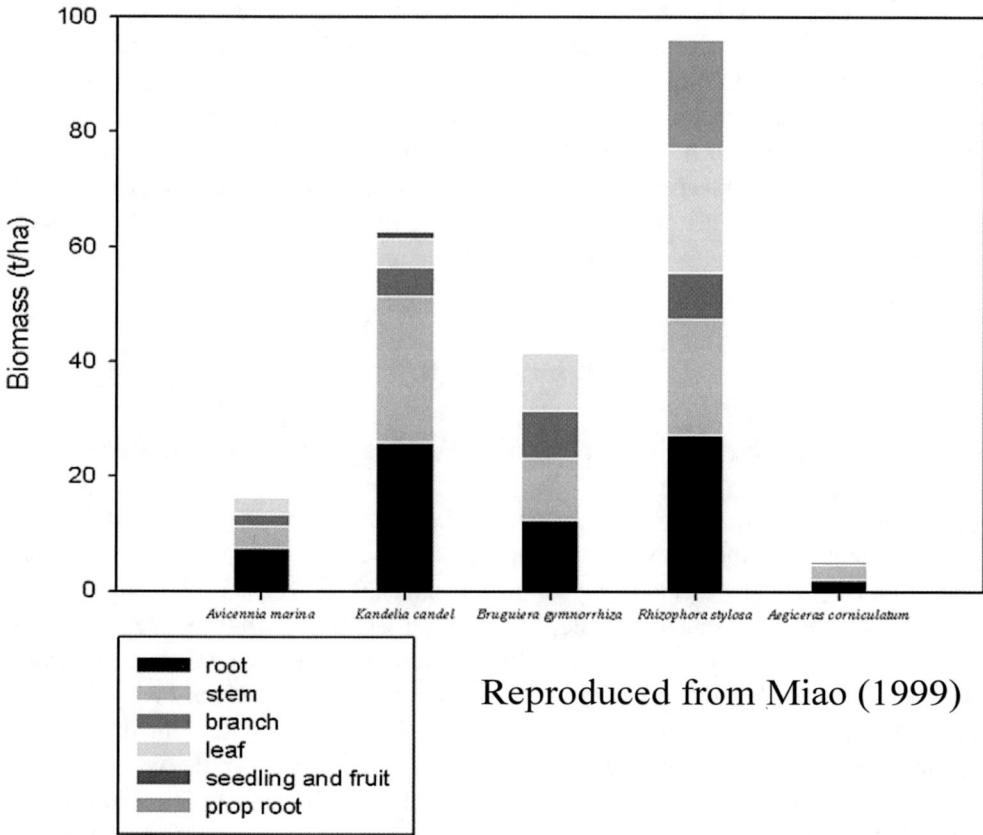

Figure 1. Population biomass. Reproduced from Miao (1999).

Species diversity (Table 1) differed among the communities since different communities were supported by different species and their different abundance. Some communities had relatively richer species diversity, lower ecological dominance and higher evenness, such as *Exoccaria agallocha* + *Bruguiera gymnorrhiza* + *Clerodendrum inerme* community, *Kandelia candel* + *Aegiceras corniculatum* community, *B. gymnorrhiza* + *A. corniculatum* community and *Rhizophora stylosa* + *B. gymnorrhiza* community. The two other communities *A. corniculatum* community and *Avicennia marina* + *A. corniculatum* community had relatively poorer species diversity, higher ecological dominance and lower evenness.

3.2. Population Biomass

Figure 1 shows that Rhizophora stylosa had the greatest total biomass, followed by Kandelia candel, Bruguiera gymnorrhiza, Avicennia marina and Aegiceras corniculatum. Biomass of A. marina and A. corniculatum, especially A. corniculatum, were much lower than the other three populations. For example, biomass of R. stylosa was 17.32 times higher than that of A. corniculatum. Biomass of different organs of different populations and their ratios to the total varied greatly. Biomass of different organs of A. marina and B. gymnorrhiza followed the order: root > stem > leaf > branch. Biomass of different organs of K. candel followed the order: root > stem > branch > leaf > fruit. Biomass of different organs of R. stylosa followed the order: root > leaf > stem > prop root > branch. Biomass of different organs of A. corniculatum followed the order: stem > root > branch > leaf. Ratios of root biomass to total biomass of A. marina, A. corniculatum and K. candel were relatively higher, more than 35%. Ratios of leaf biomass to total biomass of B. gymnorrhiza, R. stylosa and A. marina were relatively higher, more than 19%. Ratios of stem biomass to total biomass of A. corniculatum and K. Candel were relatively higher, more than 40%.

4. DISCUSSION

4.1. Common Species, Main Community Types and the Soil Attributes

The answer of how many true mangrove Species distribute in China differs from researchers' understanding and varies with time. Lin concluded there were 27 species and 1 variety (Lin, 2001). Wang reported only 19 (Wang et al., 2003). The latest number by Science Press of China was 24 (Wang and Wang, 2007). In this study we identified the number to be 24. All the above statistics didn't include Sonneratia apetala because it's an exotic species. Thus 9 naturally distributed true mangrove species accounted for 37.5% of the total in China. Miao (1999) pointed out 7 true mangrove species naturally distributed in the Zhanjiang Mangrove National Nature Reserve (the Gaoqiao mangrove forest is the core area of the reserve), but the table in his study showed it should be 10. Compared with 9 in this study, Miao took Heritiera littoralis as a true mangrove species. Due to H. littoralis is not restricted to the intertidal zone; it's identified as a semi-mangrove species. Yingluo Port, where this study was located is divided into two parts of Guangdong and Guangxi. Liang (1996) studied the flora of mangrove community of Guangxi part. True mangrove species in both parts were Aegiceras corniculatum, Kandelia candel, Bruguiera gymnorrhiza, Rhizophora stylosa, Acanthus ilicifolius, Avicennia marina, Excoecaria agallocha and Acrostichum aureum. Guangdong had additional species Acanthus ebracteatus, and Guangxi had additional species Lumnitzera racemosa.

Five main communities of the Gaoqiao mangrove forest were Aegiceras corniculatum community, Kandelia candel + Aegiceras corniculatum community, Bruguiera gymnorrhiza + Aegiceras corniculatum community, Avicennia marina + Aegiceras corniculatum community and Rhizophora stylosa+ Bruguiera gymnorrhiza community (Miao, 1999). As mentioned above, Gaoqiao has adjacent mangrove community of Guangxi part. Liang (1996) showed that there were 11 communities in Guangxi part. They were B. gymnorrhiza community, R. stylosa + B. gymnorrhiza community, R. stylosa community, B. gymnorrhiza + K. candel community, K. candel community, K. candel + A. corniculatum community, K. candel + A. marina community, A. corniculatum community, A. marina community, A. marina + A. corniculatum community and Excoccaria agallocha community. Compared with investigation of Miao (1999), E. agallocha + B. gymnorrhiza + C. inerme community came into being after 6 years because the community didn't appear in 1999. This implies that the Gaoqiao mangrove forest tends to form new late stage succession community. Compared with results of Liang (1996), Gaoqiao and the adjacent Guangxi part contained almost the same common species, but formed different community types. Both presented similar two species dominated communities, but Guangxi part presented additional single species dominated communities. Obviously Guangxi part showed more integrated succession, and perhaps after 9 years succession single species dominated communities were replaced by the communities composed of previous dominant species.

The mangrove soil parameters are the result of complex interactions of physical factors (tides, microtopography, etc.) and biological factors (microbial activity, physiological activity of plants, etc.). This study showed that mangrove succession resulted in soil acidification, increased organic matter content and salinity, but total N, P content didn't change regularly with different succession stages. *Excoecaria agallocha* + *Bruguiera gymnorrhiza* + *Clerodendrum inerme* community, partly composed of mangrove associate species, apparently showed different soil properties compared to the other communities that were completely composed of true mangrove species mainly due to distribution of close or on land margin, and are not subject to large tide fluxes.

4.2. Community Composition

Aegiceras corniculatum covered 4 communities, which implied that both as pioneer species, *A. corniculatum* had better ability than *Avicennia marina* to suit the changing environmental condition. During succession, as a community developed, inter-species and intra-species competition for environmental resources might result in decline in population density. This could explain the decreased number of individual mangroves at the late succession stage. The height data indicated that all communities presented no more than two canopy layers. Much larger values of average height and DBH showed that *Rhizophora stylosa* + *Bruguiera gymnorrhiza* community perhaps represented a climax community in Gaoqiao. The result partly matched that there was a well-preserved pristine *R.stylosa* + *B. gymnorrhiza* mangrove community aged more than 80 years with a height of 6 m (Zhang et al., 2002). However, in this study the average height for *R. stylosa* and *B. gymnorrhiza* was 4.30 m and 4.10 m, respectively. Maybe the previous data were overestimated.

4.3. Species Diversity

From Table 1, we deduced canopy layer had great influence on values of the species diversity in different communities in Gaoqiao. The differences in the species average heights were 0.1 m, 0.45 m, 0.7 m, 0.2 m, 1.3 m for *Avicennia marina* + *Aegiceras corniculatum* community, *Kandelia candel* + *A. corniculatum* community, *Bruguiera gymnorrhiza* + *A. corniculatum* community, *Rhizophora stylosa* + *B. gymnorrhiza* community and *Excoecaria agallocha* + *B. gymnorrhiza* + *Clerodendrum inerme* community, respectively. Those with larger differences between species height (no less than 0.2 m) appeared as two canopy layers, and had higher species diversity. The other two communities with species average height less than 0.1 m showed only one canopy layer, and their species diversity was relatively lower.

4.4. Succession

From the distribution of six main communities, we deduced the dynamics of mangrove community succession (Figure 2). The successional process began with the formation of the pioneer community. *Avicennia marina* and *Aegiceras corniculatum* were the first two species on the arenaceous barren site. They contributed to form abundant sludge by trapping powdery remnants. Strong root systems helped them to grow well near the low tide line with large tidal fluxes. Mud continued to accumulate as new silt began to form. The salt excretion glands characteristic of *A. marina* and *A. corniculatum* (Liang, 1996) enabled them to extend seaward. As we have said, *A. corniculatum* had better ability than *A. marina* to suit changing environmental condition. Gradually the *A. marina* + *A. corniculatum* community was replaced by pure *A. corniculatum* community. This was during early succession stage.

A substrate providing adequate sludge was needed for *Kandelia candel* and *Bruguiera gymnorrhiza* to germinate. With sediments accumulating, soil salinity increased, and habitat became suitable for *K. candel* and *B. gymnorrhiza* to grow. Consequently *K. candel* + *Aegiceras corniculatum* community and *B. gymnorrhiza* + *A. corniculatum* community formed. After all *A. corniculatum* is a pioneer species, and the increasing high salinity environment was unsuitable for it gradually. As a result, the establishment of pure *K. candel* and *B. gymnorrhiza* communities were possible, which accounted for the middle succession stage.

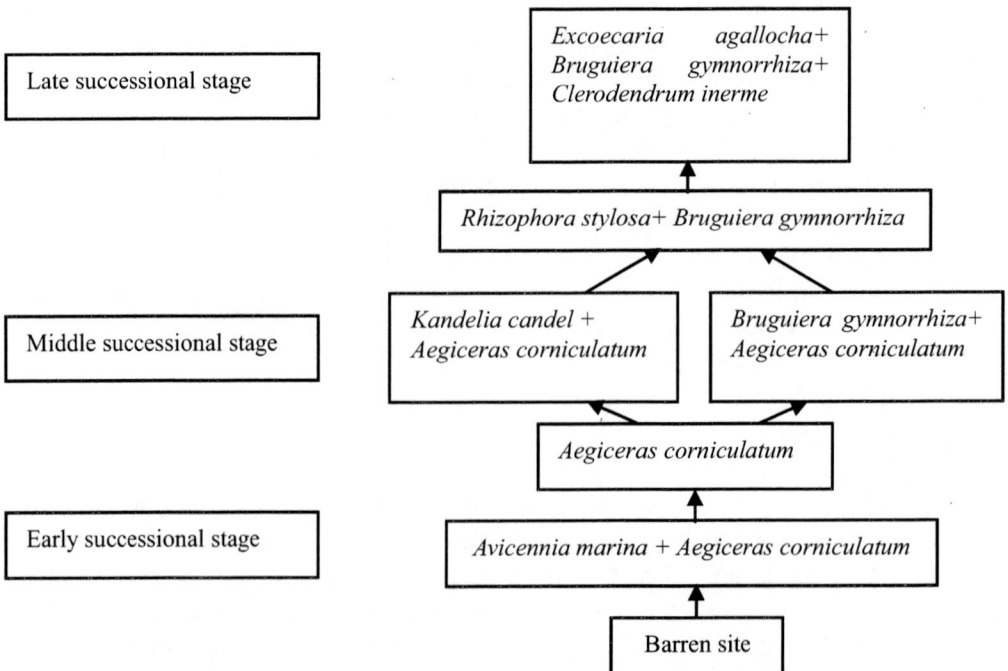

Figure 2. Mangrove succession.

Rhizophora stylosa possessed substantial prop roots, which enabled the plant to succeed in muddy hard soil. The continued accumulation of mud led to the formation of a muddy flat, and the frequency of floods reduced. The mud became firm, with the loam and clay content increased, and the environment supported the growth of *R. stylosa*. *Kandelia candel* community could not grow in a muddy hard soil with little tide influence, *Bruguiera gymnorrhiza* could still live in this environment. Gradually the *R. stylosa* + *B. gymnorrhiza* community is formed. With little tide influence, the muddy hard soil environment would turn into hard soil. This new habitat would be suitable for *Excoccaria agallocha.* Some mangrove associate species such as *Clerodendrum inerme* also occurred in this habitat. With soil becoming harder and harder over time, *R. stylosa* could not adapt to this soil property, but *B. gymnorrhiza* could still survive. Finally *E. agallocha* + *B. gymnorrhiza* + *C. inerme* community replaced the *R. stylosa* + *B. gymnorrhiza* community. These formed the late succession stage.

In a word, the mangrove substrate properties changed from sandy, semi-sandy, muddy silt to hard and firm mud, and salinity increased with community succession. During the succession, mangrove expanded both seaward and landward with continuing shifts in the dominant species over time (Lin, 1981; Liang, 1996; Wang, 2002).

4.5. Biomass

Biomass of different species varied with ages and growth rates. Older mangroves and faster growth rates resulted in larger biomass. Ratios of organic biomass to the total varied largely among different species. Ratios of root biomass to total biomass of *Avicennia marina*, *Aegiceras corniculatum* and *Kandelia candel* were relatively larger, which indicated the roots possessed stronger support systems. *A. marina* and *A. corniculatum* were pioneer species, and larger ratios of below ground part to total helped them stand firmly in loose soil substrate with large tidal fluxes. High root biomass proportion of *K. candel* was due to its giant buttress roots. The comparatively higher ratios of leaf biomass/total biomass of *Bruguiera gymnorrhiza* and *Rhizophora stylosa* indicated the two species had larger photosynthetic areas, which would enhance population development.

5. Conclusion

It's well known that mangroves have irreplaceable ecological importance. As a representative of pristine mangroves in China, the Gaoqiao mangrove forest deserves to be assessed of its ecological characteristics. The flora of Gaoqiao differed slightly to the adjacent area. Six main mangrove communities presented early, middle and late succession series. Soil acidification, increased organic matter content and salinity occurred with succession, but nutrition content did not change regularly within the different succession series. *Rhizophora stylosa* + *Bruguiera gymnorrhiza* community perhaps represented local climax community. It seemed that community canopy layer was a main factor to determine species diversity value. Succession series were established, deduced from the distribution of the main communities. Biomass of different organs of the main populations can be used as a reference to support the future comparison with other restoring mangroves.

The national reserve, where the Gaoqiao mangrove forest belongs to, was established 10 years ago. But the associated policy and management has not yet been fully implemented. Currently, insufficient personnel and financial support limit the progress of the reserve in reaching their conservation goals. Of primary importance is to ensure the following-up conservation of this core mangrove forest.

ACKNOWLEDGEMENTS

We thank Dr Wang Ruijiang at the South China Botanical Garden and all staffs of the Administration of Zhanjiang Mangrove National Nature Reserve for field assistance. This study was funded by the Knowledge Innovation Program of the Chinese Academy of Sciences (KSCX2-SW-132).

REFERENCES

Adeel, Z., Pomeroy, R., 2002. Assessment and management of mangrove ecosystems in developing countries. *Trees.* 16, 235-238.

Ashton, E.C., Macintosh D.J., 2002. Preliminary assessment of the plant diversity and community ecology of the Sematan mangrove forest, Sarawak, Malaysia. *For. Ecol. Manage.* 166, 111-129.

Benfield, S.L., Guzman, H.M., Mair, J.M., 2005. Temporal mangrove dynamics in relation to coastal development in Pacific Panama. *J. Environ. Manage.* 76, 263-276.

Cruz-Escalona, V.H. et al., 2006. Analysis of the ecosystem structure of Laguna Alvarado, western Gulf of Mexico, by mean of a mass balance model. *Estuar. Coast. Shelf Sci.* doi:10.1016/j.ecss.2006.10.013

FAO, 2003. Status and trends in mangrove area extent worldwide. By Wilkie, M.L. and Fortuna, S. *Forest Resources Assessment Working Paper No. 63.* Forest Resources Division. FAO, Rome. (Unpublished)

Fernandes, M.E.B., 1999. Phenological patterns of *Rhizophora L., Avicennia L. and Laguncularia Gaertn. F.* in Amazonian mangrove swamps. *Hydrobiologia.* 413, 53-62.

Ferwerda, J.G., Ketner, P., McGuinness, K.A., 2007. Differences in regeneration between hurricane damaged and clear-cut mangrove stands 25 years after clearing. *Hydrobiologia.* 591, 35-45.

Gunawardena, M., Rowan, J.S., 2005. Economic valuation of a mangrove ecosystem threatened by shrimp aquaculture in Sri Lanka. *Environ. Manage.* 36, 535-550.

He, B.Y., L, T.H., Fan, H.Q., Wang, W.Q., Zheng, H.L., 2007. Comparison of flooding-tolerance in four mangrove species in a diurnal tidal zone in the Beibu Gulf. *Estuar. Coast. Shelf Sci.* 74, 254-262.

Jayatissa, L.P., Dahdou-Guebas, F., Koedam, N., 2002. A review of the floral composition and distribution of mangroves in Sri Lanka. *Bot. J. Linnean Soc.* 138, 29-43.

Kathiresan, K., Bingham, B.L., 2001. Biology of mangroves and mangrove ecosystems. *Advances in Marine Biology.* 40, 81-251.

Liang, S.C., 1996. Studies on the mangrove communities in Yingluo Bay of Guangxi. *Acta Phytoecologica Sinica.* 20, 310-321. (Written in Chinese with English abstract)

Liang, S.C., Fan, H.Q., 1993. Plant population distribution pattern of *Rhizophora stylosa* community in Yinluo Bay, Guangxi. *Journal of Guangxi Academy of Sciences.* 9, 8-12. (written in Chinese with English abstract)

Lin, P., 1981. Subtropical mangrove forest research in Fujian. *Acta Phytoecologica Sinica.* 5, 177-186. (written in Chinese with English abstract)

Lin, P., 2001. A review on the mangrove research in China. *Journal of Xiamen University.* 40, 592-603. (Written in Chinese with English abstract)

Macintosh, D.J., Ashton, E.C., Havanon, S., 2002. Mangrove rehabilitation and intertidal biodiversity: a study in the Ranong mangrove ecosystem, Thailand. *Estuar. Coast. Shelf Sci.* 55, 331-345.

Méndez Linares, A.P. et al., 2006. The mangrove communities in the Arroyo Seco deltaic fan, Jalisco, Mexico, and their relation with the geomorphic and physical-geographic zonation. *Catena.* doi:10.1016/j.catena.2006.05.010.

Miao, S.Y., 1999. Ecological study on the mangrove forest in Zhanjiang Nature Reserve, Guangdong. *Journal of Guangzhou Normal University.* 21, 65-69. (Written in Chinese with English abstract)

Murray, M.R., Zisman, S.A., Furley, P.A., Munro, D.M., Gibson, J., Ratter, J., Bridgewater, S., Minty, C.D., Place, C.J., 2003. The mangroves of Belize Part 1. distribution, composition and classification. *For. Ecol. Manage.* 174, 265-279.

Primavera, J.H., 2000. Development and conservation of Philippine mangroves: institutional issues. *Ecol. Econ.* 35, 91-106.

Rasool, F., Tunio, S., Hasnain, S.A., Ahmad, E., 2002. Mangrove conservation along the coast of Sonmiani, Balochistan, Pakistan. *Trees.* 16, 213-217.

Sherman, R.E., Fahey, T.J., Martinez, P., 2003. Spatial patterns of biomass and aboveground net primary productivity in a mangrove ecosystem in the Dominican Republic. *Ecosystems.* 6, 384-398.

Shunula, J.P., 2002. Public awareness, key to mangrove management and conservation: the case of Zanzibar. *Trees.* 16, 209-212.

Ukpong, I.E., 1997. Mangrove swamp at a saline fresh water interface near Creek Town, Southeastern Nigeria. *Catena.* 29, 61-71.

Wang, B.S., Liang, S.C., Zhang, W.Y., Zan, Q.J., 2003. Mangrove flora of the world. *Acta Bot. Sin.* 45, 644-653.

Wang, B.S., Liao, B.W., Wang, Y.J., Zan, Q.J., 2002. Mangrove forest ecosystem and its sustainable development in Shenzhen Bay. *Beijing: Science Press,* pp. 3-150. (Written in Chinese)

Wang, W.Q., Wang, M., 2007. The mangroves of China. *Beijing: Science Press*, pp. 6-146. (Written in Chinese)

Ward, G.A., Smith, T.J., Whelan, K.R.T., Doyle, T.W., 2006. Regional processes in mangrove ecosystems: spatial scaling relationships, biomass, and turnover rates following catastrophic disturbance. *Hydrobiologia.* 569, 517-527.

Zhang, L., Lai, K.H., Chen, Z.T., 2002. The present situation and solution of the mangrove national nature reserve construction in Gaoqiao, Lianjiang. *Ocean information.* 1, 18-20. (Written in Chinese)

In: International Wetlands: Ecology, Conservation…   ISBN: 978-1-60456-999-5
Editor: Jose R. Herrara   © 2009 Nova Science Publishers, Inc.

Chapter 9

WETLANDS IN THE SOUTHERN KURIL ISLANDS: ORIGIN, AGE, EVOLUTION

N. G. Razzhigaeva, L. A. Ganzey, T. A. Grebennikova, L. M. Mokhova, N. I. Belyanina

Pacific Institute of Geography, Far East Branch, Russian Academy of Science,
Vladivostok, Russia

N. A. Eremenko

Reserve "Kurilskiy", Yuzhno-Kurilsk, Russia

T. A. Kopoteva

Institute of Aquatic and Ecological Problems, Far East Branch, Russian Academy of
Science, Khabarovsk, Russia

ABSTRACT

Wetlands are widespread on the Southern Kuril Islands. Wetland development is associated with late Pleistocene-Holocene climate warming. There are several types of wetlands. Blanket peat bogs are more ancient and have existed on flattened surfaces since the Postglacial warming (about 12 ka) and the accompanying increase in precipitation. Wetlands in coastal zones were formed in bays after the Holocene transgression maximum and the age of the oldest peat bogs are less than 6-4 ka. The oldest peat bogs are more than 7 ka and are found in coastal zones of volcanic islands; local conditions determined their formation. Wetland environments attained peak expansion in the Late Holocene when isthmus areas arose, coastal plains formed and surfaces turned into swamps. Wetlands in the lower portions of river valleys are also typical. Wetlands were formed in different ecological environments and during the Holocene had different peat buildup rates. Factors modulating wetland development are discussed. Wetland evolution during Holocene climatic changes is considered. Reconstructions are based on pollen, diatom, botanic analysis of peat and radiocarbon data. Sea-level changes played a key role in coastal wetland development. Specific local factors influencing wetland development on the islands are volcanic activity, varied neo-tectonic movements and tsunamis.

INTRODUCTION

Wetlands are a characteristic element of the landscapes on the Kuril Islands. They are distinguished by their floristic composition, origin, age and history of formation as well as by their association with certain relief types. Wetland development [22, 24, 27] and on the Japanese Islands [12, 17, 18, 41, 42], is closely tied to Holocene warm and wet climate phases [37]. Peat bogs, whose development stretched across almost the entire Holocene, are found on relatively small islands. Here they have limited distribution and their formation is a result of unique factors. Interest in wetland environments continues to grow since they are among the most important and informative sites for paleogeographic reconstructions, for wetlands can aid in establishing global and regional climate change patterns. This article discusses the development of wetlands as climate changed in the Holocene, a period affected by changes in sea level. We also analyze the factors underlying wetland formation on small islands with temperate zone, ocean climate conditions. This article studies the Southern Kuril Islands: Iturup, Kunashir and the islands of the Lesser Kuril Ridge. These islands not only vary in size, they also have varied relief, volcanic activity and neo-tectonic movements. The key features of peat bogs found on different relief types are studied in detail (Figure 1). Extensive drilling was carried out in wetlands on coastal lowlands and in the lower portions of river valleys to study for shoreline profiles from the coastal zone to deep into the landmass; elevation was determined line leveling. Spore and pollen data, botanic and diatomic analysis are the basis for paleoreconstructions. Sample age is based in radiocarbon dating undertaken at the Institute of Geology, RAN, Moscow (Lab. Index GIN) and at St. Petersburg University (Lab. Index LU) and in data on tephrostratigraphy. The calibrations of ^{14}C dates were made using the calibration program OxCal 3.9 [8, 44].

REGIONAL SETTING

Study area are located on south Kuril Arc, stretching from South Kamchatka to Hokkaido Island and includes Iturup, Kunashir and the islands of Lesser Kuril Ridge (Figure 1). The Pacific Ocean borders the islands on the east and the Sea of Okhotsk on the west. Lesser Kurils are located parallel Great Kuril Arc and divided from Kunashir Island by South Kuril Strait (48 km wide, depth up to 200 m). The islands are separated one by another by narrow shallow straits (1.9-5.6 km wide, up to 54 m depth), from Nemuro Peninsula – by strait with 0.7 km wide, depth up to 89 m. The islands located on south of the arc are small, less than 60 km^2 in area, maximal length is 11.4 km, width – 9.2 km (Zeleniy Island). Shikotan Island is largest and has 182 km^2 in area, 29 km long and 9 km wide.

Iturup and Kunashir Islands exhibit mountain relief (up to 500-1800 m) resulting from some volcanic groups, separated by low isthmuses (from 3-4 m to 60 m high). Shikotan has dissected low mountain relief (maximal elevation 412 m). Small islands located to the south from Shikotan have flat relief with elevation from 10-25 to 30-40 m. Zeleniy and Tanfiliev Islands are relics of Last Interglacial marine terraces.

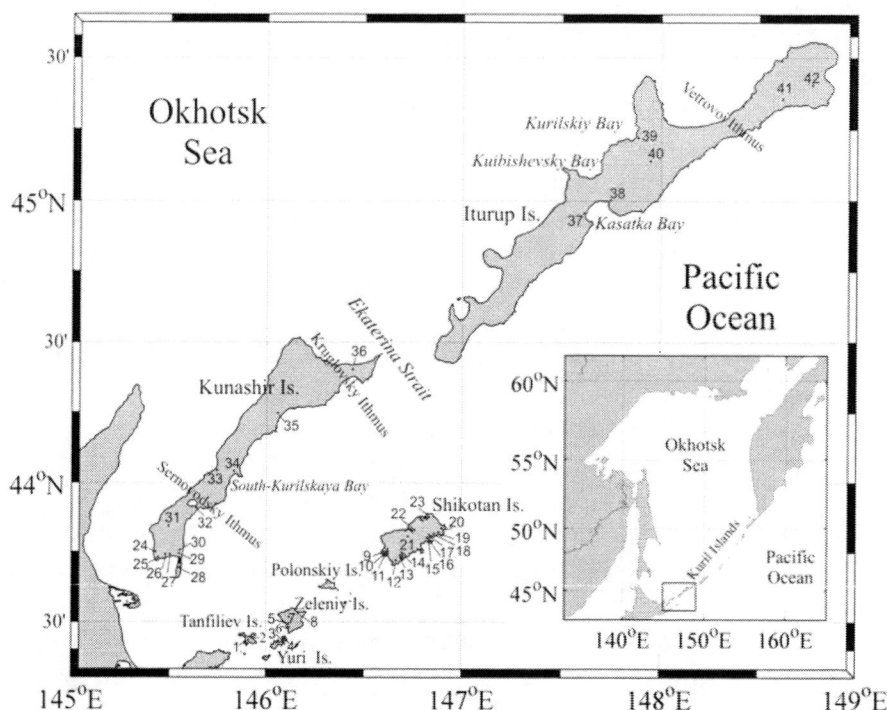

Figure 1. Location of the study area.

The South Kuril Island has numerous wetlands due to relatively warm climate with abundant precipitation under the influence of a temperate monsoon. The islands have oceanic climate with small annual temperature amplitude, warm winter and cool summer. During the winter the north-west air masses from Asia produce severe cold and snow. Winters are cold with mean temperatures below 0°C. February is the coldest month, lowest temperature is observed on Iturup Island (Table 1). During the summer moist cool Pacific air masses move south or east across the Kuril Islands toward the Asian low-pressure area, bringing with them extensive rainfall, fog and typhoons in August-September. The region has relatively low temperature in summer due to dense fog from the Pacific Ocean. The warmest month is August, mean temperature ranges from 16.0-16.6°C. Maximal temperature an rise above 30°C. Annual mean temperature vary from 4.2 on Iturup to 5.2°C on Lesser Kurils. Annual rainfall is high (1020-1255 mm). The island regularity experiences heavy rain events during the summer cyclones and typhoons. The islands have large amount of days with strong winds (101-106 days with wind > 15 m/c) and low snow cover (35 cm on Shikotan up to 68 cm on Iturup). Snow cover on small flat islands is unstable due to strong winds. N and NW winds prevail at October-March shifting to S at April-September. Amount of fog days are about 83, cloudy days – from 211 on Iturup up to 162 on Zeleniy Island [38, 39]. High humidity (82-84%) and low temperatures during the summer due to sea fog are important factors for wetland development.

Ocean currents are particularly important in influencing the regional climate of Southern Kuril Islands. The Pacific Ocean with the Oyashio cold current borders the island on the east and the Sea of Okhotsk on the west. The warm Soya current penetrates up to Kurilsky Bay on

Okhotsk side of the Iturup Island. The warm Soya current is most important, as it has a warming effect on the southern part of the Sea of Okhotsk. The Oyashio cold current brings cold water from north to south in the Kuril region.

Marine currents, mountain relief, and hot springs define a wide range of microclimatic conditions and a high diversity of ecotypes [1, 5]. Vegetation period within the Sea of Okhotsk side is warmer than Pacific coast. Phenological data show that plants in Okhotsk side develop early than in Pacific side, the same phases of plant development begin on 8-10 days earlier [5]. This is explained by two causes: influence of warm Soya current and barrier role of mountain relief, protected from fog and cold winters forming by cold water of the Oyashio current.

South Kurils are located in boreal zone. Vegetation of north part of Iturup Island (northern from Vetrovoi Isthmus) belongs to Urup floristic district, central and southern part of Iturup Island and Kunashir are divided to South Kuril floristic district and Lesser Kurils are separated in Lesser Kuril floristic district [5]. South Kuril floristic district has the most diversity flora with high contents of heat-loving plants. Cool-temperate broadleaf and mixed coniferous-broadleaf forests occupy the southern part of the Kunashir Island, boreal coniferous forests – central and north Kunashir and south Iturup. Larch and birch forests are wide spread on central and northern Iturup. Coniferous, birch and mixed coniferous-birch forests with nemoral elements are situated on Shikotan Island. Tree vegetation is absence on small islands on south Lesser Kuril Arc. The climatic condition of South Kurils facilitated the formation of a variety of wetland types.

Table 1. Climatic data for South Kuril Region.

Climatic parameters	Iturup Island	Kunashir Island	Shikotan Island	Zeleniy Island
Annual temperature, °C	+4.3	+4.7	+5.0	+5.2
Mean temperature in February, °C	-7.2	-6.7	-5.3	-5.9
Minimal temperature, ^{0}C	-26	-23	-22	-20
Mean temperature in August, °C	+16.0	+16.6	+16.3	+16.1
Maximal temperature, °C	+31	+31	+32	+28
Summa of active temperatures, (>10°C)	1451	1593	1562	1563
Annual precipitation, mm	1040	1255	1240	1020
Mean precipitation during Aprile-October, mm	595	915	870	770
Snow depth (minimal/maximal), cm	21/68	10/62	35	
Mean humidity, %	82	82	83	84
Amount of bright/cloudy days	12/221	16/173	20/175	22/162
Amount of days with strong winds (rate more than 15m/s)	80.1	69.9	101.4	63.4

TYPES, DISTRIBUTION, ORIGIN AND AGE OF WETLANDS

Wetlands on the Southern Kuril Islands are divided, based on location and hydrologic regime, into blanket, oligotrophic and lowland, eutrophic forms (Table 2). In contrast to

Sakhalin Island, where the predominate peat accumulation type is blanket [47], this wetland form has limitedly distribution on the Kuril Islands and lowland peat bogs dominate [37].

Blanket peat bogs are found on the small islands of the Lesser Kuril Ridge with flattened relief where they are a surface formation (Zeleniy, Tanfiliev, Polonskiy Islands) or where they are associated with closed depressions on terrace-like surfaces (Yuri Island). Zeleniy Island, which has an isometric form with depressions in its central portions, has the highest level of wetlands (Photoes 1, 2). Similar type wetlands are distributed on Nemuro Peninsula on Hokkaido Island and on Yururi, a small island to the east of Hokkaido [17, 28, 29].

Photo 1. Wetland of central part of Tanfiliev Island, Lesser Kurile Islands.

Peat accumulation on small islands in the south of the Lesser Kuril Ridge began at the end of the Late Pleistocene (Postglacial warming) and Early Holocene – ^{14}C-data at the foot of peat bogs 12200±150 BP, GIN-13457; 11400±100 BP, GIN-13456; 8870±110 BP, GIN-12550 (Table 3). Peat bogs sit on bluish-gray clay having a volcanogenic-aqueous origin that blocks water and that aids wetland formation on the islands. Grasslands are developed on drained portions of the island covered in yellow-brown loam. Volcanic ash seams with light-gray, mixed grain sand and a low K_2O (0.73%) content are found at the base of the Holocene portion of the peat bogs; ^{14}C-data 10360±60 BP, GIN-13455 were obtained below these seams. This is likely to be ash from Ma-g volcano Mashu on Hokkaido Island [11]. A volcanic ash seam consisting of coarse grain sand and with low K_2O (0.76%) content (^{14}C-data from underlying peat 7270±50 BP, GIN-13454; 6930±50 BP, GIN-13000; 6100±40 BP, GIN-13012) that corresponds to volcanic ash from Ma-f volcano Mashu is found in the Middle Holocene portion of peat bogs (Photo 3). Volcanic ash consisting of small grain sand with a low K_2O (0.68%) content (^{14}C-data from underlying peat 3420±40 BP, GIN-13006), whose origin is unknown, is found in the profile's Late Holocene portion. Volcanic ash composed by yellow silt and that radiocarbon dates to ash from Ta-c Tarumai Volcano that erupted between 2300-2500 years ago is located higher up in the profile. A light peat layer seam rests higher up; it formed around 1000 years ago and includes fine volcanic glass with high and low K_2O content and may be the result of two volcanic eruptions. Volcanic ash from Ko-c2 Komagatake Volcano (1694 AD) and Ta-a Tarumai Volcano (1739 AD) is often encountered in the cover of the peat bogs. These ash layers are wide distributed in peat and soil sections of Eastern Hokkaido [3, 11, 30, 31].

Table 2. Age, altitude and geomorphological setting of wetlands in Southern Kuriles

Wetland name	Number (Fig. 1)	Age (^{14}C yr BP)	Altitude,m	Peat depth (m)	Geomorphology
Tanfiliev Island					
Wetland in central part	1	12200	10-15	2	Last Interglacial marine terrace
Wetland near the lake	2	5700	1.5-6	2.70	Stream valley
Yuri Island					
Central part	3	4600	4-13	5.5	Stream valley
Blanket wetland in north part	4	5000-6000	30-40	0.8	30-40 m terrace surface
Zeleniy Island					
Wetland near Utinoe lake	5	3300	1-9	2.50	Lake coast, wetland formed due to regression and stream valley
South-western part	6	6900	5-8	1.2	Stream valley
Wetland in central part	7	8900-12000	10-15	2.0	Last Interglacial marine terrace
Northern part	8	6000	2-6	2.1	Stream valley
Shikotan Island					
Wetland near Cape Zvezdniy	9	3300	2-6	2.0	Lowland, wetland formed due to regression
Blanket wetland, southern part	10	3000	30-40	0.5	Terrace surface
Delphin Bay coast	11	2000	1-1.5	2.0	Lowland, wetland formed due to regression
Blanket wetland, south-eastern part	12	10000	60-70	1.4	Terrace surface

Wetland name	Number (Fig. 1)	Age (^{14}C yr BP)	Altitude,m	Peat depth (m)	Geomorphology
Malaya Tserkovnaya Bay coast	13	5000-6000	1-10	5.0	River valley
Tserkovnaya Bay coast	14	1200	2	0.70	Lowland, wetland formed due to regression
Dimitrov Bay coast, south	15	3300	3	1.4	Lowland, wetland formed due to regression
Dimitrov Bay coast	16	7800-8500	3-6	5.0	Stream valley
Dimitrov Bay coast, centre	17	4100-4500	2-6	4.5	Stream valley
Dimitrov Bay coast, north	18	2000-2100	1-3	2.0	Coastal lowland, wetland formed on place of paleolake
Dimitrov Bay coast, north	19	3300	2-8	3.30	Costal lowland and stream valley
Mayachnaya Bay coast	20	3000		1.0	Lowland and stream valley
Gorobets River	21	10000	8	4.0-5.1	River plain
Krabovaya Bay coast	22	2000-3000	2	2.0-2.9	Coastal lowland, wetland formed on place of paleobay
Malokurilskaya Bay coast	23	2700	2-3	2.0	Coastal lowland, wetland formed on place of paleolake
Kunashir Island					
Fedyashin Cape	24	7900	3	2.0	Coastal lowland
Izmena Bay coast	25	3500	2-4	1.2	River valley
Izmena Bay coast, near Khlebnikov Stream mouth	26	4600	2-3	1.3	Coastal lowland, wetland formed due to regression
Izmena Bay coast, near Golovnin River mouth	27	5000-6400	2-5	3.30	River valley
Veslovskii Peninsula	28	800-900	1-2	0.30	Coastal lowland
South-eastern coast	29	900	2-3	0.30	Lowland, wetland formed due to transgression
Belozerskaya River	30	3800	2-3	2.60	River valley, wetland formed due to transgression
Golovnin Volcano	31	?	120	?	Caldera bottom

Table 2. (Continued)

Wetland name	Number (Fig. 1)	Age (^{14}C yr BP)	Altitude,m	Peat depth (m)	Geomorphology
Sernovodsky Isthmus	32	2200	2-6	2.40	wetland formed due to regression
Maloe Lake coast	33	500	3-4	0.60	Coastal lowland, wetland formed on place of lake
Yuzhno-Kurilskaya Bay coast	34	4800	3-7	4	Coastal lowland, wetland formed due to regression
Coast near Rogachevka River mouth	35	2400	3-4	1.2	Coastal lowland, wetland formed due to regression
Kruglovsky Isthmus mire	36	1200	2	0.90	Lowland wetland formed due to regression
Iturup Island					
Kasatka Bay, southern coast	37	1000	3-4	0.40	Lowland wetland formed due to regression
Kasatka Bay, northern coast	38	3300	2-4	1.8	Coastal lowland, wetland formed on place of paleolake
Kurilsky Bay coast	39	1000	3-5	1.0	Coastal lowland, wetland formed on place of paleolake
near Gniloe Lake	40	4900	400	0.4	Caldera bottom
Tsirk Caldera	41	?	100	?	Caldera bottom
Medvezhiya Caldera	42	?	150-160	?	Caldera bottom

Blanket peat bogs are also found in the south of Shikotan Island. A surface wetland with shrub-moss vegetation and lichen is located among a sparse larch forest on a terrace-like surface at an elevation of 60-70 m to the south of Tserkovnaya Bay (Photo 4). Small areas of surface herbaceous and shrub herbaceous wetland are found on flattened surfaces near Cape Zvezdnyi and Delfin Bay.

In contrast to blanket peat bogs on Sakhalin Island, peat bogs on the Kuril Islands are, as a rule, a 1-1.5 m thick; they are also highly decomposed. The thickest peat uncovered in the central wetland area of Zeleniy and Tanfiliev Islands is 2 m (Figure 2), and on Shikotan – 1.5 m. The peat accumulation rate in these wetlands is less than 0.16-0.25 mm/year. The low accumulation and high decomposition rates are apparently conditioned by precipitation levels in different seasons. Even following the heavy rains that accompany typhoons, most water has drained in several days and water levels drop sharply. Strong winter winds blow snow cover off the land and reduce the role that thawing winter precipitation plays in the water balance of wetlands and this also has an affect on the hydrologic regime of wetlands.

Photo 2. Wetland of eastern part of Tanfiliev Island, Lesser Kurile Islands.

Photo 3. Peat bog with ash layers, Zeleniy Island, Lesser Kurile Islands.

Table 3. 14C-dates from peat bogs, Southern Kurile Islands

Sample no.	Depth (m)	^{14}C-age (yr BP)	Calibrated age (2 σ)	Lab. no.
Kunashir Island				
3/13003	0.37-0.40	2660±40	900BC(95.4%) 790BC	GIN-13023
4/9001	1.10-1.15	4530±110	3550BC(95.4%)2900BC	GIN-11924
5/13003	3.09-3.19	5250±90	4350BC(95.4%)3800BC	GIN-12549
5/6395	2.00-2.20	7910±140	7200BC(95.4%)6450BC	GIN-8950
2/1202	0.85-0.90	1240±40	680AD(95.4%)890AD	GIN-12223
2/5503	1.33-1.38	4800±40	3660BC(92.7%)3510BC 3410BC(2.7%)3380BC	GIN-12694
3/3196	0.35-0.40	4810±60	3710BC(86.2%)3490BC 3440BC(9.2%)3370BC	GIN-8971
3/1496	1.0-1.10	1360±280	0AD(95.4%)1300AD	GIN-9651
4/5936	2.10-2.20	1770±60	120AD(95.4%)410AD	GIN-8644
Zeleniy Island				
6/16803	0.67-0.70	8870±110	8300BC(95.4%)7650	GIN-12550
1/1904	1.08-1.10	6930±50	5980BC(2.4%)5950BC 5920BC(93%)5710BC	GIN-13000
Tanfiliev Island				
2/28405	1.35-1.40	7270±50	6230DC(95.4%)6020BC	GIN-13454
3/28405	1.67-1.72	10360±60	1090BC(95.4%)9800BC	GIN-13455
4/28405	1.87-1.92	11400±100	1190BC(12.1%)11650BC 11600(83.3%)11050BC	GIN-13456
5/28405	2.00-2.08	12200±150	13500BC(41.4%)12600BC 12500BC(54.0%)11800BC	GIN-13457
Yuri Island				
1/14204	0.60-0.62	6100±40	5210BC(95.4%)4850DC	GIN-13012
4/11604	0.93-0.95	3420±40	1880BC(95.4%)1610BC	GIN-13006
1/16004	5.45-5.50	4590±90	3650BC(95.4%)3000BC	GIN-13013
Shikotan Island				
6/21404	3.45-3.50	9420±90	9150BC(95.4%)8300BC	GIN-13020
10/13405	4.18-4.23	9750±80	9400BC(95.4%)8800BC	GIN-13473
2/15005	3.09-3.14	4200±120	3100BC(95.4%)2459BC	LU-5591
1/14405	1.31-1.36	1230±100	640AD(95.4%)1000AD	LU-5585
1/11006	1.40-1.45	2760±100	1220BC(95.4%)750BC	LU-5742
Iturup Island				
1/9798	0.65-0.70	970±60	970AD(95.4%)1220AD	GIN-10726
2/9898	0.96-1.0	1000±40	970AD(95.4%)1160AD	GIN-10490
3/9898	1.46-1.50	1180±60	680AD(95.4%)990AD	GIN-10491
2/1095	0.20-0.30	4980±90	3970BC(95.4%)3630BC	GIN-8947
Kasatka Bay Peatland [9].		3300±250	2300BC(95.4%)900BC	GIN-6516

Photo 4. Blanket wetland on South-Eastern Shikotan Island.

Photo 5. *Myrica tomentosa,* Tanfiliev Island, Lesser Kurile Islands.

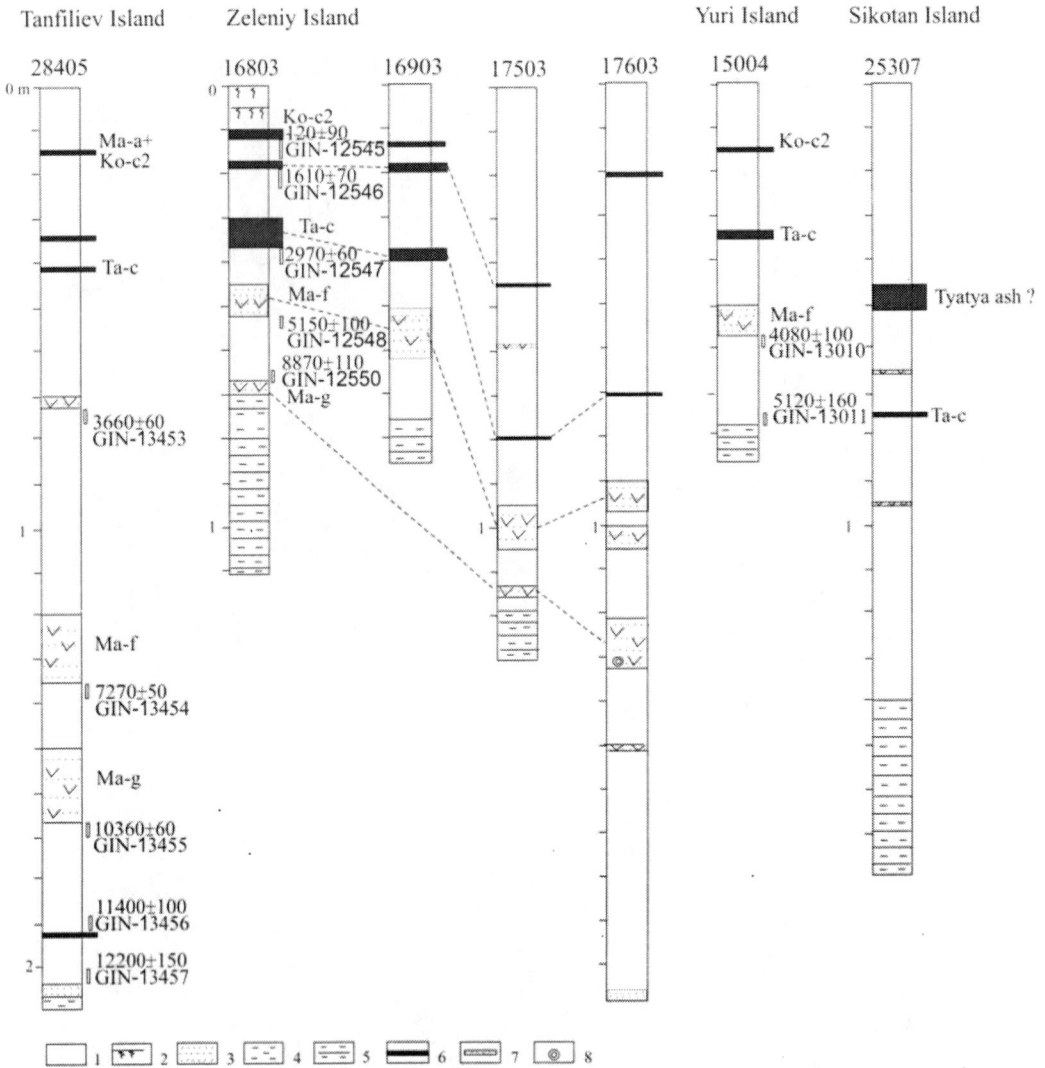

Figure 2. Blanket peat bog sections of Lesser Kuriles.
1-peat; 2-soil; 3-sand; 4-silt; 5-clay; 6-ash layer (silt composition); 7 - ash layer (sand composition); 8-wood.

Herbaceous plant debris is the primary vegetation source in peat bog formation, while dwarf low shrubs play a role in certain regions. Modern vegetation associations in different parts of a wetland are greatly dependent on the hydrologic regime. Heavily water flooded areas with active peat accumulation are associated with depressions in the central portion of wetland areas and where ground water reaches the surface. No micro-relief exists here. Where ground water reaches the surface along linear zones in the upper portions of valley streams, hummock micro-relief is characteristic. The thickness of peat declines sharply and decomposition rate increases.

Phragmites australis, Carex limosa, C. schnidtii, Lysichiton camtschatcense, Eleocharis wichurae, Juncus papillosus, Polygonum thunbergii, Parnassia palustris, Lysimachia davurica, Lobelia sessilifolia and others are representative of the floristic complex on Zeleniy

Island [5, 10]. *Myrica tomentosa, Ledum hypoleucum, Empetrum sibiricum* are among the dwarf low shrubs and *Vaccinium ulignosum, Sorbus sambucifolia* are found (Photo 5). Trees are totally absent.

Pollen analysis provides data on changes in the composition of wetland communities [35]. When peat bog formation began at the end of the Late Pleistocene and Early Holocene, sedge apparently played a large role in plant and peat formations. Dwarf low shrubs from the family Ericaceae were distributed widely in the vegetation cover structure of the Middle Holocene. *Myrica tomentosa* developed into a mass form and its pollen dominates arboreal pollen (up to 96%) in the second half of the Late Holocene. *Myrica* pollen is abundant in peat bogs on Nemuro Peninsula of Hokkaido Island [17, 28]. Throughout this period of peat bog development, as a rule *Sphagnum* spores dominate, this expressed by its high moss productivity. However, this does not entirely explain the massive development of these plants. Some seams show a high volume of Polypodiaceae and *Lycopodium* spores.

Eutrophic wetlands are divided into caldera, valley, coastal lowland and isthmus forms; these developed in inundated bays and straits existing in deflation basins in dune fields during the transgression phases of the Middle and Late Holocene.

Mountain mires on South Kuril Island are very rare. Wetlands are found on Iturup and Kunashir Islands in ancient calderas. An example is the broad, poorly studied wetlands in the ancient calderas of Medvezhya and Tsirk Volcanos in the northern portion of Iturup Island. The wetland within Medvezhya Caldera is located has elevation 150-160 m, has 4 km in length, and 2 km in wide (Photos 6, 7). There are many small pools and water level is high. In Tsirk Caldera the wetland is situated on elevation about 100 m, occupy all caldera bottom and has size 2x2.4 km. There are many drainage channels.

Photo 6. Wetland in Medvezhya Caldera, Iturup Island (Photo of A.V. Rybin).

Photo 7. Wetland in Medvezhya Caldera, Iturup Island (Photo of A.V. Rybin).

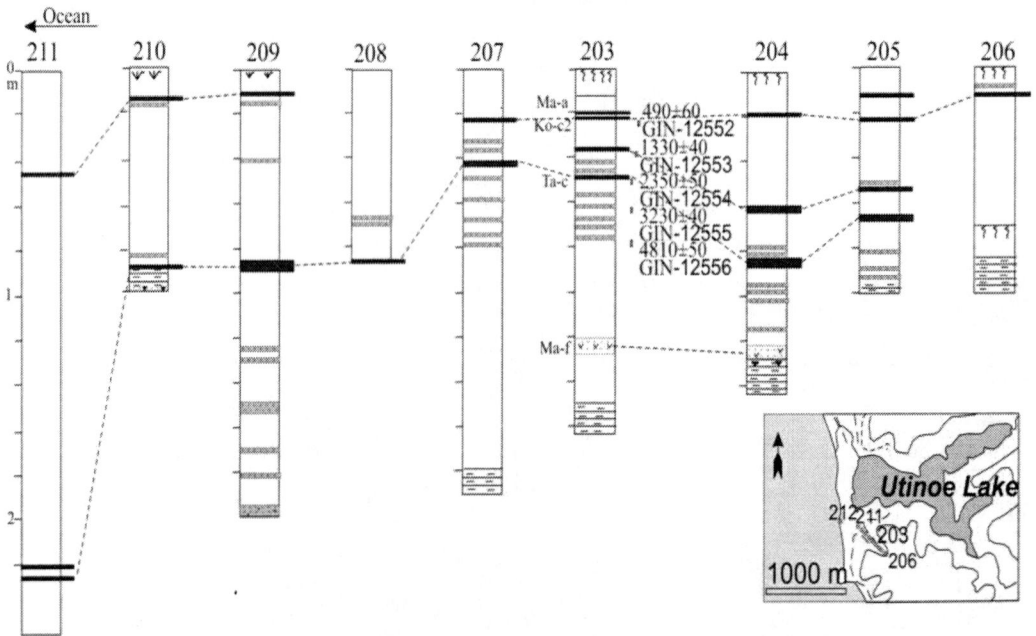

Figure 3. Peat bog sections within stream valley near Utinoe Lake, Zeleniy Island.

Photo 8. Wetland on lowland and within river valley, Dimitrov Bay coast, Shikotan Island.

A small wetland in the central portion of Iturup Island is found in an ancient caldera near Gniloe Lake at an elevation of 400 a.s.l. At a minimum, peat accumulation began in the Middle Holocene (^{14}C-data from lower portion of the peat bog 4980±90 BP, GIN-8947). The thickness of peat at the edge of the caldera is not great (up to 0.4-0.5 m) and the decomposition rate is high. Small wetlands (total size 810x110 m) with little shrubs and mosses are found in the caldera of Golovnin Volcano near Goryachee Lake on Kunashir Island at an elevation of 120 a.s.l. Not only is the saucer like shape of a caldera a factor in wetland formation, metasomatic change of the bedrock also plays a role near hot springs where it is turned into clay that acts as a water barrier.

All islands have wetlands in river valleys (Photo 8, 9). The thickest peat is in valley peat bogs with tectonic subsidence: islands of the Lesser Kuril Ridge and the southern portion of Kunashir Island. Peat bogs on small islands in the Lesser Kuril Ridge cover not only valley floors but also their sides. Small streams cut narrow channels through the peat forming vertical walls that lack any trace of alluvia. Peat accumulation on all the islands is localized in valley depressions and the foot of a peat bog can rest below sea level (Figure 3). For example, the foot of a 5.5 m thick peat bog on Yuri Island is 1.1-1.3 m below sea level. ^{14}C-data 4590±90 BP, GIN-13013 for the foot the peat bog demonstrates that peat began to accumulate when sea levels dropped during a climate cooling and regression period between the Middle and Late Holocene. Marine depositions belonging to later transgressions are not found in this profile or in profiles for peat bogs on other small islands in the Lesser Kuril Ridge. A 3.3 m thick peat bog was discovered at the outlet for Golovnin River in the south of Kunashir Island and whose foot rests 1 m below sea level (^{14}C-data from the base of the profile 5250±90 BP, GIN-12549).

Maximum thicknesses are found 200-400 m from the shore line of valley river outlets that are beyond the zone inundated during the Middle and Late Holocene transgressions (Figures 4, 5). Micro-relief in areas of active peat accumulation is not expressed and large hummock micro-relief is characteristic of areas close to stream channels. Stretches of the lower stream drainage meander intensively. In this case, the thickness of peat bogs at river valley outlets is not great and no continuous peat profiles are found.

Photo 9. Wetland within river valley, Dimitrov Bay coast, Shikotan Island.

Peat with low to moderate decomposition rates accumulates in river valleys with active wetland formation processes; accumulation rates reach 1.16 mm/year. On small islands where peat forms a continuous sheet along the sides of valleys, accumulation rates in upper valleys decrease sharply and peat bogs shift to peat soils. In several instances active peat accumulation occurs in the middle reaches of valleys where clay promotes wetland development. A wetland in the middle drainage of the Gorobets River on Shikotan Island has 5.1 m thick layer of peat (Figure 6; Photo 10). This is one of the oldest peat bogs in the Southern Kuril Islands and a profile ^{14}C-data 9750±80 BP, GIN-13473 was obtained at its base [35].

Microclimatic features promote peat accumulation. An example is a peat bog in the stream valley of Malaya Tserkovnaya Bay on Shikhotan Island that has been protected from direct ocean impacts by the small island Aivazovskiy. Peat at this location is more than 5 m thick (Figure 7). Partitioned valley peat bogs are multi-layered with terrigene material having different origins. Layers of sediments from tsunamis are found in peat bogs in the outlet portions of valleys [36].

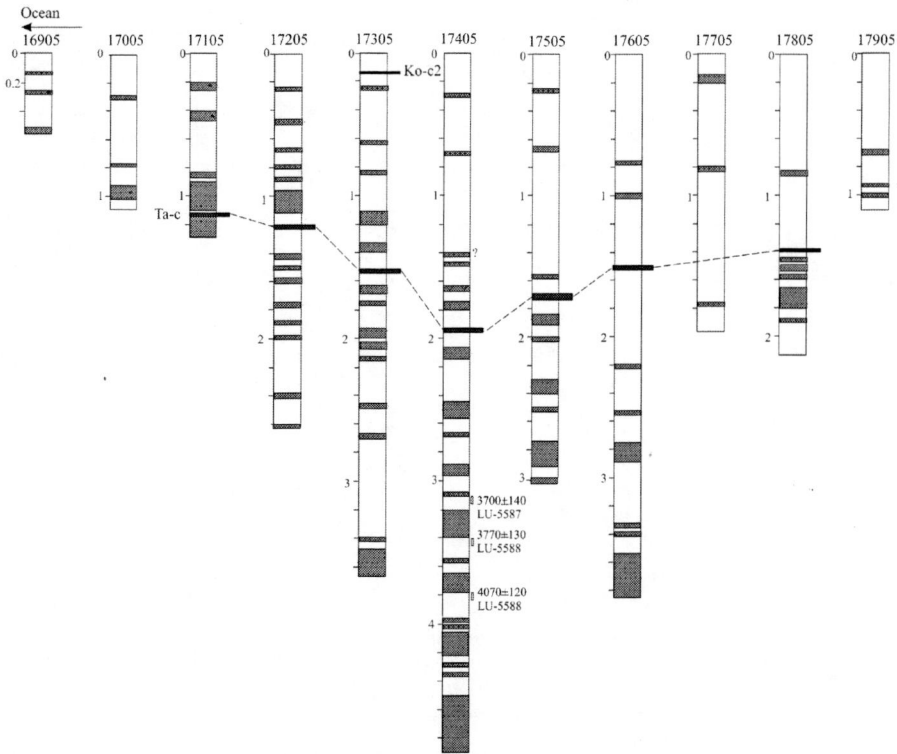

Figure 4. Peat bog sections within stream valley, Dimitrov Bay coast, Shikotan Island.

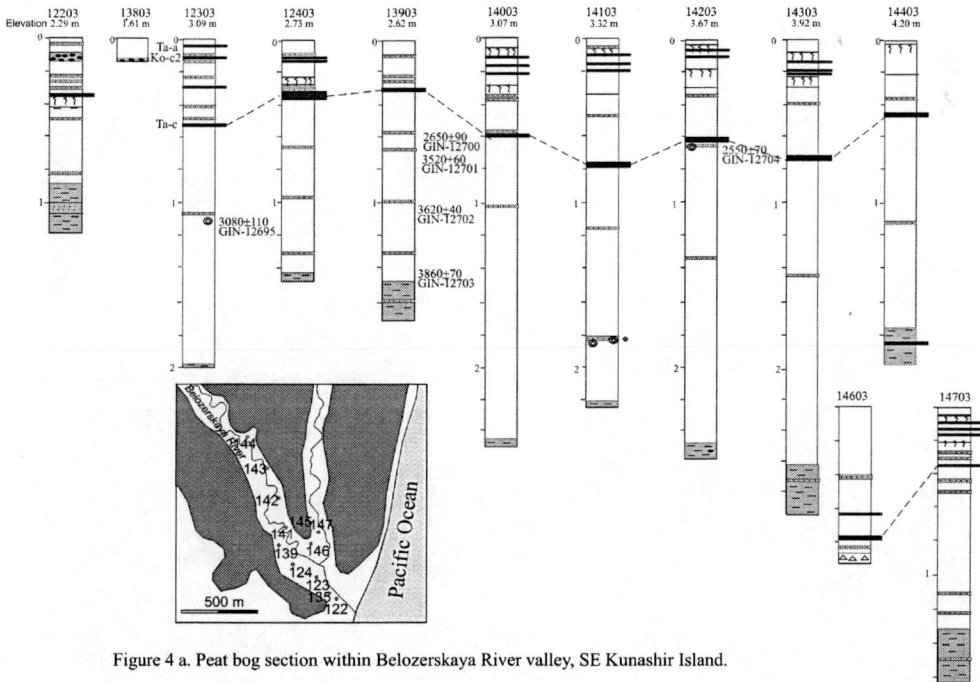

Figure 4 a. Peat bog section within Belozerskaya River valley, SE Kunashir Island.

Figure 5. Peat bog section within Belozerskaya River valley, SE Kunashir Island.

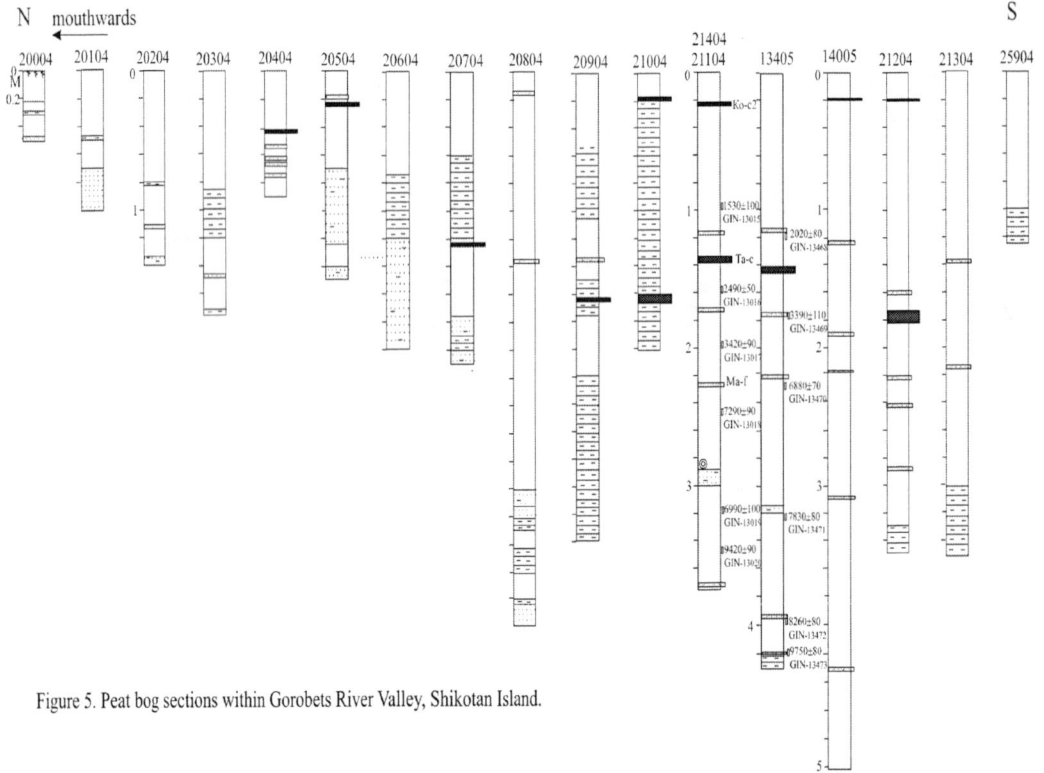

Figure 5. Peat bog sections within Gorobets River Valley, Shikotan Island.

Figure 6. Peat bog sections within Gorobets River Valley, Shikotan Island.

Photo 10. Wetland within valley of Gorobets river, Shikotan Island.

The botanic composition of peat shows frequent shifts in vegetation-peat forming elements, with herbaceous vegetation providing most of the debris (Figures 8, 9). An accumulation of lowland woody peat began on Shikotan Island in the Gorobets River valley in the Early Holocene; it has larch debris (up to 65%) and a small volume of alder. *Carex* spp. and *Calamagrostis* spp make up to 40% of the herbaceous debris. The mosses are *Bryales* spp. and *Sphagnum* spp is sporadic. The peat is highly decomposed (80-90%). Climate warming and increased humidity (^{14}C-data 9420±90 BP, GIN-13020) changed wetland associations and herbaceous peat began to accumulate. *Carex* spp., *Equisetum* spp., *Phragmites* spp. making up to 70%, dominate the debris. Woody species and shrub debris do not exceed 20%. *Larix, Spirea* are encountered. *Sphagnum* makes up to 10% of the moss layer, with *Bryales* (*Paludella squarrosa*), Hepaticopsida totaling up to 30%. Dramatic changes in ecological conditions led to *Larix* (up to 70%) dominating layers. Peat layers of mostly moss (up to 70%) are found, with *Bryales* (*Paludella squarrosa*), Hepaticae dominating. Peat decomposition is the lowest (30-40%) in such layers. Herbaceous peat began to accumulate in the Middle Holocene, is highly decomposed (60-80%) and consists mostly of *Carex* spp. There are large volumes of *Phragmites* spp debris. Ferns are found sporadically. *Chamaedaphne calyculata* is found among the dwarf low shrub debris. The content of *Calamagrostis* spp. increases in the layer of black, very well decomposed peat. Late Holocene herbaceous peat consists primarily of sedge. Peat decomposition is lower (40-60%). Other herbaceous plants (*Equisetum, Phragmites, Calamagrostis, Lobelia sessilifolia,* Polypodiaceae) are represented in small volumes. Shrub (cranberry and others), alder and birch debris is found in small volumes (less than 5-10%, in separate layers up to 20%). Moss debris is present in certain layers (up to 5-10%), represented by *Sphagnum magellanicum, Bryales* spp. Among the sedge, layers with *Calamagrostis* spp. and *Phragmites* spp. debris as well as *Comarum palustre, Eriophorum grasile, Glyceria aquatica* make up the upper portions of peat bogs. Shrub diversity increases and makes up 25% of the peat layer; accumulations occurred in the Little Ice Age. Moss content does not exceed 10%. Peat is highly decomposed (60-70%). Various species of *Eriophorum* (up to 70%) and *Carex* (up to 20%) dominate the surface layer.

Grasses dominate the vegetation (up to 85%) content of peat in valley peat bogs in the outlet of the Golovnin River on Kunashir Island. Grass composition is varied in the Middle Holocene portion of the profile; *Carex* spp., *Calamagrostis* spp., *Equisetum* spp., Polypodiaceae, *Typha latifolia, Phragmites* spp. debris, and in certain layers *Comarum polustre, Acorus calamus, Filipendula kamchatica, Menyanthes trifoliata* and *Scheuchzeria polustre* debris are found. Shrub and woody species debris does not exceed 20-30% Alder and cranberry are found. Moss is not abundant (less than 15%); *Sphagnum*, most often *S. teres* debris, a hydrophilic species characteristic of richly fed wetlands, dominates [47]. *S. squarrosum*, eutrophic species, is found less often. Maximum *Sphagnum* debris (up to 30%) rests in seams of sphagnum-shrub peat with low decomposition rates; these formed during cooling and retreat of the sea level between the Middle and Late Holocene (^{14}C-data 4530±110 BP, GIN 11924). *Bryales* spp. debris (up to 10%) is found in some seams. Decomposition rates are greatest in the lower portion (up to 90%), higher up in the profile the rate decreases to 50-60%, though a seam of well decomposed peat is found (80-90%) in the center of the profile. In the Late Holocene portion of the profile peat decomposition rates decrease to 30-50%. The composition of herbaceous-peat is less diverse. *Calamagrostis* spp. (up to 70%) and *Carex* spp. (up to 70%) debris dominate. Lower peat accumulation rates in the second half of the Late Holocene (^{14}C-data 2660±40 BP, GIN 13023) aided the development of soil.

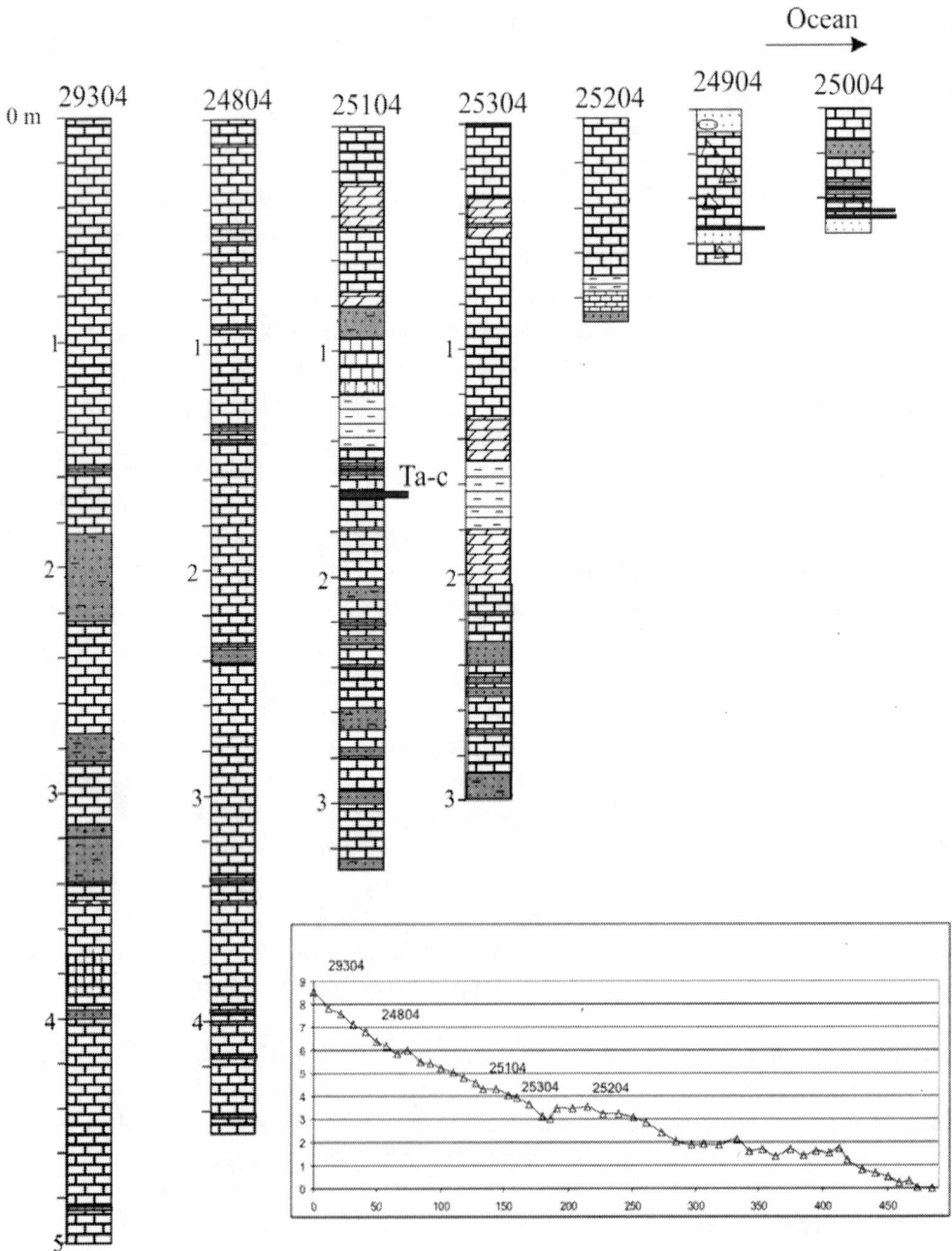

Figure 7. Peat bog sections within stream valley, Malaya Tserkovnaya Inlet coast, Shikotan Island.

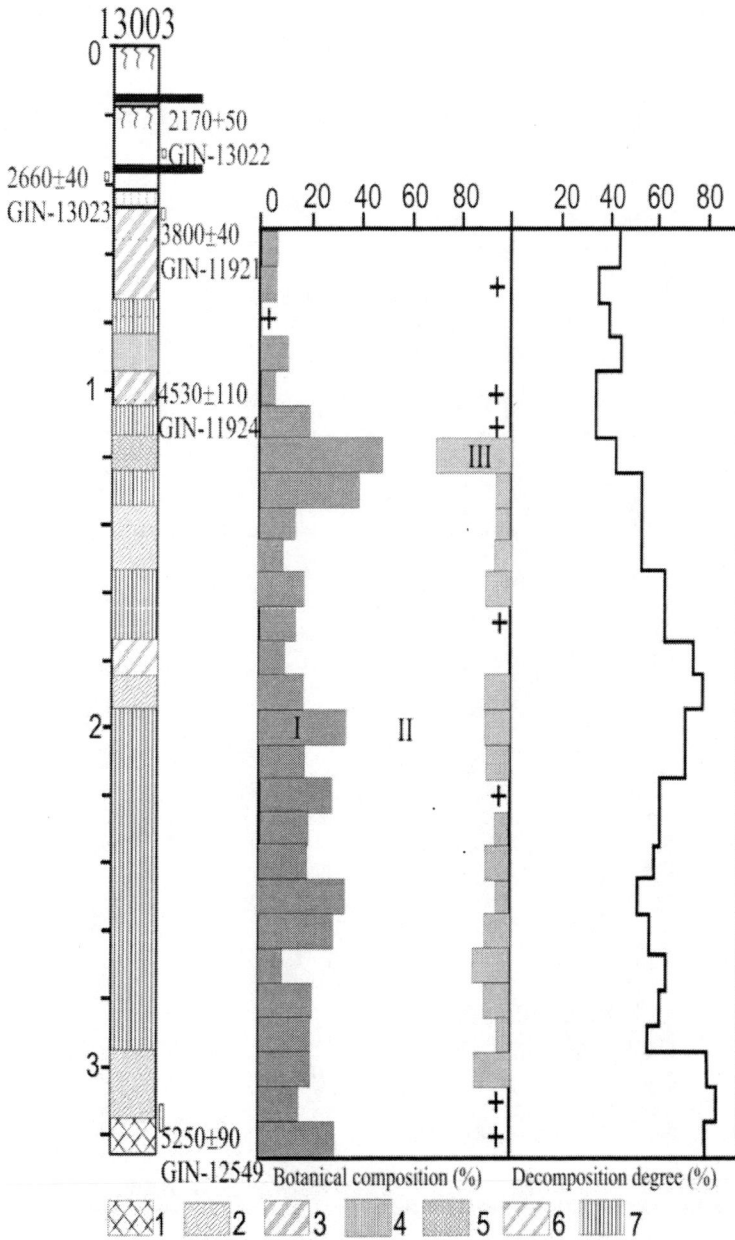

Figure 8. Botanical composition of peat bog within Golovnin River Valley, Southern Kunashir Island.
I - wood and shrubs; II – herb; III - moss.
1 - shrub-herb peat; 2 - sedge peat; 3 - *Calamagrostis*-sedge peat; 4 - *Calamagrostis* peat; 5 - *Sphagnum*-shrub peat; 6 - sedge-*Typha* peat; 7 - herb peat.

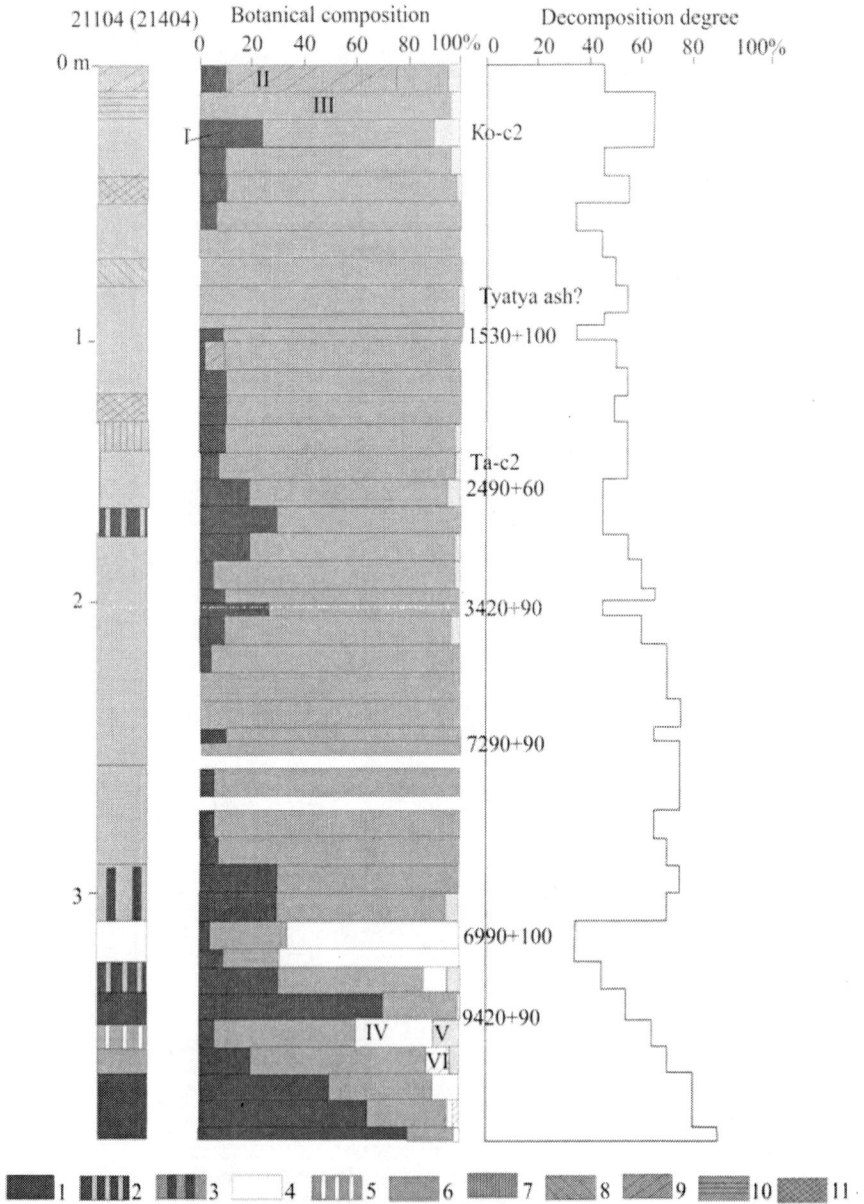

Figure 9. Botanical composition of the peat bog within Gorobets River Valley, Shikotan Island.
I-wood and shrubs; II-Eriophorum spp.; III-herb; IV-moss (Bryales+some Hepaticae); V-moss
(*Sphagnum*); VI-moss (Hepaticae).
Peat: 1-wood peat; 2-wood-sedge peat; 3-shrub-sedge peat; 4-Bryales peat; 5-moss-sedge peat; 6-sedge
peat; 7-*Calamagrostis*-sedge peat; 8-*Phragmites* peat; 9-*Eriophorum* peat; 10-*Glyceria*-sedge peat; 11-
herb peat.

Photo 11. Lowland wetland near Yuzhno-Kurilsk, Kunashir Island.

Photo 12. Lowland wetland near Olya Lake, Kunashir Island.

Photo 13. Wetland on Sernovodsky Ithmus, Kunashir Island.

Lowland eutrophic wetlands are broadly distributed in the coastal zone. Peat covers low marine terraces. Surface thickness ranges from 1-2 to 7-8 m. Shoreline walls that stretch parallel to the coastline and on which forest vegetation is often found is a feature of these wetlands. Broad lowland eutrophic wetlands are found on large islands (Iturup and Kunashir) where they are concentrated in the upper portions of bays and lowland isthmuses. Large wetlands are found on Kunashir Island near Dolgoe Lake on the Kruglovsky Isthmus, in the upper portion of Yuzhno-Kurilskaya Bay, on the Pacific Ocean side of Yuzhno-Kurilskiy Isthmus (Photoes 11, 12), on the Pacific Ocean side of Sernovodsky Isthmus (Photo 13), and on the coastal plain in the southeast portion of the island. On Shikotan Island these wetlands are found in the upper reaches of closed bays (Krabovaya and Otradnaya) where there is marine deposits under peat bogs. Wetlands are also found on the shore of Malokurilskaya Bay and on the Pacific Ocean coastline of Dimitrov Bay. Wetlands are situated on the shores of coastal lakes and on narrow lowland isthmuses on the small islands of the Lesser Kuril Ridge.

Peat accumulation began immediately after the maximum phase of the Middle Holocene transgression and the age of the oldest peat bogs is, as a rule, less than 6000 years; the majority are 4000 years [23, 33]. Peat bogs along the bay and lagoon edges are the oldest (^{14}C-data 6360±110 l.n. GIN-11924, Kunashir Island); they cut sharply into the land and accumulation began at the height of the transgression. The oldest peat bogs exceed 7000 years and are found in the coastal zone of volcanic islands where local conditions drove their formation. An example is the peat bog near Fedyashin Cliff in the southwest of Kunashir Island where peat accumulation began around 8000 years ago (7910±140 BP GIN-8950); this is on the edge of a small barrier lake in a small stream with welded tuffs and low water penetration. Most peat bogs in the coastal lowlands are associated with active accumulative formations from the Late Holocene. Active peat accumulation occurred when sea levels dropped at the end of the Middle Holocene and during the regression between the Middle and Late Holocene (4800-4500 ^{14}C yr BP). Peat accumulation began in certain areas as sea levels dropped at the beginning of the Late Holocene (3790±70, GIN-8443; 3300±250, GIN-6516) [9, 23,32, 33]. More active peat accumulation occurred on coastal lowlands beginning in the Late Holocene (2760±120, GIN-7039; 2780±60, GIN-8431). The youngest wetlands are on coastal lowlands in the lower drainage of the Kurilka River (970±60 BP, GIN-10726), on the coastline of Kasatka Bay on Iturup Island (1000±40 BP, GIN-10490), near Dlinnoe Lake on the north of Kunashir Lake (1240±40 BP, GIN-12223) and on Veslovskiy Peninsula in southeast Kunashir Island.

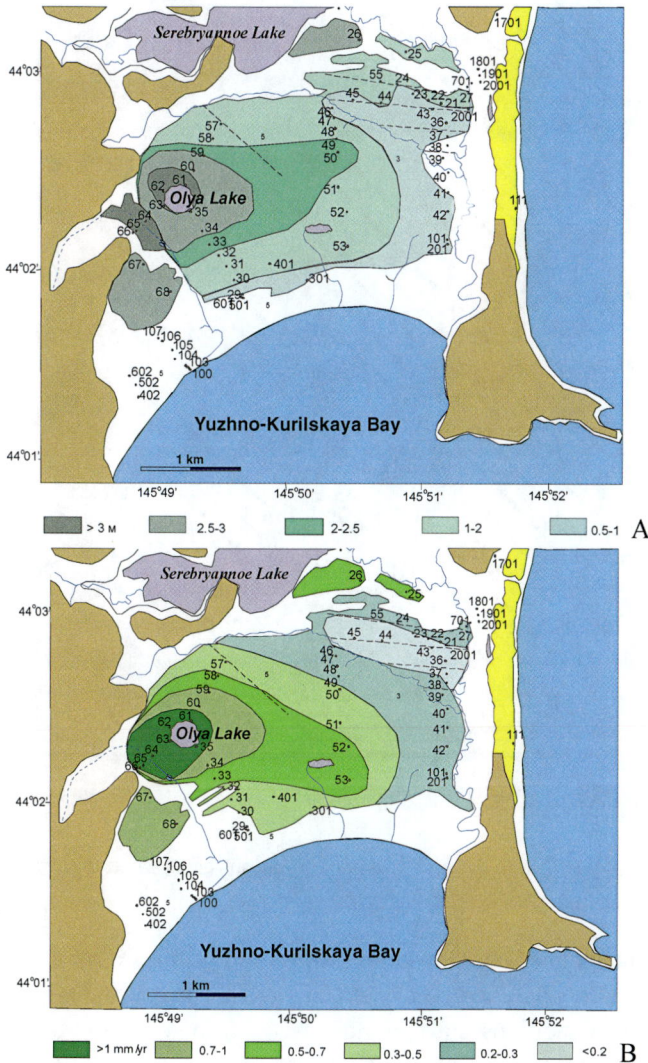

Figure 10. (Continues)

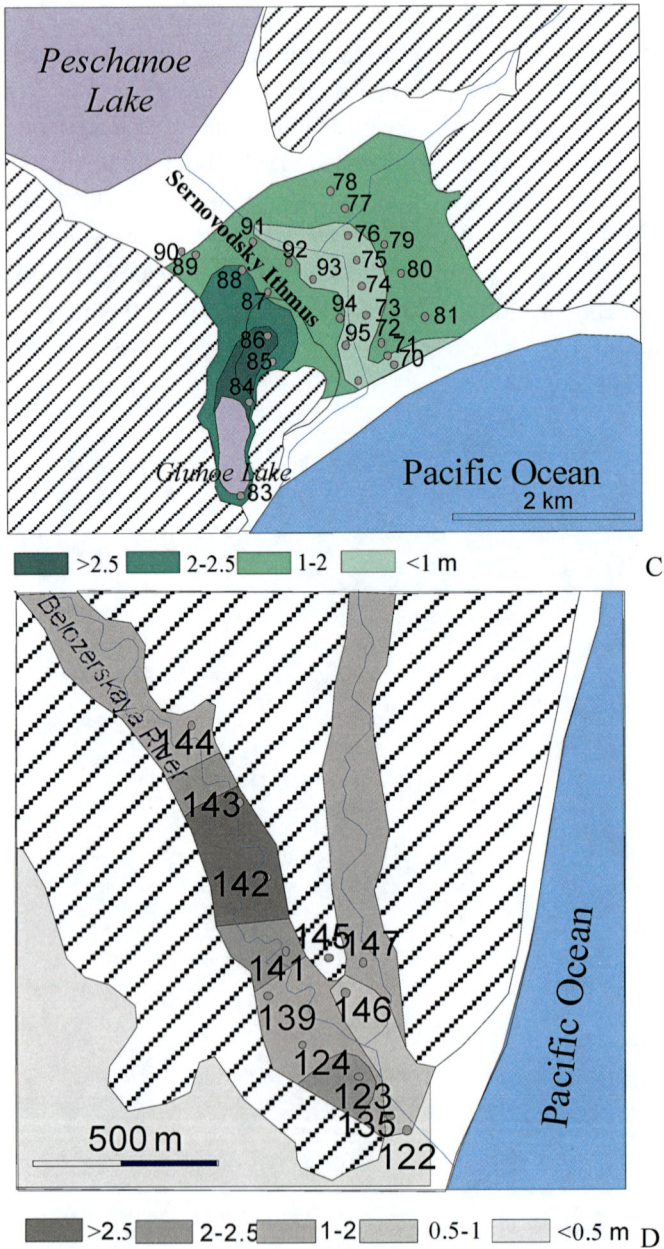

Figure 10. Thickness of peat (A, C, D) and peat accumulation rates (B) on the wetlands of Kunashir Island.

As a rule, wetlands in coastal lowlands have a thin peat layer up that is up to 2 m thick. The thickness increases where temporary streams release water and where water levels near lakes increase; these conditions aid continual watering and the rapid growth of peat bogs (Photo 14). Slightly decomposed peat accumulates and the peat accumulation rate exceeds 0.5-1 mm/year (Figure 10). Peat bog thickness in the last thousand years has been less than 1 m and rarely exceeds 0.3-0.4 m. The peat has a weak decomposition rate.

Photo 14. The wetland with maximal peat-forming rates, near Olya Lake, Kunashir Island.

Several wetlands on the coast of islands arose on the site of shallow water barrier lakes. These lakes, as a rule, underwent several stages of watering and wetland formation conditioned by climate reasons and by litho-dynamic factors in the coastal zone that in many ways are interrelated [13, 23, 33]. In the profile of a paleolake in the lower drainage of the Lesnaya River (Kunashir Island), the increase in the depth of the lake was connected with a rise in sea level and then with the level of ground water on the coast (watering phase around 6400-6200 ^{14}C yr. BP). The watering phase in the 6020-5720 ^{14}C yr. BP interval coincides with the pluvial phase identified for Japanese Islands and is probably associated with an increase in precipitation. The barrier lake Kushu on Rebun Island [43] reached its maximum depth around 4900-3100 ^{14}C yr. BP, a time that corresponds to the pluvial phase at the end of the Middle Holocene, to the regressions that occurred between the Middle and Late Holocene, and also to the transgressions in the first half of the Late Holocene.

During the watering phase terrigenous sediments accumulated in lakes and the rate of sediment accumulation increased (for example in the paleolake at the outlet of Lesnaya River the rate of increase doubled to 1.4 mm/year). As water bodies developed further, a gradual increase in sediments occurred and lakes grew shallower and wetlands began to form. The evolution of such lake environments occurred asynchronically and were heavily dependent on local conditions. The intensive and irreversible wet land development of water bodies began in the second half of the Late Holocene. For example, these processes occurred around 3300±250 BP, GIN-6516 for Kasatka Bay [9]; around 2000-2500 yr. BP for Kushiro Lake on Hokkaido Island [26] and Lake Kushu on Rebun Island [43].

The peat bogs at shallow paleolakes are, as a rule, Late Holocene. Lake sediments were studied on the east (Dimitrov Bay) and west (Malokurilskaya Bay) coasts of Shikotan Island.

The wetland on the edge of the paleolake on the north coast of Dimitrov Bay (Photo 15) existed throughout the Late Holocene (^{14}C-data 4200±120 BP, LU-5591). The lake filled in and the shifted to a wetland around 1230±100 BP, LU-5585. A shift in the diatomic complex characteristic for shallow water lakes is a key condition for wetland formation. A wetland formed at a lake on the coast of Malokurilskaya Bay around 2760±100 BP, LU-5742. The shift from lake to wetland environments is seen in the shift of diatomic complexes; different representatives of genera *Pinnularia* (up to 33%, 18 species) and *Eunotia* (up to 10%, 14 species) characteristic of wetland associations appear. The composition in the upper portion of the peat bog has diatomic species that are capable of enduring temporary drought conditions: *Pinnularia borealis, P. ignobilis, P. viridis, Hantzschia amphioxys, Eunotia praerupta*. Changes in the freshwater composition of flora indicate a gradual rise and swamping of the lake, leading to development of a wetland. A similar wetland is found on the north coast of Kasatka Bay on Iturup Island; its formation began less than 3300 ^{14}C yr. BP [9]. Coastal lowland peat bogs are distinguished by homogenous botanic composition with a predominance of herbaceous plant debris and decomposition rates (up to 80%). The botanic composition of peat bogs has been studied for wetlands located on the coast of Yuzhno-Kurilskaya Bay on Kunashir Island (Figure 11). Following the drop in sea level at the end of the Middle Holocene (^{14}C-data 4800±40 BP, GIN 12694), wetlands replaced marine environments and sedge peat environments with a high decomposition rate (75-90%) began to accumulate. *Calamagrostis* spp. debris is also found in the herb composition. Peat includes dwarf low shrub and alder debris (up to 35%). *Sphagnum* spp., *Bryales* spp. debris is found in isolation. Peat decomposition in the Late Holocene portion of the profile rises dramatically (40 to 90%). The large volume of *Scheuchzeria* spp. debris in several seams demonstrates the high watering rate. This plant is now rarely encountered. Moss in several seams increased to 20% and *Sphagnum magellanicum* and species from the genus *Bryales* are found.

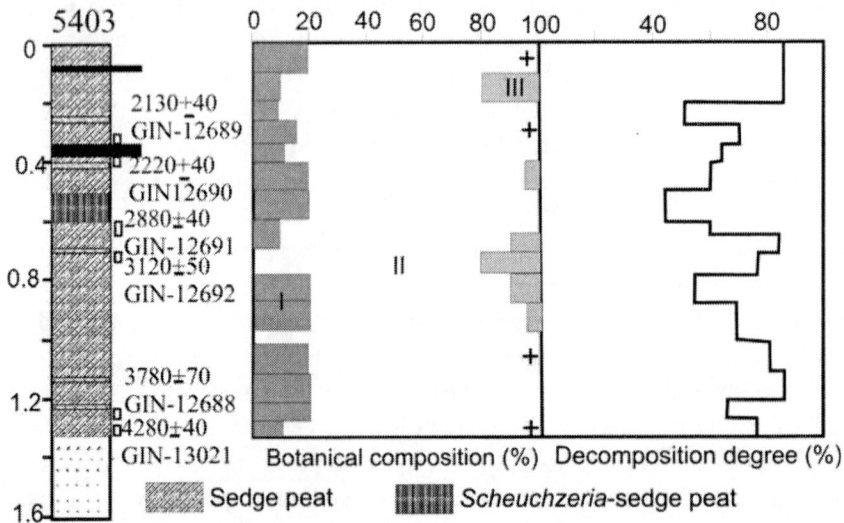

Figure 11. Botanical composition of peat bog for wetlands located on the coast of Yuzhno-Kurilskaya Bay, Kunashir Island.
I – wood and shrubs; II – herb; III-moss.

Photo 15. The wetland on the edge of the paleolake on the north coast of Dimitrov Bay, Shikotan Island.

Photo 16. *Lysichiton camtschatcense*, Shikotan Island.

Modern wetland vegetation on the Southern Kuril Islands developed along lowland lake shores, in downstream river and stream drainages, and in depressions on marine terraces with bush-sedge-moss (*Myrica tomentosa, Ledum hypoleucum, Empetrum sibiricum, Sieversia pentapetala*, very rare – *Juniperus sibirica, Chamaedaphne calyculata*) and sedge-moss associations (*Andromeda polifolia, Calamagrostis inexpansa, Trichophorum alpinum, Carex appendiculatd, C. cryptocarpa, C. cespitosa, C. gynocrates, Rhynchospora alba, Eriophorum vaginatum, Juncus yokoscensis*). There are herbaceous wetlands with *Molinia japonica, Calamagrostis langsdorfii, Hosta rectifolia, Iris laevigata, Hemerocallis esculenta, Eriophorum gracile, Juncus decipiens. Menyanthes trifoliata, Comarum palustre, Equisetum fluviatile, Carex limosa, C. koidzumii, C. cryptocarpa, Eleocharis palustris* are found in low wetlands. *Scirpus tabernaemontanii, Phragmites australis, Typha latifolia, Cicuta virosa* and *Naumburgia thyrsiflora* are found. *Carex rhynchophysa, Lysichiton camtschatcense* are found on wetland river and along stream shores (Photo 16). *Sparganium glomeratum*, less often *Acorus calamus* are present; the latter is most often encountered near hot springs [5]. Moss composition in south Kuril wetlands is poorly studied. *Sphagnum magellanicum, Sph. palustre, Sph. fallax. Sph. pulchrum,Sph. Nemoreu* are found in a sedge-sphagnum wetland near Yuzhno-Kurilsk, *Sphagnum magellanicum, S. papillosum, S. orientale, S. squarrosum* are found in a sphagnum-herbaceous wetland at Sernovodsky Isthmus on Kunashir Island. *S. japonicum, S. girgensohnii, S. quinquefarium* are near Lagunnoe Lake. *S. magellanicum, S. squarrosum, S. compactum, S. angustifolium, S. russowii* are found in a sphagnum wetland with *Pinus pumila* on Iturup Island, *S. teres, S. fimbriatum* were found in larch mires. *S. squarrosum, S. warnstorfii* have been found on Shikotan Island [4, 46]. All moss species have wide ecological amplitude, as a rule, eutrophic, rare – eutrophic-mesotrophic or mesotrophic (*Sphagnum magellanicum, S. palustre, S. russowii).* Older age wetlands exhibit much more diverse vegetation.

Small wetlands develop in deflation basins amidst dune fields. Their small size and thin peat layers that often contain sand are key features.

PALEOENVIRONMENTS AND WETLAND EVOLUTION

Wetland development on the Southern Kuril Islands as well as in other regions of the Russian Far East was first of all associated with climate change in the Holocene. The oldest existing oligotrophic wetlands on the islands in the northwestern portion of the Pacific Ocean developed in flattened divides and lava runs with slow surface flow. On large islands development took place in the lowlands of large depressions. These blanket peat bogs are characteristic for islands in temperate zones that have a clearly expressed ocean climate. Peat accumulation began in the warming period of the Early Holocene [16, 22, 24, 27, 40, 41]. The foot of peat layers in the continental portion of the southern Russian Far East are 8200-9200 years old [24].

Peat bog development on the Southern Kuril Islands began at the end of the Late Pleistocene and is associated with the warming around 12200 and 11400 ^{14}C yr BP, a trend that is also well documented in other parts of the world [7, 17]. These warming events are clearly evident in the south of the Lesser Kuril Ridge. Spore and pollen spectra from the base of blanket peat bogs show the development of birch forests. Dark coniferous forests increase and broad-leafed species appear (*Quercus, Ulmus, Juglans*) in the second warming. Sea level was lower than today and a land bridge still existed that joined the Lesser Kuril Ridge and Kunashir Island with Hokkaido Island. Wetlands began to form on terrace-like surfaces. The small islands in the south of the Lesser Kuril Ridge are a relict of these wetland territories. The wetlands of the same age (about 12000-11500 yr BP) are distributed on coastal terraces of Nemuro peninsula of Eastern Hokkaido and Yururi Island [17, 29]. The development of Lesser Kuril wetlands and Nemuro Region wetlands are very similar and the main factors are flat relief, and climate [12].

The process of wetland formation advanced in the Early Holocene, aided both by a warming climate and increased humidity. The northward movement of warm ocean current at the beginning of the Holocene caused sea fogs during the summer, low temperature and high humidity were favorable conditions for peat-formation [12]. Valley peat bogs began to form, the most ancient of which are the peat bogs in the Gorobets River valley on Shikotan Island. The Southern Kuril Islands warmed significantly around 9750-8040 ^{14}C yr PB. Birch forest expanded on the islands and dark coniferous forest with south temperate elements appeared: oak, viburnum, aralia [35]. Climate warming and high humidity aided an active wetland formation process, the expansion of peat bogs and an increase in the rate of peat accumulation. Heavy flooding in the profile for a valley peat bog on Shikotan Island is demonstrated by the presence of sand seams. The percentage of hydrophilic plants in plant communities grew and *Myrica tomentosa*, characteristic of ocean climates, appeared. Birch shrubs disappeared. Warming had a clear impact on other areas of the Russian Far East as the border of broad-leaved formations made a major shift to the north [22, 24].

The change in peat accumulation rates reflects the development and evolution of wetlands during small climate change variations in the Holocene (Figure 12). Although change in the peat accumulation rates on islands has a metachronic and in some case a asynchronic character and depends not only on climate variations but also on features of the local hydrologic region, patterns are evident. If in the Early Holocene peat accumulation rates were high, there is a noticeable decrease in the Middle Holocene. The decrease in peat accumulation rates that even led to a cessation of the process, the increase in peat decomposition rates and the appearance in the diatomic complex of a large number of species characteristic of soils demonstrates a decrease in humidity. The reason may be both a decrease in average annual precipitation rates and also an increase in evaporation, since the average annual temperature and the sum of active temperatures increased (> 10°C). Possibly the number of clear days increased. There were significant changes in the vegetation cover on Iturup, Kunashir and Shikotan Islands around 6500-5300 ^{14}C yr. BP, the role of south temperate elements increased greatly, and the area covered by broad-leafed forests expanded. The period 6500-5500 ^{14}C yr. BP on the north and east of Japanese Islands had a warm and dry climate [48]. Increased evaporation from the surface of peat bogs, an increase in temperature and a greater decomposition rate for organic material are also reasons for the reduction in peat accumulation rates [40]. This Middle Holocene decline is noted in other areas of the Russian Far East: in the blanket peat bogs in the inland portions of Hokkaido Island [40], on Sakhalin Island [22, 27] and in Primorskiy Krai [24]. The phenomenon was possibly associated with a decrease in available moisture [27, 34]. On the continental coastline of the southern Russian Far East reduced peat accumulation rates are noted in the Holocene Optimum that led to forests entirely replacing wetland environments [6, 24].

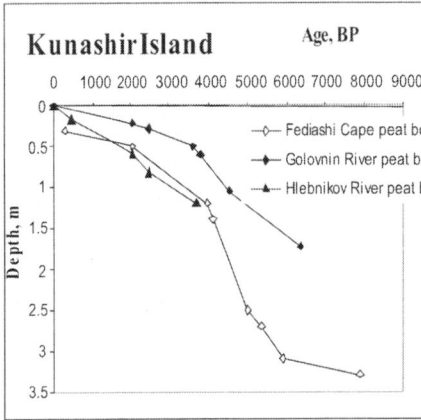

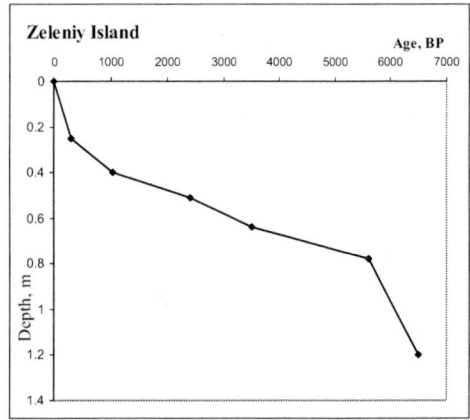

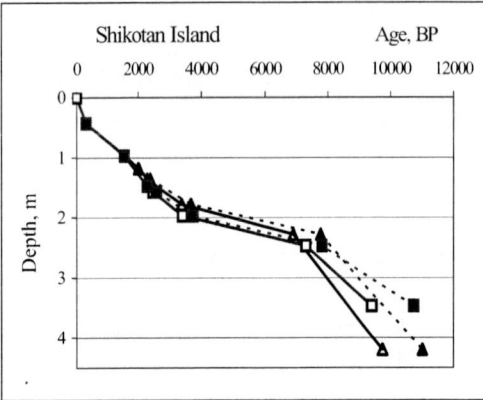

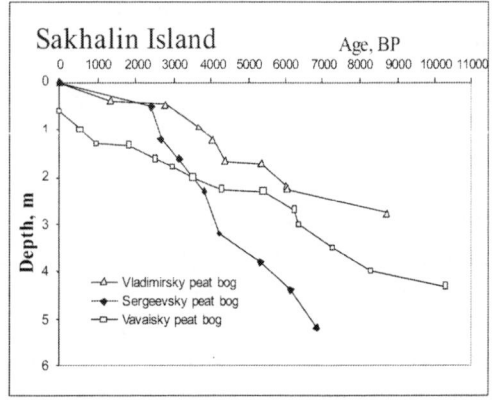

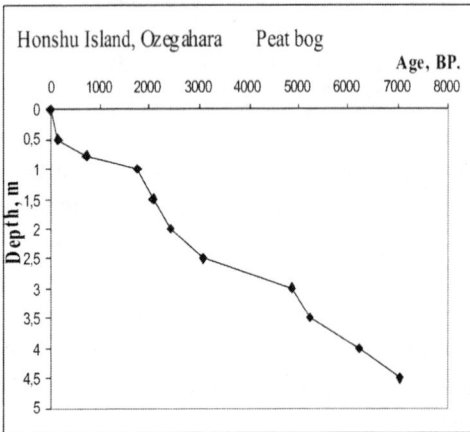

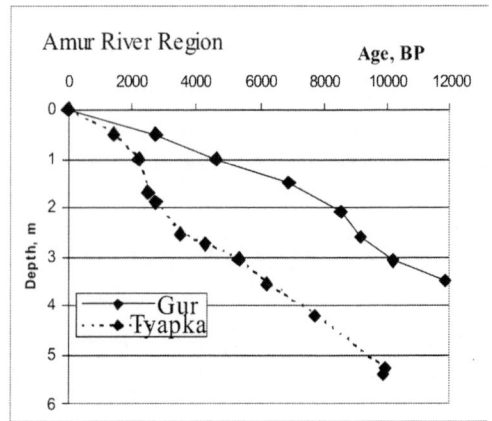

Figure 12. Position of ^{14}C-dates in peat bog sections of Southern Kurile Island, Sakhalin Island [25, 27], Honshu Island, Ozegahara Mire [40] and Amur Region [6].

In contrast to the continental coastline and large islands in the south of the Russian Far East, on Japanese islands humidity increased as temperature dropped at the end of the Middle

Holocene (after 5500 ^{14}C yr. BP) [45, 48]. This increased peat accumulation. Increased peat accumulation rates are also noted on Kunashir Island (Figure 12).

Sea level changes associated with Middle and Late Holocene climate changes had a major impact on the formation and development of wetlands on the coast and outlet portions of valleys. Increased sea level raised ground water in wetland areas. Small amplitude regressions dried lagoons and the uplands of marine bays, changing them into marshy coastline plains where on the surface of a Middle Holocene lagoon terrace synchronic peat accumulation began to occur around 4100-4300 years ago. The shift of lagoon environments to lake and wetland environments with no break in the processes demonstrates the rapid pace of the regression [23, 32].

The drop in sea level at the end of the Middle Holocene led to short term wetland environments that rapidly accumulated peat on dried out portions of shoreline slopes that were destroyed by the small amplitude transgression that followed. Fossil peat bogs (4810±60 BP, GIN-8971) with highly decomposed peat (less than 0.4 m) are now eroding away at the back of the beach.

Cooling and increased humidity in the second half of the Holocene aided the formation of broad, lowland plains on the coastlines of islands and promoted development of wetlands. High peat accumulation rates on islands are observed during the warming period in the first half of the Late Holocene, in climatic conditions that were warmer than observed today (Figure 12). Peat accumulation rates in other regions in the south of the Russian Far East reached their peak [24]. Rates on eastern Sakhalin reached 1-1.4 mm/year. As on the continent, peat accumulation patterns changed and blanket wetlands on divides and terraces were almost a universal pattern. Poorly decomposed (10-30%) dwarf low shrub-sphagnum and sphagnum-mezzo-peat and oligotrphic peat (with a predominance of *Sphagnum fuscum, S. magellanicum*) associations were the pattern. Peat accumulation possibly occurred in short intervals as is seen in seams of heavily decomposed (60-70%) mezzo-peat in the profile of a wetland in Terpeniya Bay (3660±40 BP, GIN-11141) [34]. Peat bog formation on flattened divides, slopes and lava flows in the inland portions of Japanese Islands are 3400-4000 years and occurred in heavy snowfall zones where intensive watering may have been the result of water from melting snow [16, 41].

Wetlands formed in the Late Holocene along lagoon banks and barrier lakes that were associated not only with low amplitude regression but also with the expansion of accumulative forms on coastlines. The short-term rise in sea level in the first half of the Late Holocene and the consequent rapid regression caused a dramatic shift from a marine to a wetland environment between 3600-3700 ^{14}C yr BP. Broad wetlands formed where lagoons cut deeply into the dry land mass at river outlets in the south of Kunashir Island. The low amplitude transgressions that followed did not result in wetlands. On wetland areas along lagoon edges in river outlets where peat accumulation began in the Middle Holocene, there was a sharp decline in peat accumulation rates in Late Holocene (3610±110 BP, GIN-11922). Soil developed at these peat bogs. Wetland environments that arose in early Late Holocene in outlet portions of relatively large rivers developed short term, thin peat bogs covered with floodplain facia (Kurilka, Kuibyshevka Rivers, Iturup Island).

The temperature drop in the second half of the Late Holocene decreased peat accumulation rates on islands (to 0.28-0.30 mm/year). Across the Russian Far East this period is distinguished by the lowest rates of peat accumulation in the Holocene (0.4-0.5 mm/year for divides, up to 0.1-0.18 for lagoon peat bogs, and in northern regions – up to 0.1-0.16 mm/year) [24, 40]. The decrease in peat accumulation rates was slow to move south: western Sakhalin around 2400 years ago, southern Sakhalin, the Japanese Islands and Kunashir Island – around 2000 years ago ^{14}C yr BP (Figure 12). Decreased peat bog vegetation productivity as a response to lower temperatures is why peat accumulation declined [40]. The relative siccation of peat bogs on the mainland coast may have occurred with cooling and with an intensification of continental climate features; in the coastal zone it may be associated with the drop in sea level [24].

In the second half of the Late Holocene several coastal peat bogs formed that dried out in the first half of the Late Holocene. A change in diatomic complexes that had a predominance of genera *Eunotia, Pinnularia* to associations that are typical for soil (*Pinnularia borealis, Hantzschia amphioxys, Luticola mutica, Diadesmis contenta*) demonstrates that soil making processes were underway [14]. The drop in ground water in coastal peat bogs associated with a decline in the sea level led to the growth of wetlands with shrubs, a fact established by the sharp increase in *Myrica, Alnus,* Ericaceae pollen between 2100-2000 ^{14}C yr BP (Kunashir Island).

High peat accumulation rates (up to 0.6-1 mm/year) in the Late Holocene are locally observed in areas with high ground water on lowland lake shores (the edge of Lakes Olya, Serebryanoe, Glukhoe on Kunashir Island). Here water is released near streams that are blocked by storm walls and low marine terraces have active natural springs (Cape Vostochny, Kunashir Island; coastline of Iturup Island to the south of Cape Burevestnik).

Despite the decline in peat accumulation rates in the Late Holocene, wetland environments on the islands were at their peak. The area of lowland isthmuses expanded and coastal plains formed that turned into wetlands. More active wetland processes began during the warming of the lesser Holocene optimum around 1000 ^{14}C yr BP (^{14}C-data 1000±40 BP, GIN-10491; 970±60 BP, GIN-10726). During the regression in the second half of the Late Holocene, lowland peat bogs formed on drained benches and peat began to accumulate as temperatures increased (^{14}C-data 1360±280 BP, GIN-9651). The falling sea level during Little Ice Age led to the formation of wetlands on benches and beaches of small islands of Lesser Kurils. Now the peat bogs with ash layers Ta-a, Ko-c2 are located on the beach of Tanfiliev and Zeleniy Islands and active eroded (Photo 17). Effect of regression could increased by coseismis uplift at XVII-century, which was reconstructed for shore of Eastern Hokkaido (Akkeshi-ko and Furen-ko area), the uplifted are extended at least 50 km along the southern Kurile Tranch [3].

Andesite-basaltic composition volcanic ash with small grain size and poor permeability aided wetland formation. The surface of low marine terraces (coastline from Cape Burevestnik to the outlet of Gruntovaya River on Iturup Island) are heavily waterlogged coastline areas close to active volcanoes. Peat bogs top pebble terraces in the south of Kasatka Bay have the same origin and peat accumulation began after andesite ash from the volcanic eruptions near Burevestnik ridge covered the soil on terraces around 1180±60, GIN-10491 (Figure 13). A peat bog profile (thickness 0.75 m) was made in a dune near Lake Maloe, Kuibishevsky Bay on Iturup Island (Figure 14). The peat layer is exposed in the upper portion of the dune at around 15 m. Ferric hydrate accumulations with a concretion formation resulting from the oxidation of bivalent forms of iron into trivalent occurs at the contact of peat with aeolian sand in the geochemical barrier zone. Peat accumulation began during the sub-Atlantic cooling (1770±60 BP, GIN-8644), possibly as a result of thawing snowfields that filled adjacent, deep deflation basins. The loom-like nature of basaltic volcanic ash seams appearing at the foot of the peat bogs aided wetland development. Peat is covered by aeolian sands from the Little Ice Age. Given the mix of volcanic material, peat bogs on the Kuril Islands have an elevated ash content (up to 25-27%).

Tsunamis affect wetlands on lowland coasts and in lowland valley areas. In wetlands that exposed to to the Pacific Ocean tsunami have left numerous sand layers, some times hundreds of meters inland (Photo 18). The largest run-up and inundation areas are observed on the Pacific Ocean side of islands [21]. Studies of paleotsunami show peat bog profiles for the Pacific Ocean coast of islands to have the largest volume of sand [19, 36]. Studies of diatomic complexes in peat bog profiles in the extreme portions of coastal lowlands show that acid-base conditions in peat bogs change following a tsunami [14]. The diatomic flora typical for acidic wetland conditions (acidophils and acidobionts) are replaced by diatoms characteristic for neutral and base environments following a tsunami. In the diatomic assemblages circumneutral, acidophilous and acidobiontic species dominate. These species continue to develop for a period following a tsunami. Changes in pH, Eh and ground water salinity are major impacts on wetland vegetation [2, 15,20, 31].

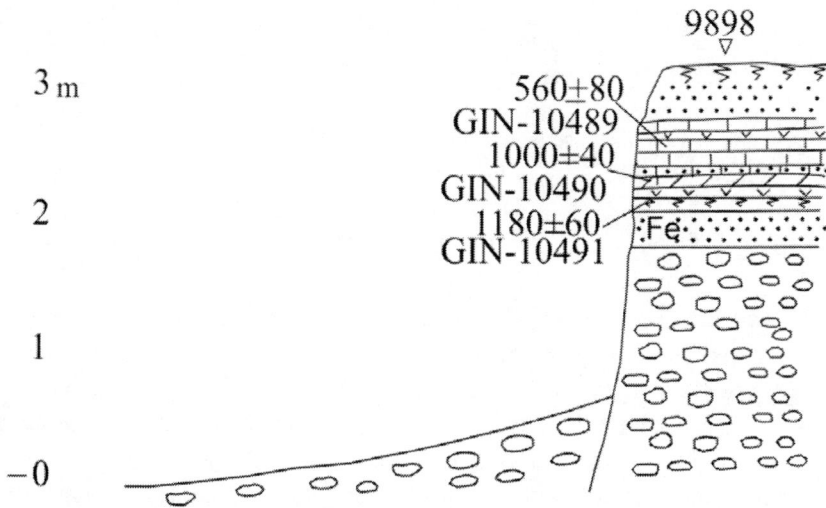

Figure 13. Section of the peat bog on Kasatka Bay coast, Iturup Island.

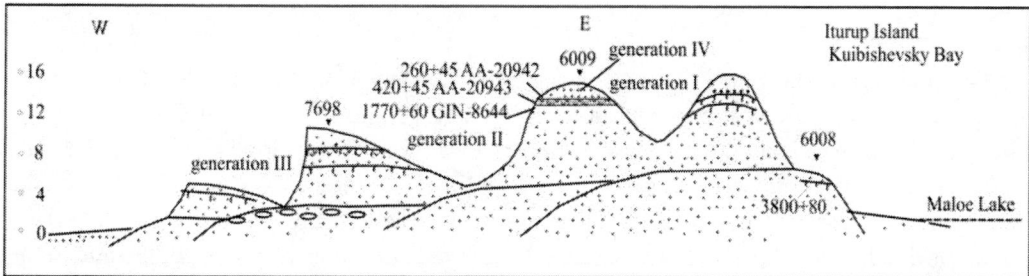

Figure 14. Cross-section of dune field with peat layers near Lake Maloe, Kuibishevsky Bay, Iturup Island.

Photo 17. The peat bog of Little Ice Age on Tanfiliev Island coast, Lesser Kurile Islands.

CONCLUSION

Multi-age wetlands are broadly distributed on the Southern Kuril Islands where conditions for the formation, development and stability are closely tied to climate change, changes in sea level and neo-tectonic movements. Wetlands are found in various relief settings. On large islands with hilly relief wetlands are distributed on terrace-like surfaces, on coastal lowlands and in lowland portions of river valleys. Valley peat bogs are characteristic of islands that have experienced tectonic subsidence in the Holocene: Lesser Kuril Ridge and the southern portion of Kunashir Island. Blanket, oligotrophic peat bogs have limited distribution. They are developed on small island in the south of the Lesser Kuril Ridge that have flattened relief and are encountered on terrace-like surfaces in the southern part of Shikotan Island.

Wetland formation began in warming periods at the end of the Late Pleistocene-Early Holocene. Today the oldest wetlands are on islands of the Lesser Kuril Ridge (Zeleniy, Tanfiliev, Shikotan Islands). The age of wetlands located in coastal lowlands are mid to late Holocene. Sea level fluctuations associated with climate change played a major role in the development of these wetlands. Iturup and Kunashir Islands have the largest wetlands on coastal lowlands. The process of peat accumulation was more intense in the Middle and at the beginning of the Late Holocene. Wetland environments, however, experienced their greatest development during the second half of the Late Holocene.

Photo 18. The peat bog with tsunami sands and ash layers, southern coast of Dimitrov Bay, Shikotan Island.

Volcanic ash with varied grain size changes the hydrologic and hydro-chemical regimes and affects biota and is one factor aiding wetland development on the islands. Tsunamis dramatically alter conditions, including acid-base conditions, and affect the natural trends in wetland development in the Kuril Islands, especially on the Pacific Ocean coastline. Tectonic subsidence is important factor, controlled wetland development within coastal zone and river valleys, especially on Lesser Kuril Islands.

ACKNOWLEDGEMENTS

We would like to thank L. D. Sulerzhitskii, M. M. Pevzner (Institute of Geology, RAN, Moscow), and H. A. Arslanov (St. Petersburg University) for providing data on radiocarbon dating. We are grateful to thank N. P. Domra (Institute of Biology and Soil Sciences FEB RAS, Vladivostok) for help in pollen analysis and A. A. Kharlamov (Institute of Oceanology

RAS, Moscow) for assistance with field work. The study was supported by RFBR, projects 06-05-64033, and field grants of FEB RAS.

REFERENCES

[1]   Alekseeva, L.M. Flora of Kunashir Island. Vladivostok: Far East Scientific Centre Publication; 1983.

[2]   Asado, T. Vegetation gradients in relation to temporal fluctuation of environmental factors in Bekanbeushi peatland, Hokkaido, Japan. *Ecological Research,* 2002, 17. 505-518.

[3]   Atwater, B.F.; Furukawa, R.; Hemphii-Haley, E.; Ikeda, Y.; Kashima, K; Kawase, K.; Kesley H.M.; Moore, A.L.; Nanayama, F.; Nishimura, Y.; Odagiri, S.; Ota, Y.; Park, S.-C.; Satake, K.; Sawai, Y.; Shimokawa K. Seventeenth-century uplift in eastern Hokkaido, Japan. *The Holocene,* 2004, 14. 487-501.

[4]   Bardunov L.V., Cherdantseva V.Ya. Materials about mosses flora of South Kurile Islands In: Cherdantseva V,Ya, editor. *Systematical-Floristic Researches of Spore Plants of Far East.* Vladivostok: Far East Scientific Centre of USSR Academy of Science; 1984; 34-53.

[5]   Barkalov, V.Yu.; Eremenko, N.A. Flora of Natural Reserve "Kurilsky" and Reserve "Minor Kurils" (Sakhalin District). Vladivostok, Dalnauka; 2003.

[6]   Bazarova, V.V.; Klimin, M.A.; Orlova, L.A.; Mokhova, L.M.; Gvozdeva, I.G. Holocene climatic changes recorded in Gursky and Tyapka peat sections (Low Priamurie) In: Kasyanov V.L.; Kostina E.E.; Lutaenko K.A., editors. *Report of the Workshop on the Global Changes Studies in the Far East.* Vladivostok: Dalnauka; 2003; 99-110.

[7]   Borzenkova, I.I. Climatic Changes at Cenozoic. St.-Peterburg: Gidrometeoizdat. 1992.

[8]   Bronk Ramsey, C. Development of the radiocarbon Program OxCal. *Radiocarbon,* 2001, 43. 2A. 355-363.

[9]   Bulgakov, R. Reconstruction of Quaternary History of Southern Kuril Islands. *Journal of Coastal Research,* 1996, 12. 930-939.

[10]   Chernyaeva, A.M. Flora of Zeleniy Island (Lesser Kuril Arc). *Botanical Journal,* 1977, 62 (11), 1672-1682.

[11]   Furukawa, R.; Nanayama, F. Holocene Fall Deposits along the Pacific Coastal Region of Eastern Hokkaido. *Bulletin of Volkanological Society of Japan,* 2006, 51. N 6. 351-371.

[12]   Fujita, H.; Igarashi, Y.; Hotes, S.; Takadam M.; Inoue, T.; Kaneko, M. An inventory of the mires of Hokkaido, Japan – their development, classification, decline, and conservation. *Plant Ecology,* 2008. In press.

[13]   Grebennikova, T.A.; Razjigaeva, N.G.; Mokhova, L.M.; Bazarova, V.B., Ganzey L.A. changes records in Quaternary lacustrine sequences of Kunashir Island, Kuril Islands. In: Kasyanov V.L.; Lutaenko K.A., editors. *Abstracts of Workshop «Global Change Studies at the Far East».* Vladivostok: Dalnauka; 1997. 12-15.

[14] Grebennikova, T.A.; Razjigaeva, N.G.; Il'ev, A.Ya.; Kaistrenko, V.M. Diatom data as indicator of the paleotsunami deposits. In: Levin B.W., Nosov M.A., editors. *Local Tsunami Warning and Mitigation.* Moscow: Janus-K; 2002; 32-43

[15] Hotes, S.; Poschlod, P.; Sakai, H.; Inoue, T. Vegetation, hydrology and development of the coastal mire in Hokkaido, Japan, affected by flooding and tephra deposition. *Canadian Journal Botany,* 2001, 79. 341-361.

[16] Igarashi, Y. History of environmental change in Hokkaido from the viewpoint of polynological research In: Higashi S., Osawa A., Kanagawa K., editors. *Biodiversity and Ecology in the Northernmost Japan.* Sapporo: Hokkaido Univ. Press; 1993; 2-19.

[17] Igarashi, Ya.; Igarashi, T.; Endo, K.; Yamada, O.; Nakagawa, M.; Sumita, M. Vegetation history since the Late Glacial of Habomai Bog and Ochiishi Cape Bog, Nemuro Peninsula, eastern Hokkaido, north Japan. *Japanese Journal Historical Botany,* 2001, 10. 67-79.

[18] Igarashi, Y. Vegetation History of Bekanbeushi Mire, Eastern Hokkaido. In: Tsujii T., Tachibana H., editors. *Wetlands of Hokkaido.* Sapporo: Hokkaido University Press; 2002; 43-50

[19] Iliev, A.Ya.; Kaistrenko, V.M.; Gretskaya, E.V.; Tikhonchuk, E.A.; Razjigaeva, N.G.; Grebennikova, T.A.; Ganzey, L.A.; Kharlamov, A.A. Holocene Tsunami Traces on Kunashir Island, Kuril Subduction Zone. In: Satake, K., editor. *Tsunamis: Case Studies and Recent Development,* Netherlands: Springer; 2005; 171-192.

[20] Iyobe, T.; Haraguchi, A.; Nishijima, H.; Tomizawa, H.; Nishio, F. Effect of fog on sea salt deposition on peat soil in boreal *Picea glehnii* forests in Ochiishi, eastern Hokkaido. *Japan Ecological Research,* 2003, 18. 587-597.

[21] Kaistrenko, V.M.; Gusyakov V.K., Dzhumagaliev V.A., Dykhan G.S., Ivaschenko A.I., Yeh G., Kato L.N., Klochkov A.A., Pelinovsky E.N., Predtechensky G.S., Sasorova E.V., Titov V.V., Kharlamov A.A., Shelting E.V. Tsunami manifestation of 4 October 1994 in Shikotan Island In: Sergeev K. F., Kaistrenko V.M., editors. *Geodynamics of tectonosphere of the Pacific-Eurasia conjunction zone.* Yuzhno-Salhalinsk: Institute of Marine Geology and Geophysics FAEB RAS; 1997; VIII. 55–73.

[22] Khotinsky, N.A. Holocene of Northern Euroasia. Moscow: Nauka; 1977.

[23] Korotky, A.M.; Razjigaeva, N.G.; Grebennikova, T.A.; Ganzey, L.A.; Mokhova, L.M.; Bazarova, V.B.; Sulerzhitsky, L.D.; Lutaenko, K.A. Middle and late-Holocene environments and vegetation history of Kunashir Island, Kuril Islands, northwestern Pacific. *The Holocene,* 2000, 10. 311-331.

[24] Korotky, A.M.; Grebennikova, T.A.; Pushkar, V.S.; Razjigaeva, N.G.; Volkov, V.G.; Ganzei, L.A.; Mokhova, L.M.; Bazarova, N.B.; Makarova, T.R. Climatic changes of South Far East Area in Late Cenozoic (Miocene-Pleistocene). Vladivostok: FESU Publication; 1996.

[25] Korotky, A.M.; Pushkar, V.S.; Grebennikova, T.A.; Razjigaeva, N.G.; Karaulova, L.P.; Mokhova, L.M.; Ganzey, L.A.; Cherepanova, M.V.; Bazarova, V.B.; Volkov, V.G.; Kovalukh, N.N. Marine terraces and Quaternary history of Sakhalin Island shelf. Vladivostok: Dalnauka; 1997.

[26] Kumano, S.; Ihira, M.; Maeda Y.; Yamauchi, M.; Matsumoto, E.; Matsuda, I. Holocene sedimentary history of some coastal plains in Hokkaido, Japan IV. Diatom assemblages in the sediments from Kushiro Moor. *Ecological Reseach,* 1990, 5. 221-235.

[27] Mikishin, Yu.A.; Gvozdeva, I.G. Environment development of South-Eastern Sakhalin at Holocene. Vladivostok: FESU Press; 1996.

[28] Morita Y. Vegetation history in Nemuro Peninsula since the Late Pleistocene. *Vegetation Science.* 2001 a. 19. 39-44.

[29] Morita Y. Vegetation history of Yururi Island in easternmost Hokkaido since the Late Holocene. *Japanese Journal Histor. Botany*, 2001 b. 20. 81-89.

[30] Nakagawa, M.; Ishizuka, Y.; Kudo, T.; Yoshimoto, M.; Hirose, W.;Ishizaki, Y.; Gouchi, N.; Katsui, Y.; Solovyow, A.W.; Steinberg, G.S.; Abdurakhmanov, A.I. Tyatya Volcano, southwestern Kuril arc: Recent eruptive activity inferred from widespread tephra. *The Island Arc,* 2002, 11. 236-254.

[31] Nanayama, F.; Furukawa, R.; Shigeno, K.; Makino, A.; Soeda, Yu.; Igarashi, Ya. Nine unusually large tsunami deposits from the past 4000 years at Kiritappu marsh along thte southern Kurile Tranch. *Sedeimnetary Geology*, 2007, 200. 275-294.

[32] Razjigaeva, N.G.; Grebennikova, T.A.; Ganzey, L.A.; Mokhova, L:M.; Bazarova, V.B. The role of global and local factors in determining the middle to late Holocene environmental history of the South Kuril and Komandar Islands, northwestern Pacific. *Palaeogeography, Palaeoclimatology, Palaeoecology*, 2004. 209. 313-333.

[33] Razjigaeva, N.G.; Korotky, A.M.; Grebennikova, T.A.; Ganzey, L.A.; Mokhova, L.M.; Bazarova, V.B.; Sulerzhitsky, L.D.; Lutaenko, K. A. Holocene climatic changes and environmental history of Iturup Island, Kuril Islands, northwestern Pacific. *The Holocene,* 2002, 12. 469-480.

[34] Razjigaeva, N.G.; Mokhova, L.M.; Zaretskaya, N.E. Climatic Rhythms and the Stages of Development of Landscapes of the Coastline of Terpenie Bay (Sakhalin Island) in the Holocene. In: Kasyanov V.L.; Razjigaeva, N.G.; Lutaenko K.A., editors. *Report of the International APN-Start Global Changes Research Awareness Raising Symposium in Northeast Asia.* Vladivostok: Dalnauka; 2005; 126-144.

[35] Razzhigaeva, N.G.; Ganzey, L.A.; Belyanina, N.I.; Grebennikova, T.A.; Ganzey, K.S. Paleoenvironments and Landscape History of Minor Kuril Islands since Late Glacial. *Quaternary International,* 2008, 179. 83-89.

[36] Razzhigaeva, N.G.; Ganzey, L.A.; Grebennikova, T.A.; Harlamov A.A., Il'ev A.Ya., Kaistrenko V.M. Geological records of paleotsunami on Shikotan Island (Lesser Kuril Ridge) at Holocene. *Volcanology and Seismology,* 2008, 2. In press.

[37] Razzhigaeva, N.G.; Ganzey, L.A. Sedimentary environments on islands in Pleistocene-Holocene. Vladivostok: Dalnauka; 2006.

[38] Reference Book of USSR Climate. Leningrad: Gidrometeoizdat; 1968.

[39] Reference Book of USSR Climate. Leningrad: Gidrometeoizdat; 1970.

[40] Sakaguchi, Y. Characteristics of Ozegahaea deposits and climate changes since Lat4eglacial in Central Japan. *Bulletin of the Department of Geography University of Tokyo,* 1976, 2. 1-27.

[41] Sakaguchi, Y. Distribution and genesis of Japanese peatlands. *Bulletin of the Department of Geography University of Tokyo*, 1979, 11. 17-42.

[42] Sakaguchi, Y. Warm and cold stages in the past 7600 years in Japan and their global correlation. *Bulletin of the Department of Geography University of Tokyo,* 1983, 15. 1-31.

[43] Sato, H.; Kumano, S.; Maeda, Y.; Nakamura, T.; Matsuda, I. The Holocene development of Kushu Lake on Rebun Island in Hokkaido, Japan. *Journal of Paleolimnol,* 1998, 20. 57-69.

[44] Stuiver, M.; Reimer, P.J.; Bard, E. *et al.* INTCAL 98 Radiocarbon Age Calibration, 24000-0 cal BP. *Radiocarbon*, 1998, 40. 1041-1083.

[45] Taira, K. Environmental changes in Eastern Asia during the past 2000 years. Volcanism,   tectonism,   climate   and   palaeooceanology.   *Palaeogeography, Palaeoclimatology, Palaeoecology*, 1980, 32. 89-97.

[46] Vasileva, L.N. About Sphagnum moss flora of Kuril Islands. In: Stocenko, A.V., Vorobiev D.P., editors. *Materials about Natural Resoueces of Kamchatka and Kuril Islands.* Magadan: Magadan Book Press; 1960; 99-100.

[47] Vlastova, N.V. Peat swamps of Sakhalin. Moscow-Leningrad: Academy of Science Publication; 1960.

[48] Yasuda, Y. Climatic changes and the development of Jomon Culture in Japan. Nature and Humankind in the Age of Environmental Crisis. In: Ito S., Yasuda Y., editors. *Natural and Humankind in the Age of Environmental Crisis. Proceedings of the VIth International Symposium at the International Research Center for Japanese Studies.* 1995. 57-77.

In: International Wetlands: Ecology, Conservation...   ISBN: 978-1-60456-999-5
Editor: Jose R. Herrara   © 2009 Nova Science Publishers, Inc.

Chapter 10

BOG ECOSYSTEMS IN THE LOWER AMUR RIVER BASIN, RUSSIAN FAR EAST

V. B. Bazarova[1], M. A. Klimin[2] and T. A. Kopoteva[2]

[1]Pacific Institute of Geography FEB RAS, Vladivostok, Russia
[2]Institute of Aquatic and Ecological Problems FEB RAS, Khabarovsk, Russia

ABSTRACT

The large quantity of publications is devoted to the studies of bog ecosystems in the Lower Amur River basin. To identification of regularities of the bogs placement, development and classification in the region under study, the most significant contribution was made by Yu.S. Prozorov [16, 17, 18, 19]. In accordance with the zoning scheme of the USSR peat fund, the Lower Amur River territory was identified as the greatest in area peat basin of the Far East. In the Lower Amur River basin, the typological composition of the bog systems is the most representative as compared with other regions of the Russian Far East [22]. This work presents materials concerning the vegetation, some properties of peat, evolution and stability of the bog ecosystems in the Lower Amur River basin.

INTRODUCTION

The Lower Amur River territory belongs to the monsoon forest area [13] of which the stable predominance of precipitation over evaporation is characteristic. This situation is the basic cause of the common distribution of bogs and boggy lands in plains and lowlands if the territory under consideration and allows us to attribute the whole region to the humid zone (Table 1). The Lower Amur River basin extends in the sub-meridional direction from 48° to 54°N (about 1000 km). The bogs and boggy lands concentrate basically at lacustrine-alluvial lowlands – Amur-Amgun, Udyl-Kizinsky, Evoron-Chuckckagir and Middle-Amur. These lowlands differ essentially from each other in both natural conditions and sizes. The swamping of all lowlands reaches 50-60%. A share of bogs of total boggy area is taken to be 65-70% while that of paludal forests and meadow bogs – 30-35% [16, 17].

The detailed calculations of the swamping degree of the Lower Amur River territory are given in Table 2 [8]. The Middle-Amur lowland located within the Russian territory is a part of a larger geographical region including also the Syantszyan plain (PRC). Total area of the Syantszyan plain is 51 300 km². The area of bogs within it reaches about 11 600 km² or 22.6% while that of meadow bogs is about 12 400 km² or 24.1%. Therefore, the factor of general swamping of the Syantszyan plain reaches 24000 km² or 46.7%. On the whole, the area of the Middle-Amur lowland and Syantszyan plain territory is 116300 km², its swamping reaches 60100 km² (51.7%), bog area is 36000 km² (31%) while that of boggy lands 24100 km² (20.7%) [8].

Table 1. Modern climatic data of the Lower Amur River basin

Hydrometeorologicl station/boggy area	Annual temperature, °C	Sum of annual precipitation, mm	Coefficient of humidity	Average January temperature, °C	Average July temperature, °C
Nikilaevsk/ Amur-Amgun	-2.2	566	1.35	-23.4	16.4
P. Osipenko/ Evoron-Chugchagir	-2.8	444	1.18	-28.3	17.8
Bogorodskoe/Udyl-Kizi	-1.9	469	1.30	-25.0	17.5
Komsomolsk/ northen part of Middle Amur	-0.6	524	1.30	-25.4	20.0
Vyazemskaya/ southern part of Middle Amur	1.8	686	1.50	-21.1	21.2

Table 2. Bogging rate of the Lower Amur depressions

Boggy area	Total square, km²	Square of boggy	bogs and land	Square	of bogs	Square of land	boggy
		km²	%	km²	%	km²	%
Amur-Amgun	9000	3950	43,89	2666	29,63	1284	14,26
Evoron-Chugchagir	20000	8950	44,75	6041	30,21	2909	14,54
Udyl-Kizi	10000	4850	48,50	3274	32,74	1576	15,76
Middle Amur (Russian part)	65000	36150	55,62	24401	37,54	11749	18,08

CHARACTERISTICS OF BOG BIOCENOSES

As the basis for classification of the bog facies of the Lower Amur River lowlands, the soils which play a large part in the diagnostics of natural peculiarities of facies and biocenoses were taken [17]. Later, the characteristics of identified groups of biocenoses were given with estimation of specificity of distribution, location, feed conditions, vegetation, soils of peat deposits in the southern Far East [18].

Eight types of facies each of which includes several groups of facies kinds were identified.

Grassy Eutrophic Type of Facies of Ground Feed

In the feed of the eutrophic grass bogs, the flood and alluvial waters play the major part. In the vegetable cover, the herbaceous plants – reed grasses, sedges, bogbean and horsetails – predominate. Some species and groups of species of facies are characterized by abundance of shrubs and mosses. The soils of this facie type belong to genera of alluvial-sod-peaty and sod-peaty silted ones. Of them, a combination of the alluvial, sod and peat-forming processes is characteristic. The grass eutrophic bogs are of frequent occurrence and occupy the spacious territories at the Lower Amur River lowlands. However, not more than 25-30% falls to share of them in the flood plains. The major part of the flood plain is covered by the meadows with mineral oils. The following groups of facies species were attributed to the grass eutrophic type of facies.

The Reedgrass Group of Facies Species of the Low Flood Plain with Alluvial-Sod-peat Soils

The reedgrass bogs are located at the sections of low flood plains of Amur and Amgun rivers. Their wetting is made by alluvial and flood waters. The vegetable cover pf this group of facies forms largely the grass layer. In this layer, Calamagrostis langsdorffii (Link) Trin. and C. angustifolia Kom. predominate. Carex minuta.Franch. grows everywhere and Caltha pygmaea Nakai, Salix myrtilloides L., Stellaria radians L., Sanguisorba parviflora (Maxim.) Takeda, Lathyrus pilosus Cham., Comarum palustre L. occur more often than the others. The sedge part of the grass stage plants including the Calamagrostis langsdorffii belongs to the meadow and meadow-bog phytocenoses. There are few typical bog species. The moss layer is feeble marked or sometimes absent. Most often, Sphagnum obtusum Warnst. and Sphagnum orientale L. Savicz, and also Climacium dendroides (Hedw.) Web. et Mohr are found. Practically all of mosses fall into in the category of eutrophic ones.

Only on the Evoron-Chukchagir lowland, the reedgrass group of biocenoses with silted flood-plain sod-peat soils in quagmires was identified. As a rule, the phytocenoses are one-layered. Only Calamagrostis langsdorffii predominates. By origin, these bogs differ from the present-day reedgrass biocenoses.

The Sedge Group of Facies Species of the Flood Plain with Sod-peat Silted Soils

The sedge bogs are located largely within the flood plains of Amur and Amgun rivers and their tributaries. The feed of valley bogs is provided by alluvial and flood waters. In the vegetable cover, the plants of bush-grass layer prevail with a dominance of Carex pseudocuraica Fr. Schmidt, C. schmidtii Meinsh. and C. minuta. The major part of species falls into category of purely boggy ones. The meadow-bog plants are rather few. Nearly all of plants are eutrophic. In some sections, Spiraea salicifolia L. and Betula ovalifolia Rupr. are in small proportion found. The moss layer is only characteristic of the sphagnum-sedge phytocenoses where Sphagnum obtusum and Sp.orientale predominate.

The Shrub-sedge Group of Facies Species of High Flood Plain with Sod-peat Silted Soils

The shrub-sedge bogs are located within the high flood plains of Amur and Amgun Rivers. They occupy a belt outside of which the bogs of atmospheric and mixed feed are situated. The vegetable cover is formed by three layers. The shrub layer is made up by Salix brachypoda (Trautv. et Mey) Kom., Betula ovalifolia and Spiraea salicifolia. In the shrub-grass layer, Carex minuta prevails. In the majority of sections, Calamagrostis neglecta (Ehrh.) Beanv., Sanguisorba pafviflora, Comarum palustr, Salix myrtilloides, Chamaedaphne calyculata (L.) Moench grow. The representatives of meadow flora (Gentiana triflora Pall., Scutellaria regeliana Nakai, Stellaria radians etc.) and species characteristic of the bogs of mixed and atmospheric feed (Rubus chamaemorus L., Eriophorum vaginatum L., Pedicularis grandiflora Fisch., Smilacina trifoliata Desf., Oxicoccus quadripetalus Gilib. etc.) are found. On the whole, the shrub-grass layer is characterized by intrusion and growth of mesotrophic and oligotrophic species with wide amplitude of ecological adaptability. The moss cover is also characterized by significant diversity. The number of only sphagnous mosses species is eleven among which there are rare plants for the Far East – Sphagnum perfoliatum L. Savicz, Sp. fimbriatum Wils., Sp. subfulvum Sjoers. The dominant Sphagnum.orientale and Sp. obtusum belong to the eutrophic species. In the vegetable cover of the shrub-sedge group of facies species, the eutrophic plants prevail, meadow species persist and mesotrophic and oligotrophic species spread out.

The Buck Bean Group of Facies Species Near Lakes with Sod-peat Silted Soils Near Lakes and Small Water Bodies

The buckbean bogs are located in the abundantly wetted sections where the scraw bogs form. The vegetable cover is formed by two layers. In the fruticulose-grass layer, Menyanthes trifoliata L. predominates. In the sphagnum-buckbean association, Carex limosa L. and Carex pseudocuraica are co-dominants of the buckbean. The comparatively large group includes Carex lasiocarpa Ehrh., Iris laevigata Fisch., Angelica maximoviczii (Fr. Schmidt) Benth., Comarum palustre, Calamagrostis neglecta, Pedicularis resupinata L.. The moss layer is not distinguished by the species diversity. In all places, Sphagnum obtusum prevails.

Eutrophic Arboreal-grass Facies Species of the Ground Feed

The bogs of this type are widely distributed on the Lower Amur River lowlands.

Reed Grass Larch Group of Facies of the Drained Sections of the First Terrace above the Floodplain with Peaty Soils

The bogs are damped due to the runoff of the surface and subsurface waters. In the vegetable cover, the arboreal layer plays the major edificatory role. In the native plantations, the forest stand is formed by Larix dahurica Turcz. with addition of Betula platyphylla Sukacz. In the derivative forest stand appearing after fires, Betula platyphylla predominates. The underbrush is entirely absent or consists of single specimens of Spiraea salicifolia, Betula ovalifolia and Alnus fruticosa Rupr.. In the fruticulose-grass layer, Calamagrostis langsdorffii prevails. Along with the sterrhads, the sylvestral plants – Vaccinium vitis-idaea L., Maianthemum bifolium (L.) Schmidt., Equisetum silvaticum L., Ledum hypoleucum Kom.

are also found. In the moss layer, Sphagnum angustifolium (Russ) C. Jens. and Sp. russowii Warnst predominate, Sphagnum magellanicum Brid. is also abundant.

The identification of the *hydrophilic moss-grass group of biocenoses with the silted sod-peat soils of the floodplain and first terraces above the floodplain* should be noted This group occurs in all places of all Lower Amur River plains and lowlands, it is ecologically confined to the most damped areas with the eutrophic conditions of feed. The phytocenoses are two-layered. The Menyanthes trifoliata prevails more often than other species while predominance of Carex limosa, C. lasiocarpa Ehrh. and C. pseudocuraica occurs more rarely. Equisetum limosum L., Pedicularis resupinata or P.grandiflora typical hydrophilic plants are found in all places. In the moss layer, Sphagnum orientale and Sp. obtusum prevail. Here and there, Sphagnum balticum (Russ) C. Jens. And Sp. jensenii H.Lindb. occur often, sometimes Sphagnum magellanicum is abundant.

The Arboreal -sphahgnous Mesotrophic Type of Facies of the Mixed Feed

The bogs of this type are situated on the surfaces of first, second and higher terraces. They are widely distributed within the Lower Amur River lowlands and occupy relatively drained sections from which the runoff to the adjoining bogs is possible. At the same time, the arboreal-sphagnous facies for the most part are provided with the additional damping at the expense of the surface runoff and deluvial waters. The forest stand contains Larix dahurica with not large addition of Betula platyfylla. Occasionally, Pinus koraiensis Sieb. et Zucc. is found in some sections. The edificatory role of the arboreal layer decreases gradually in connection with bog formation. In the cover above the soil surface, the heather undershrubs and sphagnum mosses predominate. The bog formation contributes to decrease in the number of sylvestral plants and overgrowth of sterrhads – Ledum palustre, Chamaedaphne calyculata, Sphagnum magellanicum, Sp. rubellum Wils.), Sp. fuscum.

The Larch Sphagnum-ledum Group of Facies Species of Terraces above the Floodplains with Peaty Soils

The larch and sphagnum-ledum paludal forests are of limited occurrence and do not form large woodlands. The arboreal layer is an edificator of the vegetable cover. The forest stand includes Larix dahurica and Betula platyfylla. In one of sections, a single additions of Picea ajanensis Fisch. And Pinus koraiensis were found. In the fruticulose-grass layer, Ledum hypoleucum – a plant that grows basically in the upland forests – predominates. Other plants found here – Vaccinium vitis-idaea, Chamaeperiglymenum canadense (L.) Graebn., Linnaea borealis L., Maianthemum bifolium, Rosa davurica Pall. – belong also to the sylvestral ones. The number of sterrhads (Calamagrostis neglecta, Carex minuta, Chamaedaphne calyculata and Vaccinium uliginosum L. is relatively small. Among mosses, Sphagnum angustifolium and Sp. russowii are abundant. Here, the forest mosses Pleurozium Schreberi (Brid.) Mitt., Dicranum Muchlenbeckii Br., Sch. et Gmb. grow also but their occurrence is not high. In the group under consideration, the phytocenoses evolve from the forest-bog to boggy ones.

The Larch Fruticulose-sphagnum Group of Facies Species of Terraces above the Floodland with Peaty Soils

The larch fruticulose-sphagnum group of facies species occurs on the relatively drained sections of the first terrace above the floodland, on the flat and slightly inclined surface of the second and higher terraces. The bogs of this group are widely distributed on the Lower Amur River lowlands. The vegetable cover is formed by four layers (stages). The arboreal layer includes only Larix dahurica. In the fruticulose layer, Betula ovalifolia prevails. Almost in all places, Pinus pumila Regel of 2-3 m high grows. In the fruticulose-grass layer, Ledum hypoleucum and Ledum palustre are abundant. Chamaedaphne calyculata is found and there is Carex minuta within some sections. Almost in all places, Vaccinium uliginosum, Calamagrostis neglecta, Vaccinium vitis-idaea, Oxicoccus quadripetalus grow. The most projective cover is characteristic of the moss layer. The majority of sections are characterized by predominance of Sphagnum magellanicum and more rarely Spagnum Angustifolium. Spagnum fuscum and Sp. rubellum are often found. The frequent fires contribute to the dissemination of Polytrichum strictum Sm.. On the whole, the phytocenoses of this group of facies species are characterized by predominance of sterrhads falling into category of oligomesotrophic species by ecologically.

Mesotrophic Grass-sphagnum Type of Facies of Mixed Feed

The bogs of this type are presented by one *grass-sphagnum serial (in connection with a thermokarst) group of facies species of first terrace above the flood land with slightly peaty soils*. In this group of facies species, the vegetation is more or less uniform while the soil character changes. This is related to the thermokarst processes when melting bumps-outliers of permafrost with great inclusion of ice. This serial group of facies species occurs frequently near the majority of bumps but occupies very narrow belt. The phytocenoses are formed by two layers – fruticulose-grass and moss ones. The Carex limosa that can be considered as the phytocenose indicator from the ecological viewpoint is characterized by most abundance. Among the plants occurring at all times, the majority belongs to hygrophytes: Iris laevigata, Pedicularis grandiflora, Menyanthes trifoliate and Calamagrostis neglecta. The hydrophytes being no permanent components of the phytocenoses prevail also: Scheuchzeria palustris L., Comarum palustre, Caltha pygmaea and Angelica maximoviczii. The species with the wide amplitude of adaptability are also found: Chamaedaphne calyculata, Oxicoccus quadripetalus, Andromeda polifolia L. etc. The most part of species can be attributed to the category of mesotrophic ones. The species composition of mosses is poor. Sphagnum obtusum and Sp.riparium Aongstr. predominate.

Sphagnum Heterotrophic Type of Facies of Atmospheric Feed

This type of facies is characterized by the only *fruticulose-sphagnum group of facies of the first terrace above the flood land with peaty soils*. The bogs of this group are situated on the surface of the first terrace above the flood land, confined to the sections damped by only atmospheric precipitation. The fruticulose-sphagnum bogs are largely distributed in the margins of the Udyl-Kizinsky lowland. The vegetable cover of the fruticulose-sphagnum group of facies is four-layered. The arboreal layer is characterized by sparseness and insignificant height and is presented by Larix dahurica. The shrubby layer is presented by Betula ovalifolia. A dominant of micro-elevations is Chamaedaphne calyculata. To the typical representatives of micro-elevations, Ledum palustre, Oxicoccus microcarpus Turcz., Carex minuta, Eriophorum vaginatum L., Smilazina trifoliate and Drosera rotundifolia L. can be attributed. In the low places, in addition to the hydrophilic plants, such as Menyanthes trifoliata, Scheuchzeria palustris, Carex limosa and Iris laevigata the species growing on cushions - Chamaedaphne calyculata, Andromeda polifolia, Calamagrostis neglegta, Eriophorum vaginatum etc. occur here at all times. The fruticulose-grass layer is formed, as a whole, by eutrophic, mesotrophic and oligotrophic plants. In the low places, the eutrophic species are characterized by maximum occurrence while, on the cushions – oligotrophic ones. The sphagnum mosses form the moss layer. On the cushions, Sphagnum magellanicum prevails and Sphagnum lenense H. Lindb., Sp. angustifolium, Sp. rubellum mingle with it. In the low places, Sphagnum flexuosum Dozy et Molk., Sp. jensenii H. Lindb., Sp. fallax (Klinggr. emend. Isov.), Sp. obtusum predominate. In the margins of cushions and low places, the hygrophilic, eutrophic, mesotrophic and oligotrophic mosses occur.

Sphagnum Heterotrophic Type of Facies of Mixed Feed

This type includes the *fruticulosesedge-sphagnum group of facies species of the first terrace above the flood land with peaty soils*. The fruticulose-sedge-sphagnum bogs are situated on the margins of the bog areas of the first terrace above the flood land. In their feed, apart from the atmospheric precipitation, the surface-runoff and deluvial waters participate. On the Lower Amur River lowlands, these bogs are of limited occurrence and occupy quite small areas. In the vegetable cover, the fruticulose-grass and moss layers play the edificatory role. The fruticulose layer is formed by Betula ovalifolia but, within some sections, Alnus hirsuta Turzc. occurs. The fruticulose-grass layer consists of Calamagrostis neglecta, Salix myrtilloides, Vaccinium uliginosum, Sanguisorba parviflora. By the maximum abundance, Carex minuta and Chamaedaphne calyculata are distinguished. On the cushions, oligotrophic and mesotrophic species prevail while in the low places and tussocks – eutrophic and mesotrophic plants. In the moss layer, the rudiments of cushions were formed by waste of Sphagnum magellanicum and Sp. centrale C. Jens. to which the oligomesotrophic and oligotrohic species – Sp. angustifolium, Sp. rubellum, Sp. Fuscum are added. In the low places between the tussocks and cushions, Spagnum obtusum prevails. On the whole, the vegetation of the fruticulose-sedge-sphagnum group of facies is characterized by mixing of eutrophic, mesotrophic and oligotrophic species. This feature of phytocenoses is caused by formation of unequal forms of microrelief with different ecological conditions. Independently of the age, all bogs of this group formed as a result of forest bogging. On the bog bottom, well decomposed wood, wood-sedge and wood-grass peat is observed.

Sphagnum Oligotrophic Type of Facies of Atmospheric Feed

The bogs of this type are widely distributed on the Lower Amur River lowlands. They are situated on the surface of first terrace above the flood land and are damped by only atmospheric precipitation. In the vegetable cover, the oligotrophic plants predominate. The sphagnum mosses are major edificators. The fruticulose-grass layer is thinned. The sphagnum oligotrophic type includes four groups of facies species.

Cottongrass- fruticulose-sphagnum Group of Facies Species of the Margins of First Terrace above the Flood Land with Peaty Soils

The cottongrass- fruticulose-sphagnum bogs are situated ob the margins of the first terrace above the flood land where they occupy a belt of up to 1 km or more wide. They are especially numerous on the Udyl-Kizinsky lowland. The feed of bogs is atmospheric. The arboreal layer of this group of facies id presented by Larix dahurica that does not play the edificatory part in connection with a strong sparseness. The shrubby layer consists of Betula ovalifolia sometimes with addition of Pinus pumila or Alnus hirsuta. The dominants of the fruticullose-grass layer are Ledum palustre, Chamaedaphne calyculata and Eriophorum vaginatum. From among accompanying species, one can call Andromeda polifolia, Rubus chamaemorus, Smilazina trifoliata, Drosera rotundifolia, Oxicoccus microcarpus. The plants listed are characteristic of the flora of oligotrophic phytocenoses. Among mosses, Sphagnum magellanicum is distinguished by abundance. In many sections, other species of oligotrophic mosses were also found: Spagnum lenense, Sp. fuscum, Sp. balticum Sp. Lenense, Sp. balticum and Sp. fuscum. Nearly in all places, Polytrichum strictum and Aulacomnium palustre (Hedw.) Schwaegr. occur.

Sphagnum Group of Facies of the First Terrace over the Flood Land with Peaty Soils

The sphagnum bogs are situated in different parts of the Lower Amur River lowlands on the surface of the first terrace over the flood land but concentrate largely on the lowlands margins. They occur also in the central parts of the bog areas. These bogs are damped by only atmospheric precipitation. The hydrological regime is stagnant. The moss cover is characterized by relatively plane surface with small elevations and depressions. In the vegetable cover, the moss layer is critical edificatory. Sometimes, Larix dahurica of dwarf ecological form is observed in some sections. The shrubs - Betula ovalifolia with addition of Pinus pumila, and, rarely, Alnus hirsuta grow in all places. The composition of the fruticullose-grass layer plants is characterized by certain stability. The oligotrophic plants – Chamaedaphne calyculata, Andromeda polifolia, Oxicoccus microcarpus, Ledum palustre, Rubus chamaemorus, Carex middendorffii, Smilazina trifoliata, Drosera rotundifolia grow in all places. Rarely, the eutrophic and mesotrophic species (Menyanthes trifoliata, Carex minuta, Salix myrtilloides etc.) are found. The mosses form a continuous cover. Sphagnum magellanicum prevails more often than other species while Sp. Fuscum is less abundant and, only in one section under study, Sp. Lenense is a dominant. All of three mosses belong to oligotrophic species but Sp. magellanicum is distinguished by wide ecological adaptability. Spagnum Angustifolium occurs almost in all places. As for green mosses, Polytrichum strictum is of greatest abundance. On the whole, the phytocenoses of the sphagnum group are characterized by maximum presence of oligotrophic plants and maximum oligotrophy.

Hydrophilic-sphagnum (Ridge-hollow) Group of Facies Species of the Central Parts of First Terrace above the Flood Land with Complex

The gydrophilic-sphagnum, ridge-hollow bogs are most widespread on the Lower Amur River lowlands. They are confined to the central parts of the first terrace above the flood land where they are damped by only atmospheric precipitation. The water is on the surface of hollows for the whole vegetation period. To maintenance of high dampness, the large moisture reserves in the peat deposits, long conservation of seasonally frozen ground and spacious lakes with areas of up to several square kilometers located between low-land bogs and playing the water-regulatory role contribute. Of the vegetation pf this group of facies, a complexity is peculiar. On the ridges characterized by the relatively drained conditions, four layers are inherent in the phytocenoses. The arboreal layer is presented by single specimens of Larix dahurica of dwarf ecological form. The shrubby layer consists of Betula ovalifolia and Pinus pumila. The dwarf shrubs and grasses grow well on the ridges. Here, Chamaedaphne calyculata and Ledum palustre predominate. Almost everywhere, Andromeda polifolia, Oxicoccus microcarpus, Rubus chamaemorus, Eriophorum vaginatum, Drosera rotundifolia, Carex middendorffii, Smilazina trifoliate are found. In the moss layer, the oligotrophic mosses – Sphagnum magellanicum, Sp. Lenense, Sp. Fuscum are abundant while, here and there, lichens of Cladonia sp.. As for the vegetation, the hollows are distinguished from ridges by absence of trees and shrubs as well as by great sparseness of the fruticulose-grass layer. To the typical representatives of associations, Rhynchospora alba (L.) Vabl., Carex limosa, Drosera anglica Huds., Iris laevigata and other hydrophilic species can be attributed. The mosses form, as a rule, a continuous cover. The composition of mosses in hollows is quite diverse. More often, Sphagnum papillosum Lindb. predominates while Sp. balticum to a lesser degree. These plants are typical hydrophytes.

Hydrophilic-sphagnum (Pond-ridge-hollow) Group of Facies Species in the Central Parts of the First Terrace above the Flood Land with Complex Soils

The pond-ridge-hollow bogs are very similar with the ridge-hollow in vegetation, soils and structure of peat deposits. The vegetation is presented by the association complexes which enter also into the composition of hydrophilic-sphagnum, ridge-hollow group, however, their number is much less. On the ridges, only Sphagnum magellanicum is a dominant of the moss layer. Spafnum fuscum and Sp. lenense are universal in occurrence.

Arboreal, Naturally Drained Type of Facies of Atmospheric Feed

The bogs of this type are quite widely distributed on the Amur-Amgun lowland. The bogs are located in bumps and outliers of the permafrost surrounded by the hydrophilic-sphagnum ridge-hollow bogs. The feed is provided at the expense of atmospheric precipitation. This type of facies includes two groups of facies: *larch one on the permafrost bumps-outliers of the first terrace above the flood land with residual-peaty soils* and *birch derivative on the permafrost bumps- outliers with residual-peaty soils*. In the first group, the mountain-pine association is the age stage of development of the larch one. The considerable crown density of the forest and dense underbrush prevent from the larch production and, as the overmature larches fall, the mountain pine spreads out. A destruction of underbrush and outburst of renewal processes are observed after fires causing, sometimes, initiation of lasting derivative birch plantations. The composition of derivative birch associations varies from purely birch to aspen-birch ones with admixture of larch [17, 19].

EVOLUTION OF BOG SYSTEMS

Using the submeridional profile of up to 1000 km in length (48-54° N) extended along the Amur River course several bog systems located within three vegetation zones were studied [10]. In these systems, the peatbog profiles were investigated using the sporo-pollen and radiocarbon analyses and the botanic compositions of peat, decomposition degree and ash content were determined.

Within zone of light coniferous forests, the bog systems Tyapka (Sakhalin Bay basin, Okhotsk Sea) and Chlya (within the Amur-Amgun lowland) are located. In zone of coniferous forests, there is the bog system Dudi (within the Udyl-Kizinsky lowland). Within zone of the mixed coniferous-broad-leaved forests, the bog systems Gur and Kur (north-eastern Middle-Amur lowland), Bastak (south-western Middle-Amur lowland), Kiya, Kamen-Matai and Pashino (south-eastern Middle-Amur lowland) (Fig. 1) are situated.

When studying the geosystems, a need arise always to analyze their capabilities to withstand any disturbances. Such characteristic of the geosystem quality is called stability. The stability is a capability of systems to maintain the values of their parameters and properties not exceeding the specified critical quantities. These quantities are expressed through the structural and functional characteristics of the geosystems [7]. The stability, from the aspect of vertical and horizontal structures of the geosystem, reflects the form of the object constancy which is specified by the appropriate invariant. The peatlands are admissible natural objects to which the information of the past states of the bog systems was stored and this information allows us to study their temporal and spatial development. The persistence of the plant association of the bog system is a factor of its stability. The type of the plant association is determined by hydrological conditions of the bog system. As long as the variations of these conditions are within the stable state of associations, change of associations does not occur. If the unfavorable environmental conditions, for example, a reduction of climate humidity, influences on the hydrological conditions, then the system will prove to be instable. Changes in the botanic composition of peat reflect the changes of the plant associations at different stages of developing the bog systems. For reconstruction of the hydrological conditions of bog systems, the concept of "humidity index" was proposed. At the basis of the humidity index calculation method, the ratio of peat-forming plants to moisture lies. To each peat-forming plant or their ecological group, the numerical values in points are attributed. The hydrophilic-psychrophilic plants received the least point, hydrophilic-subpsychrophilic one – middle point while the hyperhydrophilic plants – highest point [23].

The bog system Tyapka is located in the interfluve of Tyvlina and Tyvlina-Tyapka rivers, within a zone of light coniferous forests, on the Sakhalin Bay coast, Okhotsk Sea (Foto 1).

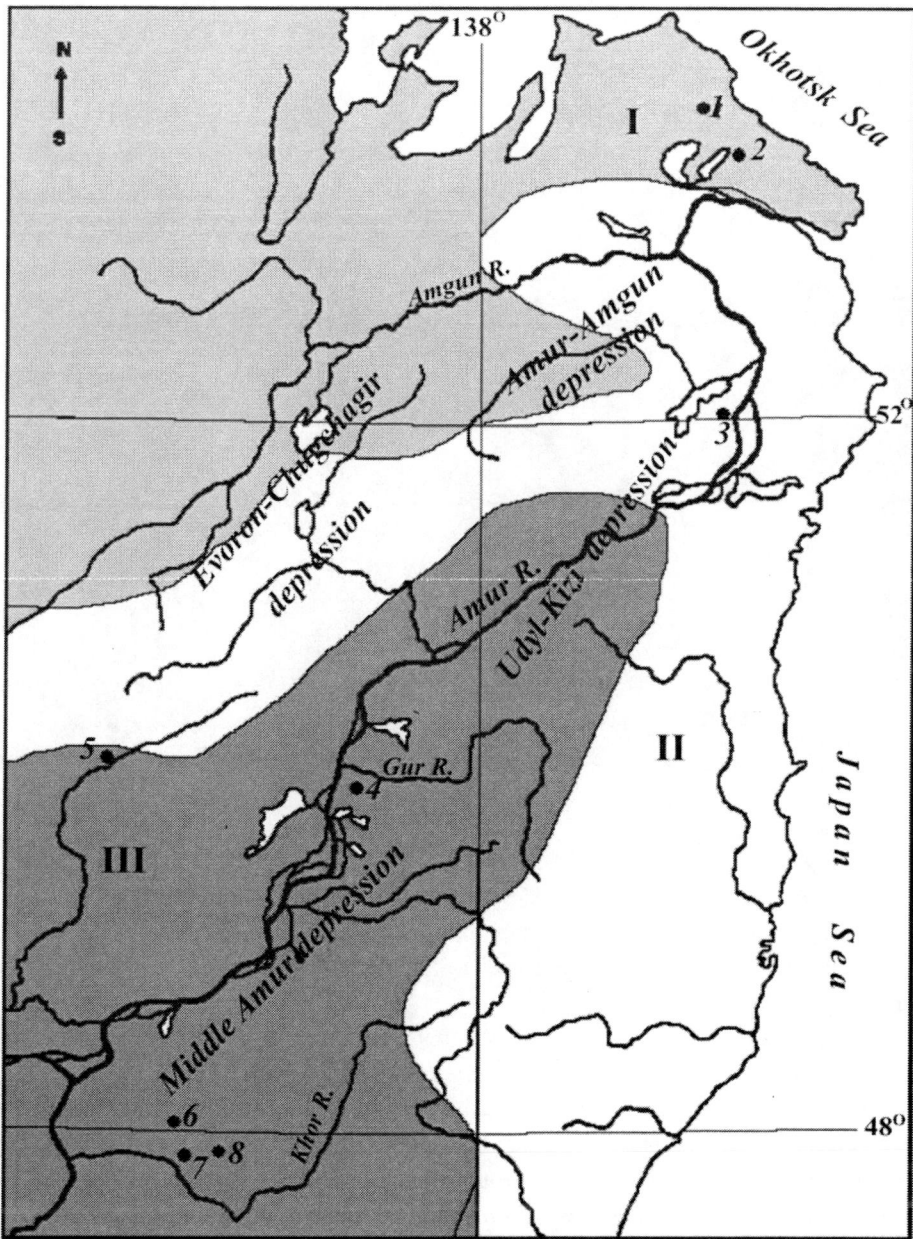

Fig. 1. Map of the Lower Amur River basin and peaty section location.
Vegetation zones (Kolesnikov, 1969): I - light coniferous forests;
II - coniferous forests; III - mixed coniferous and broad leaved forests.
Peaty sections: 1 - Tyapka, 2 - Chlya, 3 - Dudi, 4 - Gur, 5 - Kur, 6 - Kiya,
7 - Pashino, 8 - Kamen-Matai.

Foto 1. Tyapka peat bog

The peat accumulation began 9975±120 yrs BP (SOAN-4025). The diatomaceous complex of argillaceous deposits underlying the peatland includes the extinct during Pliocene freshwater species *Aulacoseira praegranulata, A. praedistans, A. praeislandica, Melosira areolata*.

The participation in the complex of diatoms characteristic of bogged surfaces is quite low, they are presented by species of *Eunotia* and *Pinnularia* genera. Such composition of diatoms is characteristic of the alluvial deposits which have, most likely, Neogene age. Here, single redeposited marine forms extinct during Miocene and Pliocene (*Coscinodiscus marginatus var. fossilis, Cosmiodiscus intersectus, C. Insignis, Pyxidicula zabelin* и *Thalassiosira gravida var. fossilis*) were found. In the profile, nearly all deposit was formed by sphagnum peat of high-moor type with not high degree of decomposition (20-30%) except the layers of peat accumulated during the following intervals: 4.80-4.50 m, 4.40-4.00 m, 3.20-2.90 m, 2.80-2.60 m and 0.40-0.20 m. In these intervals, a presence of the wood and grass components and peat decomposition degree increase (40-65%). Such changes in the botanic composition of peat allow us to suggest the variations in the hydrological conditions of the bog system related to bog water level lowering.

The bog system Chlya is situated within the Amur-Amgun lowland, in a zone of light coniferous forests (Fig. 1). A beginning of its development is related to the retreat of the Chlya lake (Foto 2). The lake bottom freeing from water began to turn into swamp. In the underlying argillaceous deposits, a rich complex of diatoms characteristic of the shallow boggy freshwater lake was found. The predominant complex at the depth of 280 cm includes the forms characteristic of the acid swamp waters (*Eunotia praerupta, Pinnularia viridis, P. Brevicostata* и *P. lata*). Here, the species characteristic of the running waters (*Aulacoseira granulata, A. distans, A. alpigena, Cymbella naviculiformis, Gomphonema parvulum*) are

Foto 2. Chlya peat bog

also present. The peat accumulation began 9700± 80 BP (SOAN-4719). The mesotrophic stage of the bog system development continued to the subboreal period of Holocene (~ to 4600 years ago). Within the interval of 1.80- 2.90 m, the wood-grassy component (up to 80%) predominates in the botanic composition of peat while the remainder belongs to the grassy group. At the beginning of the Holocene subboreal period, the bog system passes to the oligotrophic stage of development. In the botanic composition, the wood-shrubby component decreases and the sphagnum component appears and increases (up to 60%). Beginning from the depth of 1.80 m, the high moor peat has formed. In the botanic composition of peat, the sphagnum mosses predominate. At present, the bog area surface is drained and reproduction of the sphagnum cover ended. A draining of the bog area was related to the retreat of the Chlya lake that started at the end of subboreal Holocene period [12]. The peat bog is gradually overgrown with arboreal vegetation beginning with borders while the mass center is occupied by Betula ovalifolia and heathers.

The bog system Dudi began to form in the rear part of the Amur River flood land. It is located within the Udyl-Kizinsky lowlandn in a zone of coniferous forests (Fig. 1). To the bog formation time, the radiocarbon date of 8370±115 BP (SOAN-4449) points. It is a deposit of transition type. At present, only small fragment of peat bog blowing out in the Amur River shore exposure has been retained. In the botanic peat composition of the Dudi profile, the changes in the ratio of its three components (wood-fruticulose, grassy and mossy) from bog formation beginning to the present are insignificant. The whole deposit is of transition, naturally drained type.

The most ancient of all studied by now systems is the bog system Gur located ob the north-east of the Middle Amur River lowland within a zone of mixed coniferous-broad-leaved forests (Fig. 1) [9]. At the end of the Late Pleistocene, the paleolake water discharge took place through the Gur River channel to the Amur River. A complex of diatoms from clays underlying the peat includes only benthos forms. The predominant complex is presented by acidophilic, epiphytic *Eunotia monodon* and *E. praerupta* as well as bottom *Pinnularia viridis* and *Pinnularia brevicostata*. The complex contains also the species characteristic of the alkalescent waters: *Cymbella aspera*, *Stauroneis phoenicenteron* and *Tetracyclus glans*. Such complex is characteristic of the shallow, turning into swamp freshwater reservoir. The abundance of the bottom flora suggests that the water in t6his reservoir was quite clear. On the surface freeing from water, the bog system began to arise. In its evolution, three major stages can be identified. The first stage began 12120±75 BP (AA-36273) in local depressions with accumulation of valley peat with thickness of 130 cm. A feed of emergent bog system was provided by the surface-runoff and deluvial waters. At lower part of this layer, hypnum mosses predominate initially in the botanic composition (up to 50%) and remaining 50% contain wood-shrubby, grassy and sphagnum components in approximately equal quantities. Beginning with the depth of 2.60 m, the hypnum mosses disappear giving way to the wood-shrubby and grassy components. This stage of peat bog development has terminated at the end of boreal period, approximately 8000 years ago. The next stage was related to the activation of the bog-forming processes and quick extension of boundaries of primary waterlogging sites beginning 8000 years ago. Beginning with this time, the hydrophilic species were practically displaced from the vegetation composition by shrubs. This stage has terminated comparatively not long ago (~ 1000 years ago). The upper 30-cm layer of peat has formed for last ~1000 years (Fig. 2).

The bog system Kur is on the border between zones of coniferous and coniferous-broad-leaved forests at altitude of 413 m (Fig.1). It is the youngest bog system of all studied ones and the age of the lower peat layer is 3465±50 BP (SOAN-4717). In the development of the peat deposit, one can identify two stages. The lower 30-cm layer of valley peat was accumulated during the subboreal Holocene period. In the botanic composition, the wood component predominates (up to 79%) and the peat decomposition degree is high (up to 65%). Beginning with the end of subboreal Holocene period (~ 2600 years ago), the bog system has developed at the mesotrophic stage. The botanic composition of peat has changed – the quantity of sphagnum mosses has increased and the hypnum mosses have appeared. The peat decomposition degree has decreased (25-35%). At present, the bog system is naturally drained and its development continues at the mesotrophic stage (Fig. 2).

The bog system Bastak is located in the south-western Middle Amur River lowland, within a zone of mixed coniferous-broad-leaved forests (Fig. 1). The thickness of the peat layer is 2 m. The development of bog system started from the eutrophic stage. The lower peat layer is woody-shrubby-grassy and characterized by a high decomposition degree (up to 90%). In the upper 50 cm layer of peat, the woody-shrubby-grassy component is replaced by the sphagnum mosses and the peat decomposition degree has reduced (to 5-7% in the upper part of profile). At present, the development is at the eutrophic-mesotrophic stage (Fig. 2).

The bog system Kamen-Matai is located in the interfluve of Kamen and Matai rivers, on the south-east of the Middle Amur river lowland, within a zone of mixed coniferous-broad-leaved forests (Fig. 1). The thickness of the peat deposit in the profile reaches 2.60 m. In the interval of 2.60-0.60 m, the peat deposit is of valley type. In the botanic composition of peat, the woody (arboreal) component predominates (50-70%) and residues of grasses and sphagnum mosses are also present. The peat decomposition degree is high (up to 95%). The presence of alluvium in peat within interval of 2.50-2.40 m suggests a temporary flooding of the bog area with flood waters. Within interval of 2.50-2.20 m, there are the hypnum mosses in small quantities which are indicators of well watering. Here, the high values of humidity index were also obtained while the peat decomposition degree reduces (less than 50%). The upper 60-cm layer of peat contains sphagnum mosses (up to 80%) with insignificant presence of woody and grassy residues and the decomposition degree is high (up to 80%). The development of the bog system entered the mesotrophic phase.

The bog system Pashino is located on the first terrace above the flood land in the interfluve of Golda and Podkhorenok rivers, on the south-east of the Middle Amur River lowland within a zone of mixed coniferous-broad-leaved forests (Fig. 1). The thickness of the peat deposit is 3.5 m. The peat deposit is of valley type. In the botanic composition, the large quantities of sphagnum mosses are found in the profile roof. The quantity of hypnum mosses is also significant. The peat accumulation has occurred under stable hydrological conditions. The peat decomposition degree is stably high (~90%) and reduces to 50% upstream of the profile. The presence of alluvium in the botanic composition within the interval of 3.00-2.10 m suggests a flooding of the bog area by river waters. In the upper 30-cm layer, the composite peat has accumulated and the bog system entered the mesotrophic phase of development. At present, the development of the bog system continues in the mesotrophic stage.

The bog system Kiya is located on the first terrace above the floodland of Kiya River on the south-east of the Middle Amur River lowland, within a zone of mixed coniferous-broad-leaved forests (Fig. 1). The thickness of peat deposit is 2.10 m. From material of the peat bog bottom, the radiocarbon data of 8890±95 BP (SOAN-4454) was obtained. The diatomaceous flora from clays underlying the peat bog includes the species of accretion, bottom and plankton forms. The dominant complex is presented by acidophicil *Eunotia faba, E. Incissa* and *E. pectinalis*. Here, diverse species characteristic of running waters are also found among which plankton *Aulacoseira italica*, epiphytic *Fragilaria construens var. venter, F. pinnata, Cymbella turgida, Gomphonema trunctatum* etc. are most notable. Such composition of diatoms is characteristic of the dead channel conditions. When drying-up the dead channel, the valley bog system began to form. Almost all deposit consists of valley peat formed by residues of woody-shrubby and grassy vegetation. Beginning with the profile foot and to the depth of 0.5 m, in the botanic composition, the woody (arboreal)-fruticulose component prevails (up to 70-80%) and remaining 20-30% belong to grass residues. Within interval of 1.00-1.15 n, the sphagnum mosses are found to a small extent (up to 5%). They appear also in the peat composition at the depth of 0.5 m. In the upper 20-cm layer, their content increases to 50%. On the while, the bog system has developed under quiet hydrological conditions. The peat decomposition degree was high practically for the whole profile with decreasing in the upper profile portion.

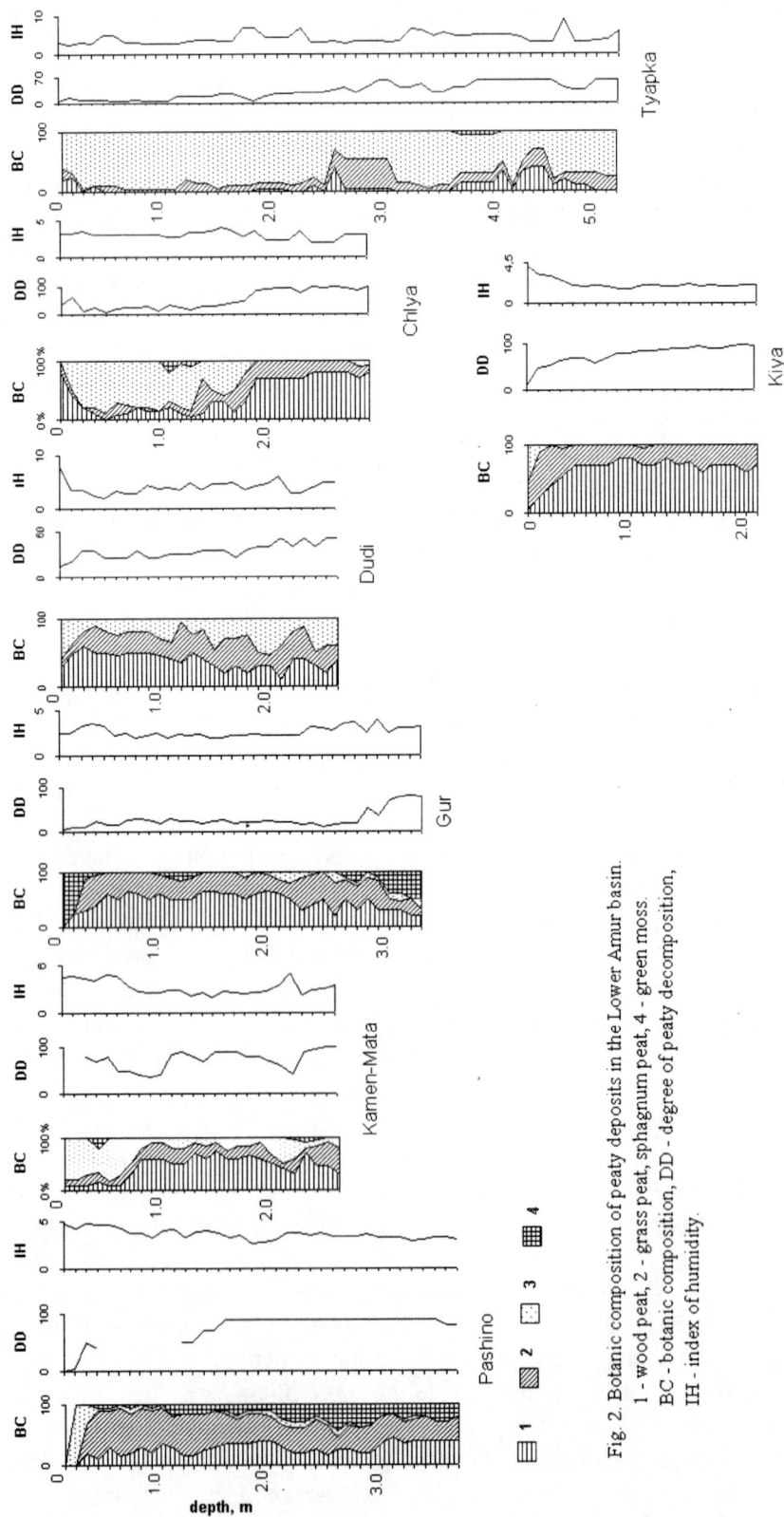

Fig. 2. Botanic composition of peaty deposits in the Lower Amur basin.
1 - wood peat, 2 - grass peat, sphagnum peat, 4 - green moss.
BC - botanic composition, DD - degree of peaty decomposition,
IH - index of humidity.

STABILITY OF BOG SYSTEMS

When the stability of bog systems is considered, the inertia conception is most applicable to them which is one of the kinds of stability of geosystems. The inertia is considered as the system's capability to maintain major qualitative features of its structure under varying environmental conditions [7]. According to response of the bog systems to hydro-climatic variations, they were divided into two groups – inert and weakly inert ones [3].

For a group of bog systems situated near the mouth of Amur River, the amplitude of variations of the humidity indices is high (from 2 to 9, Table 3). This may suggest the significant rises and drops of water level in the bog systems. Nevertheless, the differences are also observed in this group. In the profiles of bog systems Chlya and Dudi having the deposits of transitional and high-moor pet, approximately 3-4 significant drops of water level were identified for the whole period of their development. The significant drops of bog water level in the bog system Dudi were observed at time boundaries of ~ 8000, 7600, 7200 and 3500 years ago while in the bog system Chlya - ~7200 and 6400-7000 years ago. As for the bog system Tyapka containing the high-moor peat, the sharp variations of the bog water level took place up to 6 times. In this profile, the water level drops have occurred at time boundaries of 9500, ~ 8000, 5300, 4000-3500, 2700, 1400 years ago (Fig. 2). The climatic conditions of a zone of the light coniferous forests characterizing by a higher humidity and lesser evaporation contributed to accumulation of peat of transitional and high-moor types.

Table 3. Index of humidity in peaty sections of Lower Amur basin

Depth, cm	Section						
	Tyapka	Chlya	Dudi	Gur	Kiya	Kamen-Matai	Pashino
10,0	3,2	3,2	7,8	2,5	4,1	4,5	4,8
20,0	2,6	3,5	3,5	2,5	3,2	4,6	4,2
30,0	3	3	3,5	3,3	3	4,4	4,8
40,0	2,8	3	2,5	3,5	2,5	4	4,6
50,0	5	3	2	3,3	2	4,8	4,6
60,0	5	3	3,3	2,2	1,9	4,7	4,3
70,0	3	3	2,8	2,5	2	3,4	3,8
80,0	3	3	3	2	1,9	2,7	3,8
90,0	2,8	3	4,5	2,2	1,6	2,5	3,3
100,0	2,8	2,8	3,8	2,5	1,6	2,5	4
110,0	2,8	2,7	4	2	2	2,8	4,1
120,0	2,8	3,4	3,5	2,4	2	2,8	3,3
130,0	3,3	3,4	5	2,3	1,8	2,1	3,9
140,0	3,9	3,4	3,5	2,4	1,9	2,6	4
150,0	3,9	3,4	4,6	2	2,1	2	3,8
160,0	3,5	3,9	4,6	2	1,9	2,7	3,3
170,0	3,5	5,4	5	2,2	2	2,5	3,5
180,0	7	2,7	3,5	2,2	2	2,4	2,6

Table 3. (Continued)

Depth, cm	Section						
	Tyapka	Chlya	Dudi	Gur	Kiya	Kamen-Matai	Pashino
190,0	7	2	4,3	2,4	2	2,5	2,8
200,0	4,5	1,8	4,6	2,2	1,9	2,7	3
210,0	4,5	2,2	3,2	2,3	2	3,5	3,8
220,0	4,5	2,8	6	2,2	-	5	3,8
230,0	7	2,8	4	2,3	-	2,3	3,5
240,0	3,2	-	5	3,2	-	2,8	3,8
250,0	3	-	5	3	-	3	3,4
260,0	3,4	-	5	2,7	-	3,4	3,4
270,0	2,9	-	-	3,5	-	-	3,4
280,0	3,4	-	-	3,7	-	-	3,6
290,0	3,4	-	-	2,5	-	-	3,2
300,0	3,4	-	-	4	-	-	3,2
310,0	3,1	-	-	2,5	-	-	3,3
320,0	3,4	-	-	3	-	-	2,9
330,0	6,6	-	-	3	-	-	3
340,0	6,2	-	-	3,2	-	-	3,2
350,0	4,6	-	-	-	-	-	3,2
360,0	5,7	-	-	-	-	-	3
370,0	4,6	-	-	-	-	-	-
380,0	4,6	-	-	-	-	-	-
390,0	4,9	-	-	-	-	-	-
400,0	4,9	-	-	-	-	-	-
410,0	4,9	-	-	-	-	-	-
420,0	4,9	-	-	-	-	-	-
430,0	4,5	-	-	-	-	-	-
440,0	9	-	-	-	-	-	-
450,0	3	-	-	-	-	-	-
460,0	3,2	-	-	-	-	-	-
470,0	3,2	-	-	-	-	-	-
480,0	3,1	-	-	-	-	-	-
490,0	3	-	-	-	-	-	-
500,0	3,5	-	-	-	-	-	-
510,0	3,8	-	-	-	-	-	-
520,0	5,9	-	-	-	-	-	-
530,0	5,5	-	-	-	-	-	-
540,0	5,5	-	-	-	-	-	-

For group of bog systems being within a zone of mixed coniferous-broad-leaved forests, the amplitude of humidity index variations if insignificant. The hydrological conditions of these bog systems are more stable (Fig. 2).

In the bog systems situated near the mouth of Amur River, the hydrological conditions have changed several times and bog water level differences were essential. This reflected in termination of the sphagnum cover recovery, slowdown or termination of peat accumulation. These bog systems were attributed to weakly inert ones.

The water-mineral feed of bog systems confined to the continental part of the Lower Amur River basin is more stable. These bog systems are attributed to inert ones in accordance with their hydroclimatic variations.

FEATURES OF PEAT DEPOSITS

When studying the bog soils, the prominence was given to the reasoning of possibility of their agricultural use after the drainage engineering and examination of developed soils [1, 6, 18, 21]. In this connection, the properties of the upper part of peat deposits rather than the whole profile were studied in most cases.

We obtained materials concerning some properties of peat in the profiles made in three bog areas with different genesis. The high-moor peat is considered by example of the bog system Tyapka, transitional peat by example of system Dudi and valley one using the system Gur. A choice of just these profiles was dictated by the fact that they are most ancient (9-12 thousand years) and each of them is nearly entirely presented by the peat of the same genesis. A sampling was made layer-by-layer from exposures of natural or artificial origin. The samples of peat and underlying rocks obtained were crushed after air-drying and passed through the sieve with mesh diameter of 1 mm and, thereafter, ash content, pH of aqueous and salt suspensions, hydrolytic acidity by Cappen, composition of exchangeable cations $(Ca^{2+}, Mg^{2+}, Al^{3+}, H^+)$ by Karpachevsky, absorptive capacity by Aleshin, total carbon and nitrogen by Ponomareva and Plotnikova were determined [15, 20].

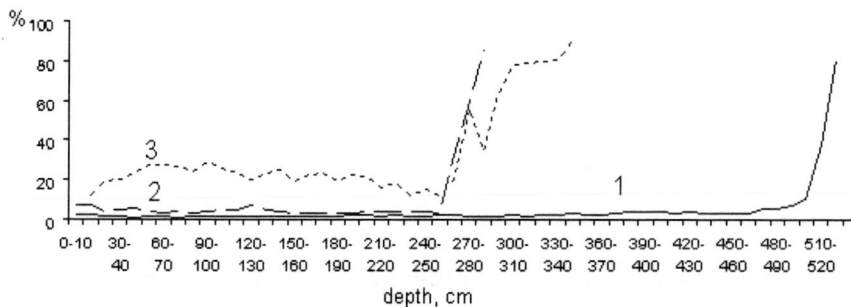

Fig. 3. Content of ash in peat (%). Peaty sections: 1 - Tyapka, 2 - Dudi, 3 - Gur

Generalizing data obtained by first researchers of the Far East peat bogs, M.I. Neishtadt [13] gives some information of the features of the bog deposits in the south part of the Lower Amur River basin. To them, he attributes the extremely high decomposition degree (55-75%), complexity of botanic composition as well as high density of peat. This suggests, on the one hand, the instability of water regime and, on the other hand, a slow accumulation of peat. The peat bogs widely distributed in the northern part of the Lower Amur River basin are characterized, in most cases, by the smaller indices of density and decomposition degree.

As in other regions of Eurasia, many peat deposits of Priamurye, especially ones of valley type, are characterized by the increased ash content as a result of silting. As a consequence,

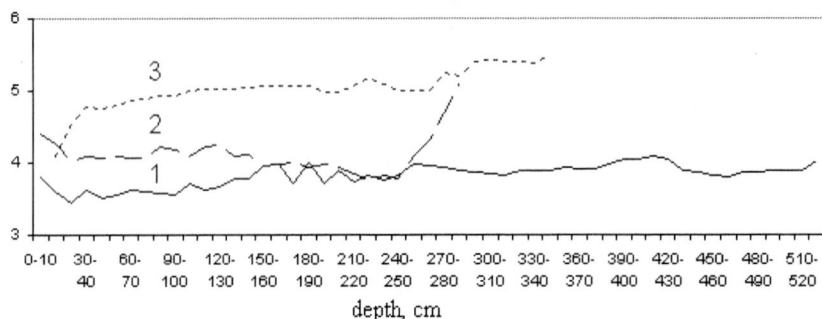

Fig. 4. Data of pH_{H2O} in peaty sections

the peat enriches with silica, iron and aluminium with deficiency in calcium, magnesium, molybdenum and zinc [20]. The profiles considered by us belong to ones with normal ash content (Fig. 3) except the Gur peat bog for which the ash content is slightly higher.

The excess of atmospheric precipitation over evaporation on the whole territory of the Lower Amur River basin has determined the major peculiarity of the peat deposits – increased acidity (Fig. 4, 5). As for the acidity level, the valley peat is close to transitional peat of Siberia and Europe. Just high acidity has determined a number of specific features of peat

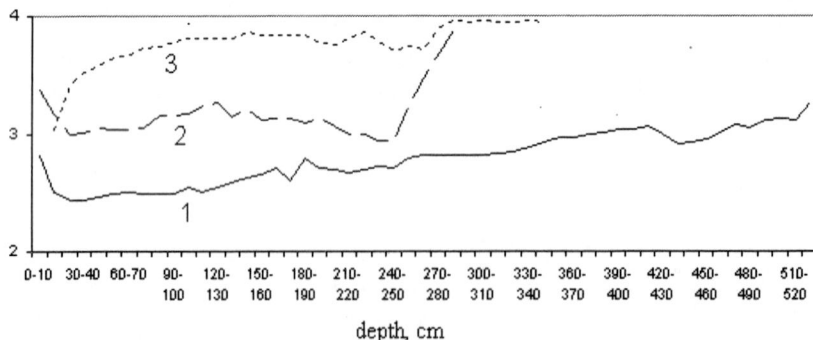

Fig. 5. Data of pH_{KCl} in peaty sections

deposits. To them, one can assign a low content in the absorbing complex of alkaline-earth bases, considerable quantity of absorbed Al^{3+} and H^+ of which the latter occupies the predominant place in the high-moor peat, high potential acidity and low degree of saturation with bases (Table 4).

In the group composition of organic matter even in valley peat, the content of waxes and resins reaches 15-17% of organic matter which is characteristic, to a larger degree, of peat bogs of transitional and high-moor types. In the fractional composition, very low content of humic acids related to calcium are observed [5] which are instable under conditions of high humidity [14].

A direct determination of total carbon (Fig. 6) indicated that although the ash content of the peat bog of high-moor type is much lower than that of two others, the content of organic carbon in it is little more than that in peat bog of valley type. It is determined by differences in both botanic composition and decomposition degree of peats of different types. The nitrogen content in all peat bogs (Fig. 7) is close to the lower limit of its content in the

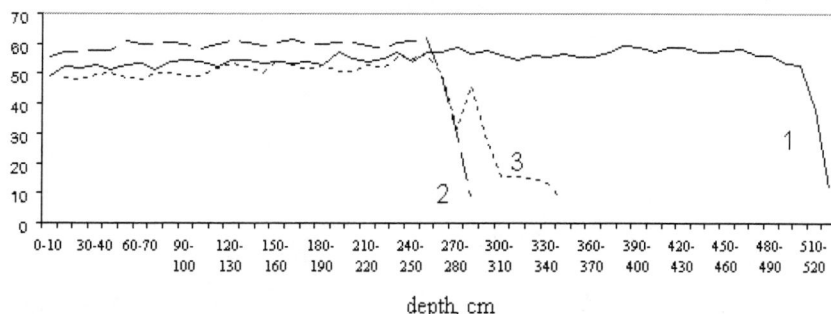

Fig. 6. Content of C total in peaty sections

European peat deposits of appropriate types [11]. As a consequence, the C:N ratio is wide (Fig. 8) which suggests very low enrichment of organic matter with nitrogen.

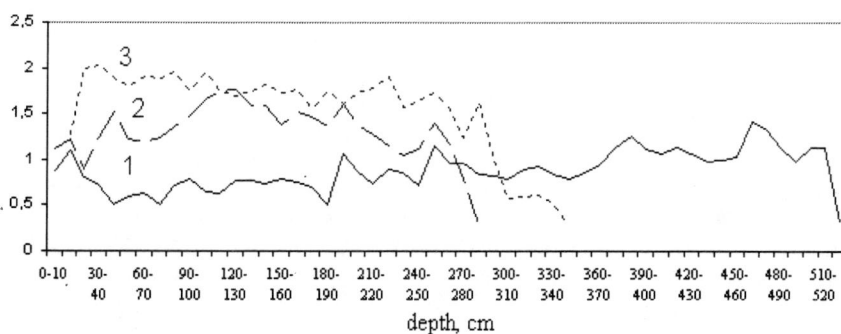

Fig. 7. Content of N in peaty sections

Table 4. Properties of different genesis peat deposits of the Lower Amur basin

Peat type	meq/100g				Potential acidity	Cation exchange capacity	Saturation percentage
	Exchangeable cations						
	Ca^{2+}	Mg^{2+}	H^+	Al^{3+}			
Eutrophic	5.48-12.2	3.33-9.15	1.89-5.94	7.54-11.76	61.33-79.96	42.52-57.32	13.6-24.0
Mesotrophic	14.18-25.50	3.46-8.27	2.31-6.11	9.36-14.65	56.80-95.16	60.22-85.56	19.9-27.0
Oligotrophic	5.18-5.86	5.77-10.94	15.12-25.08	5.43-8.52	96.37-140.67	83.59-119.15	7.5-10.7

The content of carbon in the organic mass of different types of peat (Fig. 9) varies from 51 to 68%. The minimum content was noted in the upper part of the high-moor peat bog formed by the sphagnum peat with a low decomposition degree. The maximum content is confined to the most decomposed peat in the middle part of valley peat bog.

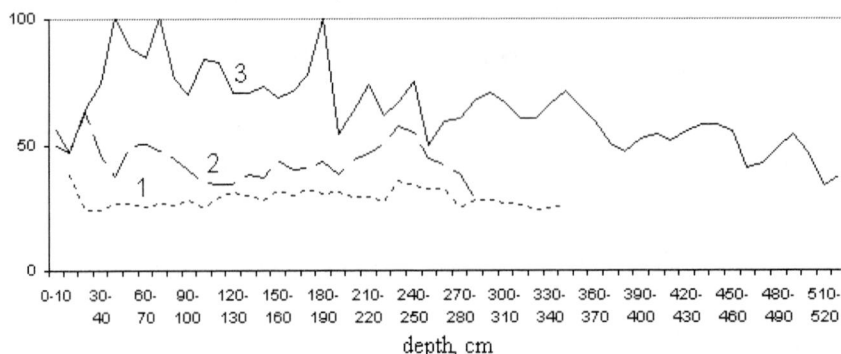

Fig. 8. Correlation of C:N in peaty sections

Thus, the biochemical features determined by the specificity of geochemical processes and biological cycle are in the high acidity of peat deposits of the Lower Amur River basin, low concentration of alkaline-earth bases and low enrichment of organic matter with nitrogen.

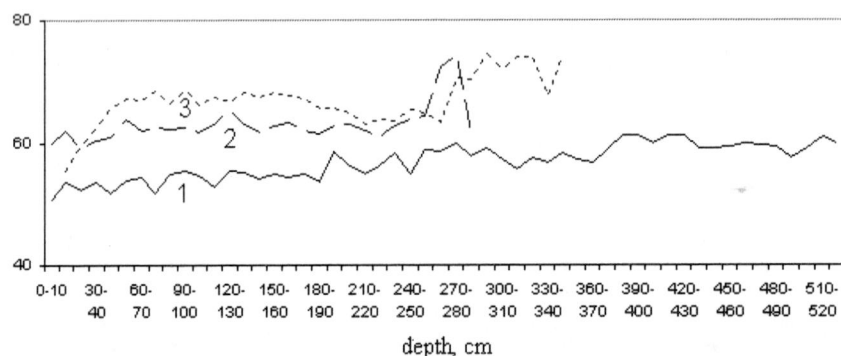

Fig. 9. C total, % from organic matter

CONCLUSION

In the Lower Amur River basin, the typological composition of bog systems is most representative as compared with other regions of the Russian Far East. Within the Middle Amur River lowland, the complex heterotrophic biogeocenoses predominate in combination with mesotrophic larch forests. The Amur-Amgun, Udyl-Kizinsky, Evoron-Chukchag lowlands are mainly occupied by oligotrophic bogs.

A zone of eutrophic grassy and heterotrophic sphagnum bogs in the coniferous-broad-leaved and dark coniferous forests is presented by the Middle Amur River area of fruticulose-sphagnum bogs. A zone of oligotrophic sphagnum bogs of the light coniferous forests corresponds to the boundaries of Udyl-Kizinsky lowland. The Chukchag district of woody (arboreal), naturally drained and heterotrophic grassy-mossy bogs and Amur-Amgun district of woody (arboreal), naturally drained and oligotrophic sphagnum bogs are intergrated into a zone of hilly bogs of the dark coniferous taiga [4, 17, 18, 19].

The intense bog formation in the Lower Amur River basin started in the Early Holocene (10-9 thousand years ago). Against the background of global paleoclimatic changes having effect on the development of bog systems of the Lower Amur River region, the regional natural factors become apparent and the strongest of them is a type of feed of bog systems. The bog systems of the Lower Amur River basin are invariant because they have maintained their vertical and horizontal structures throughout the Holocene.

Based on the long-term studies, its is proposed to set in the Lower Amur River area 10 bog districts as the objects of conservation due to the specificity of landscapes characterizing by distribution of rare flora and fauna.

At present, in the region, new approaches to exploitation of bogs with the thick peat layers consisting in the use of largely reproducible part of their resources were evident. The major advantage of this layers in much lesser change of the appearance and ecology of bog landscapes.

REFERENCES

[1]   *Agrochemical methods for soil investigations.* Moscow: A.V.Sokolov. 1975.

[2]   Anisimov V.M. Bogs of piedmont band of Middle Amur lowland and perspective of its using. In: Ed. A.M.Ivlev. *Matters of geography of Far East. Biotic components of ecosystems of the southern Far East.* Khabarovsk: Khabarovsk Complex Research Institute, FEB of RAS; 1973; p. 73-140.

[3]   Bazarova V.B. Features of formation and evolution of bog systems in the Lower Amur area. Eds. B.M.Ishmuratov, L.M.Korytny, K.N.Misevich et al. *Geography of Asian Russia at boundary of centuries.* Irkutsk: Institute of Geography SB RAS; 2001; p. 35-36 (in Russian).

[4]   Chakov V.V. Composition and biodiversity of vegetation cover of moss bogs in the Lower Amur area. Ed. by V.A.Voronov and A.N.Makhinov. *Investigations of aquatic and ecological problems of Amur area.* Vladivostok-Khabarovsk: Dalnauka; 1999; p. 175-179 (in Russian).

[5]   Chakov V.V., Klimin M.A. Bog evolution in the Middle Amur lowland and transformation of organic matter in peat. In: Ed. by P.V.Ivashov. *Biogeochemical and ecological investigations of natural and anthropogenic ecosystems of the Far East.* Vladivostok: Dalnauka; 1996; p. 126-134.

[6]   *Chernoukhov A.M. Reclamation of lands in the Priamurye.* Khabarovsk: Khabarovsk Publ. House; 1978

[7]   Diyakonov K.N., Ivanov A.N. (1991). Stability and inertia of ecosystem. *Vestnik of Moscow University,* v. 5, 28-34 (in Russian).

[8]   Klimin M.A. Genesis of peaty deposits of the Lower Amur region. In: Ed. P.V.Ivashov. *Geochemical and ecological-biogeochemical investigations in Amur area.* Vladivostok: Dalnauka; 2000; p. 170-179 (in Russian).

[9]   Klimin M.A., Kuzmin Y.V., Bazarova V.B., Mokhova L.M., Jull A.J.T. (2004). Late Glacial-Holocene environmental changes and its age in the Lower Amur River basin, Russian Far East: the Gursky peat bog case study. *Nuclear Instruments and Methods in Physics Research,* v. 223-224, 676-680.

[10]   Kolesnikov B.P. Vegetation. In: Eds. V.V.Nikolskaya and A.S.Khomentovsky. *Southern Far East.* Moscow: Nauka; 1969; p. 206-250 (in Russian).

[11] Lishtvan Yi.Yi., Korol N.T. (1975). *Primary properties of peat and methods of its determination.* Minsk, Nauka.

[12] Mikishin Yu.A. Holocene history of the Lower Amur area large lakes and tendency of its modern development. In. Ed. Yu.B.Zonov. *Data of geography and geomorphology of the Russian Far East.* Vladivostok: Far Eastern State University; 1992; p. 170-175 (in Russian).

[13] Neishtadt M.I. *History of forests and paleogeography of the USSR in the Holocene.* Ed. Ye.M.Lavrenko. Moscow: AS USSR; 1957.

[14] Petrov Ye.S., Novorotsky P.V., Lenshin V.T. (2000) *Climate of Khabarovsk Territory and Jewish Autonomous Region.* Vladivostok-Khabarovsk: Dalnauka (in Russian).

[15] Ponomareva V.V., Plotnikova T.A. *Humus and soil formation* (methods and results of study). Ed. By D.S.Orlov. Leningrad: Nauka; 1980.

[16] Prozorov Yu.S. Bogs of mari landscapes in the Middle Amur lowland. Ed. N.I.Piyavchenko. Moscow: AS USSR; 1961 (in Russian).

[17] Prozorov Yu.S. General description of bogs. In: Ed. by A.P.Muranov. *Resourses of surface waters of USSR. Far East. Lower Amur.* Leningrad: Hydrometeoizdat; 1970; p. 454-462 (in Russian).

[18] Prozorov Yu.S. *Bogs of Lower Amur depressions* Ed. A.M. Ivlev. Novosibirsk: Nauka; 1975 (in Russian).

[19] Prozorov Yu.S. Conformities of develop, classification and utilization of bog biogeocoenosis. Eds. N.I.P'yavchenko and S.E.Vompersky. Moscow: *Nauka;* 1985 (in Russian).

[20] *Theory and practice of soil chemical analysis.* Ed. L.A.Vorobiyeva. Moscow: GEOS; 2006.

[21] *Transformation of reclaimed peaty soils in the Priamurie.* Ed. P.V.Ivashov. Vladivostok-Khabarovsk: Dalnauka; 1995.

[22] *World peaty resources: reference book.* Ed. by A.S.Olenin. Moscow: Nedra; 1988.

[23] Yelina G.A., Yurkovskaya T.K. (1992) Methods of reconstruction of paleohydrological conditions as a base of objectification of succession reasons of bog vegetation. *Botanical Journal, v. 77*, 120-124 (in Russian).

In: International Wetlands: Ecology, Conservation...   ISBN: 978-1-60456-999-5
Editor: Jose R. Herrara   © 2009 Nova Science Publishers, Inc.

Chapter 11

THE EPISTEMOLOGY OF COMPLEXITY AND APPLICATIONS TO WETLAND RESEARCH

Waleska Valença Manyari[1], Luiz Pinguelli Rosa[2] and Osmar Abilio De Carvalho Junior[1]

[1]Departament of Geography, University of Brasília, Brazil
[2] Energy and Environmental Program, Federal University of Rio de Janeiro

ABSTRACT

The emergence of an epistemology of complexity is a contemporary phenomenon strongly based upon technology that has claimed new research needs and revitalize old studied themes. The study of wetlands can be considered in this context. The ability to respond in an organized manner to disturbances caused by seasonal flooding, a characteristic of wetlands ecosystems, define them as complex systems. The spatio-temporal heterogeneity of floodplain river systems is responsible for a diverse array of dynamic aquatic habitats. Thus, natural disturbances, represented by flooding and responsible for intensifying ecosystem heterogeneity, are the main factor in maintaining the ecological integrity. This requires knowledge of long-term patterns of inundation to preserve geomorphic formations, in other words, habitat diversity. Nevertheless, the processes that control water, sediment and nutrients transfer through the floodplain are not well understood. The maintenance of river-floodplain connectivity has been recognized as a central strategy in ecosystem management. In this sense, remote sensing technology for a broad-scale systematic focus is particularly relevant. Research studies in this area provide new approaches for models useful in management on a regional scale. This chapter aims therefore to illustrate this wide research field involving current concepts and methodologies in order to improve the comprehension of the wetlands ecosystems as well as establish conservation criteria.

1. INTRODUCTION

While Newtonian determinism was partially put to rout by the probabilistic approach of statistical mechanics, especially Boltzmann's contribution to the study of entropy, the uncertainty of quantum mechanics dealt it the coup de grace. In other respects, the theory of electromagnetism and the theory of relativity altered the way Newtonian physics viewed matter and the forces of nature. Similar sea changes were forthcoming in Biology with Darwin's theory of evolution in the 19th century and with Crick and Watson's discovery of DNA and the advent of molecular biology in the 20th century. In classical mechanics and in the theory of dynamic systems, predictability became dissociated from determinism, discarded along with the deterministic theory of chaos. Today, the theory of complexity provides an alternative to the role chance plays in the theory of evolution.

An approximation is now under way between the fields of Biology and Physics, with knock-on effects in other areas of knowledge. This process began when a clean break was made with the Newtonian paradigm in the internal domain of Physics. Yet "predictive" Newtonian determinism has survived to the present day as the general paradigm for other areas of knowledge. Its effective waning in the prevailing outlook has only now begun. This is partly due to its limitations in dealing with certain current issues like environmental degradation resulting from unbridled human intervention.

On the other hand, science has seen Biology come to the fore with the advent of biotechnology and genetic engineering. Its influence on other fields can be seen in computational models – neural networks, genetic algorithms, cellular automatons, and self-organised criticality.

. The path trailed has led to the search for a biology-inspired theory of complexity. With the proliferation of computers, research on a kind of experimental mathematics has made it possible to simulate the temporal evolution of dynamic systems, even when no analytical solution for the equations describing their behaviour is available. This is the case of systems governed by non-linear differential equations that have proved highly sensitive to initial conditions. The same occurs in Newton's mechanics with regard to non-linear problems, as Poincaré perceived at the turn of the 19th to the 20th centuries in the problem of three bodies in gravitational interaction among themselves (Campbell, 1983; Poincaré, 1894). Therein lay the origin of the study of deterministic chaos for, despite the fact that the system's future was theoretically determined by an equation, its future is actually unpredictable since minimal initial variations produce huge differences after a certain elapse of time. A system can switch from being well behaved to a state of chaos. At the frontier between order and chaos emerges what has come to be called complexity, characterised by the possibility of something new or unexpected arising. Complexity is associated with the phenomenon of life, thus inspiring a new general paradigm for worldview influenced by Biology.

An historical process can be observed here that admits a dialectic interpretation harking back to Marx, Hegel and Plato. Two arguments sustain this interpretation. The new biological theories vying to take the place of Physics as the general paradigm for science have received a major boost from Molecular Biology grounded in Physics. Moreover, although Molecular Biology has gained ground with DNA, the historical turning point for the growing influence of Biology dates back to Darwin's theory of evolution. The current theory of complexity, though, calls into question the role of chance inherent to Darwinian evolution theory.

A new angle on the evolution of life has been provided by Kauffman and what he calls the emerging science of complexity (Kauffman, 1995). He praises Newton's feat of formulating a theory for the movement of bodies with laws and equations only to declare that Darwin did not follow this deterministic view with his theory of evolution by natural selection. Man came to be seen as the result of a chain of accidental mutations haphazardly selected by capacity to adapt to the external environment. Kauffman holds that this idea is wrong, that order is not accidental. There are laws of complexity now being formulated by this emerging science according to which order is spontaneously generated in the natural world. So whereas in the physical sciences that have inspired the modern world's outlook the break with the global paradigm has progressed from quantum uncertainty through deterministic chaos to complexity, in the biological sciences we are now witnessing a break with the Darwinian theory of random evolution, which Kauffman's theory of complexity replaces by biological, albeit non-predictive, determinism.

Fertile cross-breeding is afoot: while the idea of complexity that typifies Biology penetrates and fertilises Physics, determinism invades Biology, evolving from predictive Newtonian determinism to deterministic chaos and the laws of self-organised criticality. Kauffman draws the analogy of a virus which, in a watery solution, maintains itself without using up any energy or matter until it is installed as a parasite in some living cell, much as occurs in mechanics with a ball in equilibrium at the bottom of a hole. Here is order in a system at minimum energetic balance. Here is the order of Physics. The order typical of Biology occurs conversely in systems out of balance that are dissipative, open and fed by a constant flux of matter and energy.

Such systems were studied by Prigogine (2002). Kauffman makes an analogy with computational systems in a Turing machine that executes algorithms. The aim of a theory is to furnish a reduced description of a system such that its behaviour can be predicted under given conditions. That is done by algorithms. However, irreducible systems and incomprehensible algorithms do exist: predictions cannot be made and all one can do is observe the system to see what happens.

Biological systems are of this sort. There are systems comprised by mixtures of different chemical substances that may spontaneously turn into autocatalytic systems. Molecules in these systems are formed by chemical reactions in such a way that their by-products catalyse the reactions producing them. This can take place in an intricate chain of reactions. Such systems sustain and replicate themselves. For Kauffman, life stemmed from molecular diversity in an autocatalytic system that surpassed a threshold of complexity, as if passing through a transition phase.

The emergence of order would be the bedrock of the origins of life and the evolution of organisms. To the present day biologists believe selection to be the source of order, but in this new approach selection operates at a level engendered by the laws of complexity. That implies a profound revision of Darwinism as we know it. The question now is to understand how selection combines with self-organisation and the spontaneous generation of order pursuant to laws of complexity. To a certain extent, we have here a combination of order in Physics with selection in Biology. Order requires an exchange of energy and matter with the outside world, as occurs with living organisms. They are never closed systems. What we observe in nature is order throughout, in plants, in flowers, in animals. Plants, for example, take light from the sun to transform water and carbon into complex molecules.

2. RIPARIAN WETLANDS AS COMPLEX SYSTEMS

Wetlands are actually a generic, fairly imprecise designation for quite singular ecosystems in waterlogged terrain, be it flooded or subject to flooding, generally situated between aquatic and terrestrial ecosystems. Indeed, adaptation to the strong presence of water on the surface or in the root zone is the prime hallmark shared by a wide variety of ecotones assembled under the one label (Mitsch and Gosselink, 1993). Despite this singular characteristic, criticism has been levelled at the idea of classifying these ecosystems as transition zones between land and water. This has serious implications for the management criteria and conservation strategies to be adopted. If that were the case, they should at least be recognised as unstable areas of tension that display unique patterns of variability, especially in the case of alluvial plains (Neiff, 2008).

The wealth of communities and species occupying wetland areas is another characteristic trait. Thus, classifying associations encountered in discrete units across wetland areas is a tough, even arbitrary task, given that their limits are frequently hard to determine. Moreover, conservation and recovery of these ecosystems are strongly dependent on advances in knowledge regarding the dynamics and functional organisation of such communities (Krebs, 1996).

According to systemic theory (Bertalanffy, 1975), flows of matter and energy mediate between community and non-living environment, maintaining an organised state of life. This elementary understanding is essential for grasping the way integral ecosystems in general work. In actual fact, these flows produce clearly defined biotic structures besides performing the cycling of matter among the living and non-living components. It is likewise acknowledged that abiotic structures are shaped by the importing of matter rich in energy (Odum, 1983). As a result, components, or rather physical subsystems (i.e. self-organised patterns) emerge. They establish environmental gradients directly proportional to the biodiversity encountered. In the case of wetlands, greater variation in such gradients can be interpreted by combining, for example, hydrometric levels, duration of surges, and so on, making for a diversity of landscapes and composition of species (Neiff, 2008). Ab'Saber (2002) provides a detailed description of Amazonian wetlands that highlights not only the biodiversity but especially the wealth of physical subsystems:

It is a highly diversified, labyrinthine patchwork of water and land: low-lying dykes along the flanks that were once covered by forest, immersed during the great floods, tall floodplain forests in some sectors where the plain backs up against the adjacent uplands, strands of biodiverse forest flanking dykes that run inland; meadows in scrolls of white sand following the meandering course of streams that have since run dry, herbaceous or meadowland vegetation in abandoned riverbeds, native grasses on the banks of floodplain lakes, exposed or submersed as the waters rise and fall. Finally, asymmetric ecosystems along the banks of side channels where *terra firma* forests cover uplands, steep banks or terraced embankments on the inner side of these discontinuous side channels, while on the other side a multifarious array of aquatic and sub-aquatic terrestrial ecosystems sprawls out across the alluvial plains. This veritable interspersed land-water patchwork supports a variety of ecosystems that are highly differentiated from each other although they all belong to a single family from the hydrogeomorphological standpoint.

As one type among a prolific classification (Mitsch and Gosselink, 1993), the riparian wetland ecosystems described above are a fine example of the interaction of various ecosystems producing a high degree of complexity. The description also draws attention to the fact that the evolution of living organisms is interconnected with the evolution of their habitat, there thus being a single process that could be described as co-evolution (Capra, 2006). Nevertheless, in a context of hierarchical relations between numerous subsystems, there is clearly a need for further knowledge about their basic properties and functions, for new and better data or new principles associated with abiotic factors.

It has thus been observed that hydrological characteristics fundamentally control such ecosystems. In these environments, specifically variations in discharge on time scales ranging from minutes to months and the frequency of exceptional flooding determine a river's every biological, chemical and physical attribute. Hey (1997) points out that even in unpolluted rivers biodiversity will not grow richer if hydraulic and morphological variations are limited, although the quality of the water also affects biological response.

In his chapter on the geomorphic activity of water, Sternberg (1998) identifies the structure and functioning of floodplains as the key factor – one that correlates closely with riparian vegetation – when it comes to the genesis of floodplains and other fluvial aspects:

> The importance of the waters is not confined to the fact that they periodically cover and uncover the land, nor can it be measured by the depth of the expanse of water lying above it. The river's geological activity must be taken into account. And the result of this activity, besides being subordinate to the nature of the terrain on which the water acts, depends on its properties, on the matter it carries with it and on the characteristics of its flow. Indeed, rivers that flow through alluvial plains are peculiar in that they themselves establish the discontinuous surfaces flanking and defining them. In the case under consideration, therefore, it is in the water, and in the water alone, that we should seek the key to the plains' morphology.

Indeed, fluvial hydrometric series can be acknowledged as a form of disturbance. By analysing them, one can detect recurrent phenomena (lymnophases and potamophases) whose characteristics (intensity, amplitude and seasonality) hinge on a tension peculiar to each river, section or even sector of the floodplain. Rocha et al. (2003) identify five components related to ecological processes in riparian ecosystems: magnitude, duration, periodicity, frequency and rate of alteration to the flux. As a result, fluctuations in the physical-chemical environment can be predicted on the basis of climatic and hydrosedimentological factors, without ignoring the fact that "The spatial and temporal patterns of the hydrology are in turn influenced by the topography, soils, and vegetation of the floodplain" (Mertes et al., 1995).

Hence, the river system to which the riparian wetlands belong can be conceived as a complex unit, with non-linear connections and interconnections produced by the transference of matter and energy among its component parts. In this sense, two features of complex systems present in riparian wetlands become crucial to understanding their physical structure and the way they operate: self-regulation and interconnectivity. These can be viewed as deriving from complex patterns of fluvial aspects encountered in the riparian environment that provide the key to monitoring and recovering such ecosystems.

2.1. Self-Regulation and Adjustability

In response to the adaptive processes set in motion by the flux of matter and energy, river courses freely adjust their geometric variables, i.e., depth, declivity, and width, including the development of meanders and other planimetric configurations. Knighton (1998) assesses rates the adjustment of the channel to independent variables according to four degrees of freedom: declivity of the riverbed (longitudinal profile), shape of cross-sections (lateral profile), configuration of the riverbed (vertical profile) and the channel's planimetry.

Study of the evolution of each of these four dimensions or adjustability planes comprises peculiar features, measurements and specific, very varied methodological procedures (American Society of Civil Engineers – ASCE, 1988; Babinski, 1992; Darby and Wiel, 2003; Lewin and Manton, 1975; MacCullah and Gray, 2005). Although they can be examined separately, they should not be considered independent of each other. Hence, adjustments to the internal geometry of the river system involve an array of variables, the interdependence of which is not very clear owing to the difficulty of isolating one variable's effect on the remainder.

Attention should, however, be drawn to planforms (vertical *orthographic projection*s of objects on a horizontal *plane*) as an analytical dimension. This adjustability plane refers to aerial perspectives shot from above the channel. This means that on the drawing board rivers display features and match types intrinsically associated with processes of erosion, transportation and depositing at work on the riverbed, on the banks and in the overflow areas (Amsler, 2004). The planform is thus one outcome of the adjustability of all the other planes, producing a great variety of standard forms that contribute to the evolution of riverine environments. Typology for classifying channel patterns is extensive but three main types can be identified: straight, meandering and intertwining (Simon and Castro, 2003).

On the one hand, the channel pattern can be conceived as an amalgam of responses to alterations in dependent and independent variables and to those deriving from human intervention. At the same time, though, the pattern shapes the floodplain resulting from sediments the channel deposits and from seasonal or continuous flooding by the flow of the river. The floodplain and its geomorphic features are relatively constant over time with major adjustment to hydrological surges. This differentiates them from the classical definition of ecotones as transition zones, as mentioned above. In actual fact, a wealth of ecotones can be identified, particularly in riparian wetlands.

Any natural modification to a river (e.g. climate changes) or alterations resulting from engineering work can destabilise the river system if the pattern of flow is disturbed. Profound changes to the characteristics of the channel and the floodplain will ensue and may adversely affect biodiversity and the potential economic exploration of existing resources. It should be added that flow characteristics are the most widely used and appropriate indicators for assessing the integrity of a riverine ecosystem on account of certain restrictions encountered.

Besides the fact that the morphology of the channel and the alluvial plain is shaped by fluvial processes governed by the flow of the river – high levels of flux in particular – many other abiotic characteristics are affected by flow conditions, including water temperature, distribution of the size of sediments suspended in the water and deposited on the riverbed. Likewise, contrasting with the scarcity of information available and the precarious resolution of temporal series for biological data, the availability of daily fluviometric data can provide a clear picture of the natural variability and the recent history of anthropological disturbance in rivers (Richter *et al.*, 1997, *apud* Rocha *et al.*, 2005). Attention should thus be focused on data parameters for forces acting on the edges of the channel, represented by the flux of the current and the forces resisting the flow, already extensively reported on in the literature (Bordette *et al.*, 1998; Knighton, 1998; Leopold *et al.*, 1994; Thorne *et al.*, 2003). Similarly, analysis of the channel pattern (i.e. its planform) is based on measurements that help describe and quantify active processes in the light of variables such as discharge, transportation of sediment, and stability of the channel banks, all of which are crucial parameters for wetland management.

One way of approaching river channel processes and morphology is by employing the concept of stream power (Knighton, 1999). The energy moving the river system starts out as potential energy. As the flow moves downstream, the potential energy is progressively converted into kinetic energy, thus triggering erosion and transportation. Hence, if one is to understand the behaviour of the alluvial channel when intervention occurs, one must identify the factors directly controlling the balance or imbalance between force and resistance. Generally speaking, if force and resistance are balanced, the river current will neither erode nor deposit sediment, transporting the sediment brought from upstream and thus maintaining the existing pattern and its attendant features. Such balance indicates stability conditions in the dimensions of the channel and can be mathematically expressed as the stream power proportionality (Lane, 1995; Simon and Castro, 2003):

$$QS_b \propto Q_s D_{50}$$

Where Q is discharge, S_b declivity, Q_s the material deposited on the riverbed and D_{50} average size of the material on the riverbed, indicating that 50% of the material is comprised by sediment of a particular granulometric size. This equation suggests that if the energy of the flow is increased by added discharge Q or by the gradient in the channel S_b, an excess in the proportion of energy flow relative to the discharge of sediment Q_s from the riverbed will result, resistance being proportional to the diameter of particles D_{50}. In this case, an additional quantity of riverbed material will be eroded.

Similar response can be expected to a reduction in resistance to erosion on the edges of the channel or to a diminishment in the size of the particles encountered on the riverbed D_{50} (assuming, of course, it is not cohesive). Conversely, a reduction in flux energy brought about by an increase in the size or discharge of riverbed sediment will produce agradation of the channel's bed. This will depend on where forces are most effective: gravity, should the channel walls be less resistant, or flux, should the bed be less resistant and lowering or incising of the thalweg take place.

Indeed, erosion of the banks plays a major role in controlling and shaping the migration of channels. This is especially true in sections where meanders predominate. Erosion likewise determines the way the alluvial plain develops and is maintained. Riverbank erosion poses a serious challenge for riparian environment management. Consequent loss of potential farmland and endemic species can be acute.

Stream power provenly affects many aspects of river systems, including riverbed shape, channel pattern, and processes like rate of sediment transportation and channel migration. It is also used as a criterion for explaining balance in the development of channel shapes (Knighton, 1999). It should, however, be stressed that, useful as it may be in working out geomorphic classifications, this concept is usually employed in studies dealing with smaller spatial scales, not with reference to the spatial resolution analysis of channel patterns requires.

Adjustments to the internal geometry of river systems produced by flow patterns and alterations to them actually involve a large number of variables. The interdependence of these variables is not very clear given the difficulty of isolating a single variable's effect on the remainder. In other words, flow geometry and channel geometry interact in space and time on the basis of circular feedback processes. It is these processes that account for the formation and maintenance of the floodplain and consequently of the riparian environment. Given the river system's self-regulation and mechanisms for adjusting its geomorphic features, our aim here is to acknowledge the planform as an analytical tool for monitoring riparian wetlands. Gradual evolution and sharp changes observed in the channel pattern can serve to diagnose the conditions under which the river system operates and thus to assist maintenance of its ecosystems.

2.2. Interconnection and Stability

In view of the considerations made hitherto, it is equally important to establish the way the river system's self-adjustment and the degree of interconnection between the sub-units, main channel and floodplain interact. Alterations to the flux of matter and energy between these subsystems can make for disruptions in transmission and conservation of complex information affording a wealth of components and structures (Bayley, 1995; Capra, 2007; Ward and Stanford, 1985). Besides determining processes of erosion, transportation and depositing performed by the river current, the type of flow establishes the spatial pattern of flooding in riparian ecosystems. Like the river channel itself, the floodplain also contributes water and sediment from the basin. As a result, the flow between channel and floodplain is frequently reversed in riparian ecosystems during low-water or reflux phases and periods of drought. Both subsystems are inter-linked by tributaries and secondary channels, so proving the existence of two-directional flow of matter and energy (Mertes et al., 1995; Sklar, 2000). There should thus be keen interest in observing alterations such as the lowering of the riverbed, emergence and incorporation of sandbars on the banks obstructing connecting channels that link the river to its floodplain, among others (Manyari, 2007).

Studies on the need to maintain lateral connectivity cover river systems from all continents across a wide range of biomes (Adams, 1985; Ligon, 1995; Makkaveyev, 1972). An illustrative example of a large hydrographic system in South America being examined can be found in Agostinho and Zalewski (1995), who studied the Paraná River, focussing on the importance of lateral migration of fish to the alluvial plains, essential for the reproduction of many species.

Regulation of hydrological surges caused by the installation of dams, which impairs the natural variability of discharge, similarly generates an interruption of the lateral connectivity between river channel and floodplain, be it by erosion and silting of the channel modifying its internal geometry and pattern or by the elimination of major floods. These effects engender a state of instability and degradation in both the fluvial and riparian ecosystems. It is easy to perceive that the wealth of species is associated as much with variability of physical parameters and diversity of habitats pertaining to the floodplain environments as with the course of the river determined by hydrological fluctuations (Salo, 1986). Hence, if the flow geometry describes a spatial pattern of flooding that correlates in time and space with periods of drought, flooding, high water and reflux of the river course and floodplain (Lewin and Manton, 1975), monitoring of the area flooded deserves attention as a wetland conservation measure. In that case, the natural flow regime should be taken as a paradigm for conserving and restoring river systems, as demonstrated in Neiff (1990) and Poff *et al.* (1997), who relate ecological integrity to the dynamic nature of fluvial and, by extension, riparian ecosystems.

In these riparian environments, it is clear that the function of stability – a premise for conservation – is distinct from that which applies, for instance, to climax forests, as Neiff (2008) points out. He stresses the fact that in these highly variable ecosystems, stability is determined by the system's capacity not to maintain but to recover its balance. The high degree of variability (magnitude, duration and recurrence of hydrological surges) does not make wetlands unstable. Proof of this is that they maintain a standard landscape of biotic communities and dominant physiognomies for both low-water and high-water phases. This means that if, on the one hand, multiple basic states (i.e. various states of balance) are observable, on the other, macroscopic order tends to be robust and stable in the face of modifications to the system's internal order. It should further be underscored that stability is essentially obtained by the capacity to mould assembly structures (understood as natural sets of populations) to states of change in a sinusoidal variability system, as shown in the graph.

The variability river systems display in morphology and states of balance in time and space is thus abundantly clear. According to Knighton (1998), at best, rivers attain an approximate state of equilibrium on an intermediate time scale between short-term fluctuations and long-term evolutionary trends. This holds true for maintenance of the channel's geometric regularity adjusted to independent variables. Hence, maintenance of alluvial plains – understood as a morphological subsystem – depends on maintenance of the variability of hydrological surges: the more variable their spectrum, the more varied the gradients produced by gradual variations introduced in the physical and chemical environment.

In sum, as open systems, the river system's dynamics develop on the basis of a range of possible configurations not characterised by highly precise scales of time and space. This property is referred to as self-organised criticality, and the dynamics of the systems triggering it operate on the verge of chaos. It can equally be observed that "Although the standard is optimal global (a state of minimal energy), it is pointless expecting these systems to attain it."

3. REMOTE SENSING FOR WETLANDS MAPPING

The dynamics of wetlands environment is extremely complex, given its variations in short intervals of space and time, when compared to other terrestrial environments. The spatial and temporal patterns of flood inundation have a crucial role in the wetland ecology. Thus, the need of mapping and monitoring the surface (intrinsic vegetation processes, fluvial or coastal dynamic, land-use changes and flood inundation patterns), over a range of spatial and temporal scales, is of paramount importance for assessing the status of wetland ecosystems. In this context, the remote sensing becomes a powerful tool for inventorying and monitoring those environments due to its capability of generating a large amount of information on surrounding land uses and their changes over time (vegetation phenology and wetland losses by human-induced changes).

Remote sensing data offer as advantages: (a) synoptic estimation over large areas, appropriate to analyze the scaling of wetland ecosystems; (b) a rapid, non-destructive and cost-effective inventory of the wetland landscape over varied spatial and temporal scales; (c) estimation of wetlands properties in remote and inaccessible areas; and (d) multi-temporal data (winter/spring and summer) that shows seasonal variability in water regime and vegetation status.

In the literature on satellite remote sensing of wetlands, studies have been conducted in different types of wetlands by using several satellites and classification techniques (Ozesmi and Bauer, 2002, Lulla, 1983).

Thus, satellite remote sensing has been used in order to map the following wetland environments: permanently flooded or intermittently exposed open water ponds (FGDC 1992), large wetland ecosystems (i.e. Chopra et al. 2001, Fuller et al. 1998); coastal wetlands (i.e. Hardisky et al. 1986, Hinson et al., 1994, Jensen et al., 1993); forested wetlands (i.e. Hodgson et al., 1987, Townsend and Walsh, 1998); forested wetlands (Ernst and Hoffer 1979; Ernst-Dottavio et al. 1981; Llewellyn et al. 1996; Sader et al. 1995); and, inland freshwater marshes (i.e. Ernst-Dottavio et al. 1981).

Both optical and microwave sensors are applied in wetland studies, where each technology describes advantages and disadvantages. Amongst the types of optical sensors, it can be mentioned: Landsat MSS, Landsat TM, SPOT, AVHRR, Indian Remote Sensing Satellite (IRS-1B) and Linear Imaging Self-scanning Sensor (LISS-II) (Ozesmi and Bauer, 2002; Lulla, 1983; Hardisky et al., 1986). The optical sensor is suitable for vegetation and land-cover mapping, phenology and flooding study during leaf-off periods.

However, there are some disadvantages of optical remote sensing estimation, such as: (a) the apparent upwelling radiance remotely measured is not only a function of water and vegetation properties, but also a function of data acquisition geometry and atmospheric conditions; (b) the information is limited to the surface, what results in inability to detect flooding in forested wetlands because of the presence of dense canopies; (c) the spectral, temporal and radiometric resolution can be inadequate; and, (e) the occasional presence of cloud and smoke cover.

Radar data is more appropriate for flood inundation mapping than the optical data. Microwave sensor has proved to be an effective tool for identifying differences between flooded and non-flooded regions, including the detection of the flooding beneath forest canopies (Hess et al., 1990, Hess and Melack, 1994, Townsend and Walsh, 1998). Besides that, the radar has as advantages the data collection at any time of day and under almost any weather conditions (cloud cover).

Ozesmi and Bauer (2002) reviewed the different classification techniques that have been used to identify wetlands. Amongst those it can be mentioned: visual interpretation; unsupervised classification or clustering; principal component analysis; supervised classification; hybrid classifications; vegetation indexes; mixture estimation; spectral mixture analysis; sub-pixel classification and rule-based classifiers.

Amongst the remote sensing studies in wetlands, great progress has been reported for multi-temporal analysis. The high coverage repetition of the satellite data enables seasonal and yearly monitoring of the wetland systems. There are two types of temporal remote sensing data commonly employed in wetland vegetation studies: (a) discrete snapshots with annual data to analyze spatial changes in land cover and (b) seasonal data to infer trends and dynamics of vegetation phenology (growth cycles). The former, is described by Munyati (2000) that used multi-temporal data over a ten-year period to assess a spatial reduction in area of dense green vegetation in upstream sections of the Kafue Flats floodplain wetland system in southern Zambia. The latter, is described by Johnston and Barson (1993) that considered images throughout one year to describe seasonal variability of water dynamic and vegetation pattern.

The Amazon River and its large tributaries present a floodplain region of about 300,000 km^2 (Iron et al., 1997), which represents approximately 56% of floodplain areas on the globe (Mitsch and Gosselink, 2000). The floodplain is a determinant factor for fish habitat quality (Gomes et al., 2006), occurrence of economically significant tree species (Parolin, 2000), and agricultural potential (Barrios et al., 2004). These floodplains are built by the formation of bars and the accumulation of sediment carried in diffuse overbank flows and channelized flows (Dunne et al., 1998). The periodic oscillation between terrestrial and aquatic phases on floodplains causes a constant exchange of the sediments and waters between river channels and floodplains. Floodplains affect erosion, transport and sedimentation processes in the watershed system (Junk, 1997). During residence of water on floodplains, substantial biogeochemical modifications occur under the influence of sorption and redox reactions and biotic processes (Melack et al., 2004; Richey et al., 1988, 1990; Seyler and Boaventura, 2003).

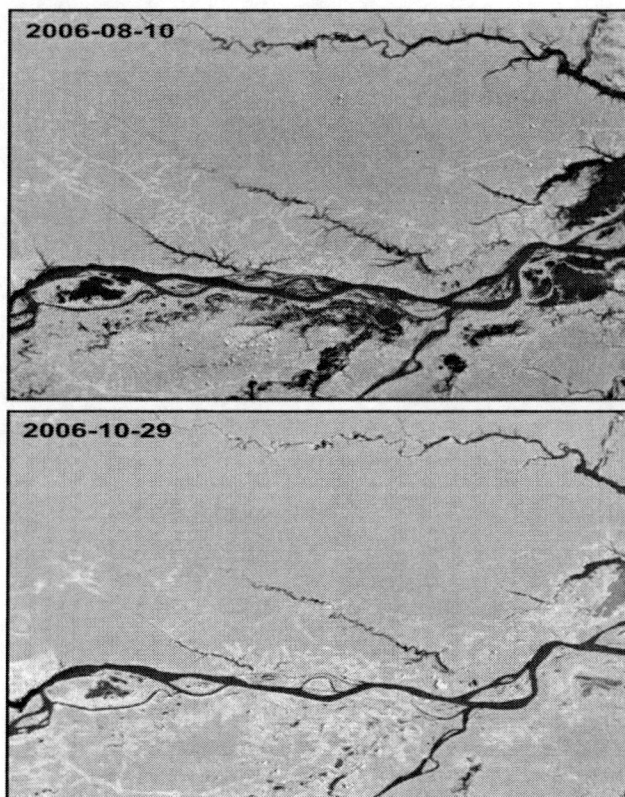

Figure 1. TM-Landsat sensor image of the annual water level variations between high and low water stages in the Amazon River.

The water storage in these wetlands and its outflow represent a significant part of the water balance in the basin (Alsdorf et al., 2001; Richey et al., 1989). Figure 1 highlights the differences between high and low water stages by using remote sensing orbital products over large Amazonian floodplain.

Sedimentation rates in the Amazonian floodplains vary considerably, both temporally and spatially (Mousinho de Meis, 1971; Filizola, 1999; Bush et al., 2000). Moreira-Turcq et al., (2004) argue that probably the sedimentation rate is directly related to the geographic location, proximity to the Amazon River channel, and duration of connection between floodplain and river.

The Amazon floodplains can be divided into nutrient-rich whitewater (*várzea*) and nutrient-poor blackwater (*igapó*) ecosystems according to the water quality of the flooding rivers (Prance, 1979). Fertile soils and nutrient-rich waters characterize the *várzea*, which becomes into a productive region within the Amazon Basin (Fiore et al., 2005).

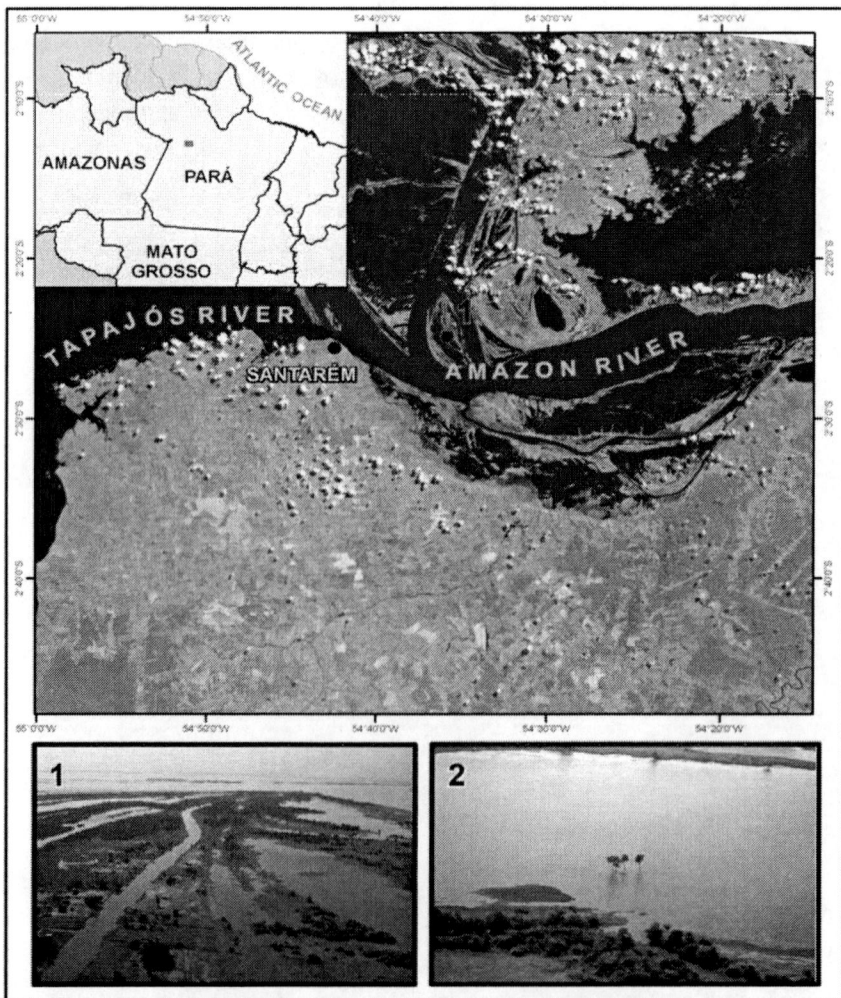

Figure 2. Satellite image from TM-Landsat sensor depicting the floodplain during flooding period in the Amazon river near Santarém city.

The Amazon floodplain monitoring at basin scale is only possible by using remote sensing data, due to the extent and inaccessibility of the inundation areas. Figure 2 presents an optical remote sensing image of the diversity of landscape units with open lakes, bogs, large meadows, alluvial forests and land forests. Another example of the central Amazon flood plain is Jarauçu river near the city of Porto de Moz presented in Figure 3. This figure shows palisade-huts of the riverine people adapted to the local hydrological floodplain conditions.

However, in the Amazon region, the SAR systems are most used because the optical sensor images are strongly affected by cloud cover and cloudless scenes are sparse (De Costa et al. 1998). Thus, several studies using SAR provide a basis for improved estimates of the contribution of wetlands to biogeochemical and hydrological processes in the Amazon.

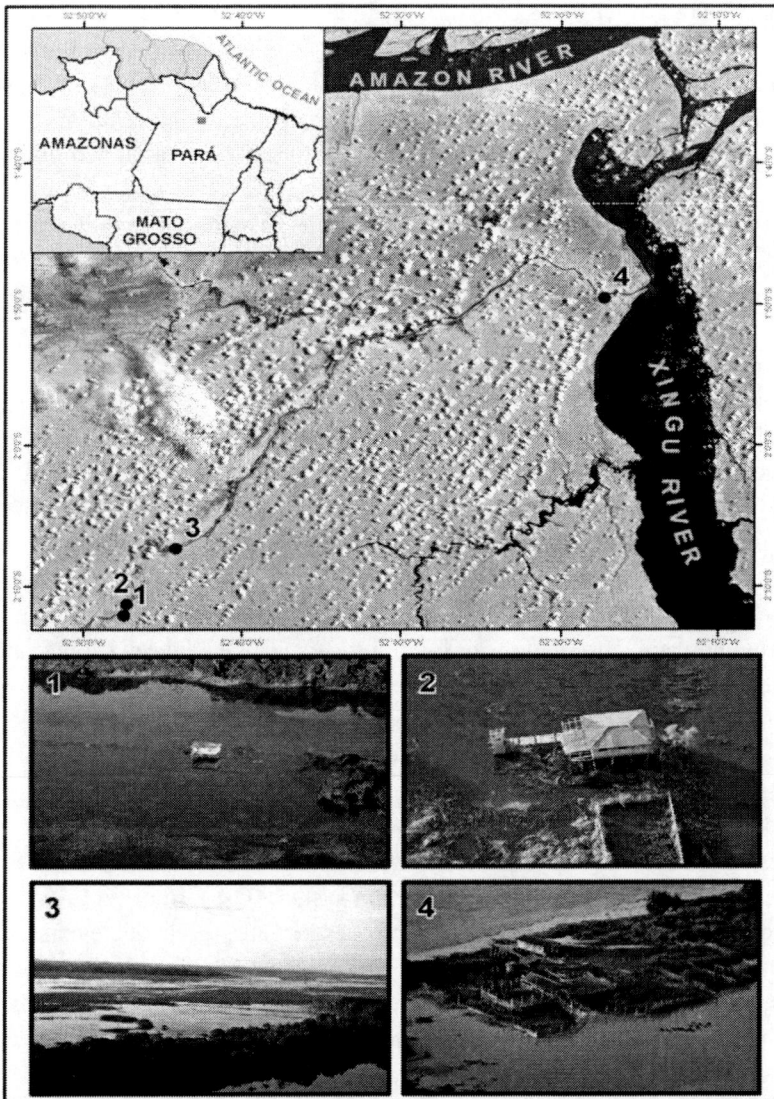

Figure 3. Satellite image from TM-Landsat sensor of Jarauçu River near the city of Porto de Moz in Pará state. The pictures show floodplain (várzeas) areas with the presence of palisade-huts of the riverine people adapted to the local hydrological conditions.

Wang et al. (1995) simulated the radar backscatter from a floodplain forest with a flooded or non-flooded ground condition at C-band (5.7 cm wavelength), L-band (24 cm wavelength) and P-band (68 cm wavelength). The flooded /non-flooded backscatter ratio was higher: (a) at HH polarization than at VV polarization; (b) at small incidence angles than at large incidence angles; and (c) at a long wavelength than at a short wavelength. The increase of the volumetric moisture from 10% to 50% in non-flooded caused the decrease of the flooded /non-flooded backscatter ratio (small at C- and L-band but large at P-band). The leaf area index (LAI) had a large effect on the simulated C-band (not L-band or P-band) flooded /non-flooded backscatter ratio.

Several studies have reported the classification of the radar images over the Amazon region to discriminate main land cover types and map flooded areas (Hess et al., 2003; Saatchi et al., 2000; Sippel, 1998; Siqueira et al., 2003). Thus, spatial variations in radar pulse amplitudes are accessed for characterization of the floodplain habitat and area of inundation (Hess et al. 1995; 2003). These studies presented the following patterns during Amazon recessional flow conditions: (a) open water yields the lowest radar returns; (b) flooded forest produces the strongest returns; and (c) flooded herbaceous vegetation and non-flooded forest produces intermediate L-band backscatter amplitudes.

Hess et al. (1995) used multi-frequency polarimetric synthetic aperture radar (SAR) data from SIR-C missions and a decision-tree classifier to map the floodplain inundation and vegetation along the Negro and Amazon rivers. Backscattering statistics indicate that both "C" and "L" band are necessary for accurate delineation of herbaceous versus woody and flooded vs. non-flooded cover types.

Hess et al. (2003) mapped wetlands in the Amazon region using L-band synthetic aperture radar (SAR) imagery from the Japanese Earth Resources Satellite-1 (JERS-1). The mapping included segmentation and clustering validated by high-resolution digital videography with an accuracy of 95%. Five cover classes: non-vegetated, herbaceous, shrub, woodland and forest combined with inundation sate (flooded or non-flooded) resulted in 10 possible covers.

Remotely sensed observations of the water surface provide an alternative to permanent gauging, providing new observations of hydrologic exchange between tributaries, floodplains, and mainstem rivers. Recent work has shown that interferometric processing and analyses of spaceborne synthetic aperture radar (SAR) described stage centimeter-scale changes for vegetation covered floodplain lakes and tributaries in flooded forests (Alsdorf et al., 2000, 2001, 2003; Birkett et al., 2002).

Alsdorf et al. (2003) examine the spatial properties of these interferometric measurements and demonstrate that flow routes across the floodplain can be automatically extracted from SAR imagery by using a geographic information system (GIS). In central Amazon floodplain localities, these authors observed one-day decreases in water levels of eleven centimeters. The spatial distribution of the water level is not homogeneous, because the flow through the floodplains includes a maze of interconnected lakes, channelized and non-channelized overbank flow, pans, and swales. Thus, water-level changes are inversely related to lengths of flow paths connecting water bodies to the main channel; that is, long distances result in small water-level fluctuations compared to short paths (Alsdorf et al., 2003).

Recently, studies have evaluated time series of SAR images in Amazon floodplain. Frappart et al., (2005) estimated both flooded area and water levels and determined the water volume stored in the floodplains of the Negro River, during the 1995–1996 seasonal cycle, by using combined observations from the Synthetic Aperture Radar (SAR) onboard the Japanese Earth Resources Satellite (JERS-1), the Topex/Poseidon (T/P) altimetry satellite, and in-situ hydrographic stations. Martinez and Toan (2007) used time series of SAR images (21 images acquired by JERS between 1993 and 1997) to map the flood temporal dynamics and the spatial distribution of landscape units (open lakes, bogs, large meadows, savannahs, alluvial forests and terra firma forest).

4. COMPLEXITY: APPLICATION TO GREENHOUSE GAS EMISSIONS STUDY BY "ARTIFICIAL WETLANDS"

It is likewise interesting to refer to the creation of "artificial wetlands" when reservoirs are formed in forested areas to generate hydroelectric energy. The emission of greenhouse gases peculiar to submersed vegetation is related to the current problems of global warming and climate change. Experimental studies have advanced considerably, supplying data that require statistical analysis involving mathematical and computational modelling if one is to understand the main processes involved in the emission of gases in Brazil's hydroelectric reservoirs, especially in the Amazon. The problem is modelling experimental measurements of gas emissions: methane (CH_4) and carbon dioxide (CO_2) issuing from reservoirs in the form of bubbles or by dispersal. What is sought is the reservoir's overall rate of emission and a total estimate for a set of reservoirs. Experimental results are available for some but not for others.

The power law[1], typical of phenomena described as self-organised criticality, whose fractal profile is generally of the intensity of the phenomenon over time, provides the number of events N per intensity I (expressed in the case we are examining in carbon mass or gas emitted per unit of area and time): $N(I) = AI^{-\lambda}$, A being a constant.

This law was applied to the statistical study of gas emissions by hydroelectric schemes (Dos Santos, 2000) for the following reasons: emissions measured are of a predominantly low intensity, some are of greater intensity and a few are very intense in terms of carbon mass per area in a unit of time ($kgC\ m^{-2}\ year^{-1}$); there are major variations in intensity of emissions in the same place at different times, from one place to another at the same time, and at each

---

[1] One type of phenomenon generally taken as a simple illustrative example of complexity is self-organised criticality, observable in a large mound of sand fed by a continuous flow of sand being poured onto it. The mound tends toward a relatively stable shape once it reaches a certain height, there being a critical value of inclination of its surface above which it loses stability and avalanches ensue. Below that degree of inclination the mound is sub-critical and all the sand added falls so as to recover that critical inclination. In other words, the system is self-organised around criticality. By relating the intensity of collapses (the quantity of sand collapsing) to the frequency with which they occur, one obtains a power law according to which the number of collapses observed (N) diminishes with intensity (I) raised to a characteristic power (small collapses occur more often than big ones) (Bak, 1995). Pressure on the grains of sand is maximal not at the centre of the base of the mound but rather in a ring some distance from the centre. The centre, paradoxically, is where least pressure occurs at the base of the mound due to a structural effect produced by highly complicated interaction between the grains of sand resting on each other in vast number. The effect is similar to that which gives arches in old buildings their structural stability. Nonetheless, the power law can be used to describe the collapse of the mound without taking account of the complicated interactions between the grains of sand that make pressure minimal at the centre of the base. Though it may be hard and painstaking work to model such interactions between vast numbers of grains, the power law makes it easy to predict the behaviour of the collapses.

point of the dam over time; a variety of factors affecting emissions in the reservoir coexist, including the time taken to complete and close the dam, the life and processes in the lake, the carbon transported from the hydrographic basin by the water, pumped from the atmosphere or removed from the soil, depth, the presence of oxygen, formation of an anaerobic layer on the bottom, the seasonal hydrological regime, temperature, wind, and discharge. It is thus difficult to obtain sufficient control to establish a causal relation, and the choice of one factor among so many others raises considerable uncertainty. The result of the statistical adjustment of the data from a set of measurements in Brazilian hydroelectric reservoirs has provided a reasonable correlation, of approximately 80%, using the simple power law (Dos Santos, 2000).

The importance of this problem is thrown into relief by the controversy surrounding studies carried out on reservoirs in the Amazon region (Rosa *et al.*, 1996). One extreme case is that of the Balbina hydroelectric scheme, whose emissions exceed those for coal-fired thermoelectric power plants producing an equivalent amount of energy (Dos Santos, 2000; Rosa *et al.*, 2004). The new result is the measurement of gas emissions downstream from the dam, immediately after the water has passed through the turbines, especially of methane (Kemenes *et al.*, 2007) found in high concentrations in water at the bottom of the reservoir (Rosa *et al.*, 2006). Kemenes *et al.* (2007) calculate the level of downstream emission of methane resulting from the turbines' ebullition flux and immediately below the Balbina dam by multiplying the discharge of turbinated water by the difference of methane gas concentration at the turbine water intake (which the authors measure at a depth of 30 metres) and the methane gas concentration 50 metres beyond the turbine outflow: $\Phi = Q (\rho_2 - \rho_1)$; Φ = ebullition flux; Q = turbine discharge; ρ_2 = concentration of CH_4 50 metres downstream from the dam; ρ_1 = concentration of CH_4 at the height of the water intake (upstream).

The authors took the maximum depth, 30 metres, as representative for calculating the discharge of methane where there is a greater concentration of methane: "*Water upstream from the dam was collected at a depth of 30m using a RUTNER bottle*" (Kemenes *et al.*, 2007). The point is that "*... it is well known that environmental conditions in deep water tropical lakes favour methanogenesis and that concentrations of dissolved methane increase significantly with depth in tropical reservoirs*" (Galy Lacaux *et al,.* 1999). Reviewing the data for the Balbina scheme, it can be observed that water intake ranges from depths of 30 metres (lower limit) to about 15 metres (upper limit) with a diameter of approximately 15 metres. So the concentration of methane to be used in calculating ebullition flux should be average concentration within this depth range and not at 30 metres, as were the measurements the authors employed. Depending on the profile, the average concentration of methane in the water varies from values close to zero at a depth of 15 metres to ~245 μmol of methane at 30 metres in the dry season; and from 5μmol at 15 metres to 125μmol at 30 metres in the rainy season. A simple calculation of the basis of these profiles provides an average concentration of methane in the water at depths between 15m and 30m in the dry season equal to about 40% of the value for 30 metres, and in the rainy season equal to about 30% of the value for the same depth. Were these averages to be applied, they would produce an average annual concentration of methane ρ_1 at the intake slightly above the average ρ_2 concentration the authors recorded 50m downstream from the dam, which is roughly half the ρ_1 concentration they measured.

Therefore, as the ebullition flux of turbining is not measured directly but rather calculated by subtracting from two sums measured with overlapping experimental errors, the result these authors obtained is highly sensitive, requiring more rigorous statistical analysis.

5. CONCLUSION

Given the foregoing, we trust it is now clear that riparian ecosystems should be interpreted on the basin-river-plain scale through long temporal series of data. On this space-time scale, it has stated that wetlands are peculiar, highly stable ecosystems. Likewise, in view of this interpretation scale, research, monitoring and conservation of these ecosystems require specific strategies, remote sensing figuring as an efficient technology for this purpose. Besides its capacity for covering large tracts of land and manipulating a large volume of data, channel planimetry combined with the prospect of taking satellite images is a significant dimension of analysis that allows fundamental properties of complex systems, self-regulation and interconnectivity to be highlighted.

ACKNOWLEDGEMENTS

The authors especially acknowledge Leonardo Figueiredo de Freitas and Arthur Manoel Cavalcante de Souza of the Laboratório de Sistemas de Informações Espaciais, for a critical and helpful review of this paper.

REFERENCES

AB'Saber, A. N. 2002, Bases para o estudo dos ecossistemas da Amazônia brasileira. *Estudos Avançados*. vol. 16 (45), 7-30.

Adams, W. M. 1985, The downstream impacts of dam construction: a case study from Nigeria". *Transactions of the Institute of British Geographers*. New Series, vol. 10 (3), 292-302.

Agostinho, A. A. and Zalewski, M. 1995, The dependence of fish community structure and dynamics on floodplain and riparian ecotone zone in Paraná Rivers, Brazil. *Hydrobiologia,* 303, 141-145.

American Society of Civil Engineers-ASCE. 1988, Task Committee on Bank Erosion and River Width Adjustment: Processes and mechanisms. *Journal of Hydraulic Engineering*, 124, vol. 1, 881-902.

Amsler, M. L. 2004, Dinâmica de processos fluviais: processos e morfologias. Curso Internacional de Pós-Graduação. 15-25 de out de 2004. 64 f. Porto Rico, Paraná. Nupélia. Universidade Estadual de Maringá.

Alsdorf, D. E. 2003, Water storage of the central Amazon floodplain measured with GIS and remote sensing imagery. *Annals of the Association of American Geographers*, 93(1), 55–66.

Alsdorf, D. E., Melack, J. M., Dunne, T., Mertes, L. A. K., Hess, L. L., and Smith, L. C. 2000, Interferometric radar measurements of water level changes on the Amazon floodplain. *Nature*, 404, 174–177.

Alsdorf, D. E., Smith, L. C., and Melack, J. M. 2001, Amazon floodplain water level changes measured with interferometric SIR-C radar. *IEEE Transactions on Geoscience and Remote Sensing*, 39(2), 423–431.

Babinski, Z. 1992, Hydromorphological consequences of regulating the lower Vistula, Poland. *Regulated Rivers: Rresearch and Management*, v. 7, pp. 337-348.

Bak, P. *How nature works, The science of self organized criticality*. University of Oxford Press, 1995.

Barrios; E.; Herrera, R. and Valles, J.L. 2004, Tropical floodplain agroforestry systems in mid-Orinoco River basin, Venezuela. *Agroforestry Systems*, 28(2): 143-157.

Bayley, P. B. 1995, Understanding large river-floodplain ecosystems. *BioSciense*, vol 45 (3), 153-158.

Bertalanffy, K. L. von. *Teoria Geral dos Sistemas*. Petrópolis vozes. 1975.

Birkett, C. M., Mertes, L. A. K., Dunne, T., Costa, M. H., and Jasinski, M. J. 2002 Surface water dynamics in the Amazon Basin: Application of satellite radar altimetry. *Journal of Geophysical Research*, 107(D20), 8059–8080.

Bordette, G. et al. 1998, Ecological complexity of wetlands within a river landscape. *Biological Conservation*, vol.85, 35-45.

Bush, M.B., Miller, M.C., Oliveira, P.E., Colinvaux, P.A., 2000, Two histories of environmental change and human disturbance in eastern lowland Amazon. *The Holocene* 10 (5), 543–553.

Campbell, D. et al. *Nonlinear phenomena*. N. Holland, Amsterdam, 1983.

Capra, F. *A teia da vida: uma nova compreensão dos sistemas vivos*. São Paulo, Ed. Cultrix. 1996.

Chopra R., Verma V.K., and Sharma P.K. 2001, Mapping, monitoring and conservation of Harike wetland ecosystem, Punjab, India, through remote sensing. *International Journal of Remote Sensing* 22: 89–98.

Darby, S. E. and Wiel, M. J. Models in Fluvial Geomorphology. In: Kondolf G. M. AND Piégay, H. (ed), *Tools in Fluvial Geomorphology*, West Sussex, England, Wiley. 2003. 503-537.

Dos Santos, M. A. *Inventário de emissões de gases de efeito estufa derivadas de hidrelétricas*. Rio de Janeiro, Tese de Doutorado, COPPE / UFRJ, p. 147, 2000.

Dunne, T., Mertes, L.A., Meade, R.H., Richey, J.E., Forsberg, B.R. 1998, Exchanges of sediment between the floodplain and channel of the Amazon River in Brazil. *GSA Bulletin* 110 (4), 450–467.

Ernst-Dottavio C.L., Hoffer R.M. and Mroczynski R.P. 1981, Spectral characteristics of wetland habitats. *Photogrammetric Engineering and Remote Sensing* 47: 223–227.

Filizola, N. 1999, *O fluxo de sedimentos em suspensão nos rios da Bacia Amazônica brasileira*. Publ. ANEEL (Agencia Nacional de Energia Eletrica), Brasilia, Brazil. 63 pp.

Fiore, M.F., Neilan, B.A., Copp, J.N., Rodrigues, J.L.M., Tsai, S.M., Lee H., Trevors, J.T. 2005, Characterization of nitrogen-fixing cyanobacteria in the Brazilian Amazon floodplain. *Water Research* 39(20): 5017-5026.

Frappart, F.; Seyler, F.; Martinez, J.M.; León, J.G.; Cazenave, A. 2005, Floodplain water storage in the Negro River basin estimated from microwave remote sensing of inundation area and water levels. *Remote Sensing of Environment* 99, 387 – 399.

Fuller R.M., Groom G.B., Mugisha S., Ipulet P., Pomeroy D., Katende A. et al. 1998, The integration of field survey and remote sensing for biodiversity assessment: a case study in the tropical forests and wetlands of Sango Bay, Uganda. *Biological Conservation* 86: 379–391.

Gay-Lacaux, C., Delmas, R., Kouadio, J., Richard, S., Gosse, P. 1999, Long-term greenhouse gas emissions from hydroelectric reservoirs in tropical forest regions, *Global Biogeochem. Cycles* 13, 503–517.

Gomes, L.C., Chagas, E.C., Martins-Junior, H., Roubach, R., Ono, E.A., Lourenço, J.N.P. 2006, Cage culture of tambaqui (Colossoma macropomum) in a central Amazon floodplain lake. *Aquaculture* 253: 374-384.

Hardisky M.A., Gross M.F. and KlemasV. 1986, Remote sensing of coastal wetlands. *BioScience* 36: 453–460.

Hey, R. D. River engineering and management in the 21st Century. In Thorne, C. et al. (ed), *Applied fluvial geomorphology for river engineering and managemet*, West Sussex, England, Wiley. 1997. 3-11.

Hess, L. L., Melack, J. M., Filoso, S., and Wang, Y. 1995, Delineation of inundated area and vegetation along the Amazon floodplain with SIR-C synthetic aperture radar. IEEE *Transactions on Geoscience and Remote Sensing*, 33, 896– 904.

Hess, L. L., Melack, J. M., Novo, E.M.L.M., Barbosa, C. C. F., and Gastil, M. 2003, Dual-season mapping of wetland inundation and vegetation for the central Amazon basin. *Remote Sensing of Environment* 87(4), 404– 428.

Hess, L.L., Melack, J.M., 1994. Mapping wetland hydrology and vegetation with synthetic aperture radar. *Internacional Journal of Ecology Environmental Sciences* 20: 197-205.

Hess, L.L., Melack, J.M., Simonett, D.S., 1990, Radar detection of flooding beneath the forest canopy: a review. *International Journal of Remote Sensing* 11 (7), 1313-1325.

Hinson J.M., German C.D. and Pulich W. Jr 1994, Accuracy assessment and validation of classified satellite imagery of Texas coastal wetlands. *Marine Technology Society Journal* 28: 4–9.

Hodgson, M.E., Jensen, J.R., Mackey, H.E. Jr., Coulter, M.C., 1987, Remote sensing of wetland habitat: a wood stork example. *Photogrammetric Engineering and Remote Sensing* 53 (8), 1075-1080.

Jensen, J. R., Cowen, D. J., Althausen, J. D., Narumalani, S. and Weatherbee, O. 1993, An evaluation of the Coast Watch change detection protocol in South Carolina. *Photogrammetric Engineering and Remote Sensing* 59 (6): 1039–1046.

Johnston, R.M. and Barson, M.M. 1993, Remote sensing of Australian wetlands: An evaluation of Landsat TM data for inventory and classification. *Australian Journal of Marine and Freshwater Research* 44(2): 235 – 252.

Kauffman, S. *At home in the universe*, Oxford Univ. Press, 1995.

Kemenes, A., Forsberg, B., Melack, J. M. Methane release below a tropical hydroelectric dam. *Geophysical Research Letters*, 34, L12809, doi:10.1029/2007GL029479.2007.

Knighton, D. *Fluvial forms and processes, a new perpective*. London, Arnold. 1998.

Knighton, D. 1999, Dowstream variation in stream power. *Geomorphology*, 29, 293-306.

Krebs, C. J. *Ecology: The Experimental Analysis of Distribution and Abundance*. Haper Collins. 4th ed. 1994.

Lane, E. W. 1995, Design of stable alluvial channels. *Transactions of the American Society of Civil Engineers*, vol. 120, n. 2776, 1234-1260.

Leopold, L. B., Wolman, M. G., e Miller, J. P. *Fluvial Processes in Geomorphology*. New York, Dover Publications. 1994.

Lewin, J. e Manton, M. M. 1975, Welsh Floodplain Studies: The Nature of Floodplain Geometry. *Journal of Hydrologyy* n. 25, 37-50.

Ligon, F. K., Dietrich, W. E., Trush, W. J. 1995, Dowstream Ecological Effects of Dams, a Geomorphic Perspective. *BioSciense* vol. 45, n. 3, 183-192.

Lulla K. 1983. The Landsat satellites and selected aspects of physical geography. *Progress in Physical Geography* 7: 1–45.

MacCullah, J., AND Gray, D. 2005, *Environmentally sensitive channel and bank protection measures. National* Cooperative Highway Research Program. Report 544. Washigton, D.C.

Makkaveyev, N. I. 1972, The impact of large water engineering projects on geomorphic processes in stream valleys. *Soviet Geography: Review and Transactions*, n.13, 387-393.

Manyari, W. V. *Impactos a jusante de hidrelétricas: o caso da Usina de Tucuruí, PA*, Rio de Janeiro, Tese de Doutorado, COPPE / UFRJ, 2007.

Melack, J.M., Hess, L.L., Gastil, M., Forsberg, B.R., Hamilton, S.K., Lima, I.B.T., Novo, E.M.L.M., 2004, Regionalization of methane emissions in the Amazon basin with microwave remote sensing. *Global Change Biology* 10, 530–544.

Mertes, L. A. K. et al. 1995, Spatial patterns of hydrology, geomorphology, and vegetation on floodplain of the Amazon River in Brazil from a remote sensing perspective. *Geomorphology*. 13, 215-232.

Mitsch, W. J. and Gosselink, J. G. *Wetlands*. 2 ed. New York, Van Nostrand Reinhold.1993.

Moreira-Turcq, P.; Jouanneau, J.M.; Turcq B.; Seyler, P.; Weber, O.; Guyot J.L. 2004, Carbon sedimentation at Lago Grande de Curuai, a floodplain lake in the low Amazon region: insights into sedimentation rates. *Palaeogeography, Palaeoclimatology, Palaeoecology* 214 (2004) 27– 40.

Mousinho de Meis, M.R., 1971. Upper Quaternary process changes of the middle Amazon area. *Geological Society of America Bulletin* 82, 1073–1087.

Munyati C. 2000. Wetland change detection on the Kafue Flats, Zambia, by classification of a multitemporal remote sensing image dataset. *International Journal of Remote Sensing*, 21(9):1787-1806.

Neiff, J. J. 1990, Ideas para la interpretacion ecologica del Paraná. *Interciencia* 15 (6) 424-441.

Neiff, J. J. 2008, Planícies de inundação são ecótonos? http://ar.geocities.com./Mneiff/

Ozesmi, S.L. and Bauer, M. E. 2002, Satellite remote sensing of wetlands. *Wetlands Ecology and Management* 10: 381–402.

Parolin, P., 2000. Seed mass in Amazonian floodplain forests with contrasting nutrient supplies. *J. Trop. Ecol.* 16, 417–428.

Poff, N. L. et al. 1997, The natural flow regime: a paradigm for river conservation and restoration. *BioScience*, v. 47, n.3, 769-784.

Poincaré, H. 1894, Sur la Théorie Cinétique des Gaz. *Révue des Sciences Pures et Apliqées*.

Prance, G.T., 1979, Notes on the vegetation of Amazonia. III. Terminology of Amazonian forest types subjected to inundation, Brittonia, New York. *Bot. Garden* 31 (1), 26–38.

Prigogine, I. *As leis do caos*. São Paulo, UNESP, 2002.

Rocha, P. C. et al. 2003, Considerações sobre a variabilidade hidrológica do alto rio Paraná. In: *X Simpósio Brasileiro de Geografia Física Aplicada, Revista GeoUERJ*, n. 1. .2022-2031, Rio de Janeiro.

Richey, J.E., Devol, A.H., Wofsy, S.C., Victoria, R., Ribeiro, M.N.G., 1988, Biogenic gases and the oxidation and reduction of carbon in Amazon River and floodplain waters. *Limnol. Oceanogr.* 33 (4), 551–561.

Richey, J.E., Hedges, J.I., Devol, A.H., Quay, P.D., Victoria, R., Martinelli, L., Forsberg, B.R., 1990, Biochemistry of carbon in the Amazon river. *Limnology Oceanography*. 35 (2), 352–371.

Rosa, L. P., Santos, M. A., Matvienko, B., Santos, E. O., Sikar, E. 2004, Greenhouse gas emissions from hydroelectric reservoirs in tropical regions. *Climatic Change*, 66, 9-21.

Rosa, L. P., Schaeffer, R., Santos, M. A. 1996, Are hydroelectric dams in the Brazilian Amazon significant sources of greenhouse gases. *Environmental Conservation*, 66, n.1, 2-6.

Saatchi, S. S., Nelson, B., Podest, E., and Holt, J. 2000, Mapping Land-cover types in the Amazon Basin using 1 km JERS-1 mosaic. *International Journal of Remote Sensing*, 21(6and7), 1201– 1234.

Salo, J. et al. 1986, River dynamics and the diversity of Amazon lowland Forest. *Nature*, vol. 322, 254-258.

Seyler, P., Boaventura, G.R. 2003, Distribution and partition of trace metals in the Amazon basin. *Hydrological Processes* 17, 1345–1361.

Simon, A AND Castro, J. Measurement and analysis of alluvial channel form. In: *Tools in fluvial Geomorphology*; Kondolf, G. M. and Piégay, H., England, Wiley. pp. 291-322. 2003.

Sippel, S. J., Hamilton, S. K., Melack, J. M., and Novo, E. M. M. 1998, Passive microwave observations of inundation area and the area/stage relation in the Amazon River floodplain. *International Journal of Remote Sensing*, 19, 3055–3074.

Sklar, L. Report on Hydrological and geochemical processes in large scale river basins, 15-19 November 1999, Manaus, Brazil. Draft 1, 10 February, 2000.

Sternberg, H. 1998, *A água e o Homem na várzea do Careiro*. 2ª. ed. Belém, PA, Museu Emilio Goeldi, Coleção Friedrich Katzer. vol.1.

Thorne, C. R. et al. *Applied fluvial Geomorphology for river engineering and management*. UK, John Wiley and Sons. 2003.

Townsend, P. A. and Walsh, S. J. 1998, Modeling floodplain inundation using an integrated GIS with radar and optical remote sensing. *Geomorphology* 21 (3–4): 295–312.

Townsend, P. A. and Walsh, S. J. 2001, Remote sensing of forested wetlands: application of multitemporal and multispectral satellite imagery to determine plant community composition and structure in southeastern USA. *Plant Ecology* 157: 129–149.

Wang, Y., Hess, L. L., Filoso, S., and Melack, J. M. 1995, Understanding the radar backscattering from flooded and non-flooded Amazonian forests: Results from canopy backscatter modelling. *Remote Sensing of Environment*, 54, 324– 332.

Ward, J. V. and Stanford, J. A. 1995, Ecological connectivity in alluvial river ecosystems and its disruption by flow regulation. *Regulated rivers research and management*, vol. 11, 105-119.

In: International Wetlands: Ecology, Conservation...   ISBN: 978-1-60456-999-5
Editor: Jose R. Herrara   © 2009 Nova Science Publishers, Inc.

Chapter 12

EFFECT OF CONSTRUCTED WETLAND ON RIVER WATER QUALITY AND GENETIC DIVERSITY OF *PHRAGMITES AUSTRALIS*

S. Zhou, M. Hosomi and M. Nishikawa

Faculty of Engineering, Tokyo University of Agriculture and Technology,
2-24-16 Naka, Koganei, Tokyo 184-8588, Japan

ABSTRACT

Lake Kasumigaura is the second largest lake in Japan and well known for its eutrophication. Surface-flow wetlands (area: 38,000 m^2; 50 m × 40 m × 19 units) planted with common reed (*Phragmites australis*) were constructed near the estuary of the Seimeigawa River, which is a nutrient polluted river flowing into Lake Kasumigaura, to treat the river water. The hydraulic loading rate, reduction in nutrient contents, plant transition, and components of sediment were investigated after construction and further surveyed periodically from 1999 to 2005. Although nutrient removal efficiencies varied after several years running, particularly from spring to summer, considerable SS, nitrogen, and phosphorus amounts were retained by the constructed wetland in 2005 after 10 years running. The average total sediment accumulation was $3.61 - 4.43$ kg m^{-2} yr^{-1}. The total phosphorus content in sediment was similar or slightly more than the amount removed from the inflow river water, whereas the total nitrogen content in sediment was significantly less than that removed, suggesting that phosphorus was removed with particle sedimentation while most nitrogen was removed by denitrification.

On the other hand, the average community diversity index of four representative constructed wetland units was 0.88 in 2004, which was significantly higher than 10 years ago, indicating greater diversity. In addition, compared with the *P. australis* population in native wetland, *P. australis* in the constructed wetland had more gene types in 2005 probably owing to the constructed wetland being in an establishment stage, whereas the native wetland had already entered a stationary stage.

Keywords: Lake Kasumigaura, Constructed wetland, Nutrient reduction, Sediment accumulation, Plant transition, Genetic diversity

1. INTRODUCTION

Deterioration of the water quality of streams, rivers, and lakes has been an important issue in many countries due to nutrients being discharged from non-point sources, which includes farm runoff. Even though streams and rivers themselves are sometimes not greatly affected by increased concentrations of nutrients, most of these waterways eventually flow to inland ponds, lakes, and sea where eutrophication has become a problem. Constructed wetlands have been used as nutrient sinks or removal systems for polluted river water (Mitsch et al., 2005a; Reilly et al., 2000; Jing et al., 2001; Zhou and Hosomi, 2008) and farm runoff (Comin et al., 1997; Fink and Mitsch, 2004). The effects of constructed wetland on water quality are described from several aspects, which include the settling of particulates, transferring oxygen into the rhizosphere, nitrification and denitrification, adsorption and absorption of phosphorus, and reduction of other pollutant concentrations (Kadlec and Knight, 1996). The most commonly used plants in constructed wetlands are common reed (*Phragmites australis*), bulrush (*Scirpus spp.*), and cattail (*Typha spp.*), among which *P. australis* is one of the most important aquatic plant communities in Japan as well being widely distributed worldwide. It is a perennial emergent aquatic plant. The annual stems develop from a perennial rhizome system, which is responsible for the rapid vegetative expansion of the species in native wetland. It grows in a variety of wetlands such as littoral zones of lakes (e.g. Lake Biwa, the largest freshwater lake in Japan, Lake Kasumigaura, and Lake Teganuma), rivers (e.g. the Arakawa River) and wetlands (e.g. Kushiro-shitsugen) in Japan. It dominates or codominates as mono-specific or mixed stands with other plants (*Typha latifolia , Typha angustifolia , Typha orientalis , Miscanthus sacchariflorus, Zizania latifolia*), playing a multi-purpose role in nature conservation and environmental protection (Karunaratne et al., 2003). It is of ecological importance as habitat for shoreline protection (Coops et al., 1996), aquatic biodiversity conservation (Kira, 1991), and nutrient retention (Kovacs et al., 1978; Brix, 1994; Yuji and Hosomi, 2002).

Nowadays, there are thousands of constructed wetlands operating as wastewater and river water purification systems in Europe and North America. In Europe, horizontal sub-surface flow wetland is the most popular type (Vymazal, 2005) while in the United States, most constructed wetlands are free water surface flow wetland (FWSF) for treating domestic wastewater, farmland runoff, and river water (Kadlec and Knight, 1996; Mitsch et al., 2002; Fink and Mitsch, 2004). For example, Bachand and Horne (2000a, b) built six macrocosms (approximately 0.13 ha each) to study nitrogen transformation processes in the adjacent Prado Basin constructed wetland and attributed nitrate removal to denitrification. In addition, two created wetlands in the Mississippi River Basin were investigated periodically for 9 years to evaluate nitrogen retention capacity (Mitsch et al., 2005a). Furthermore, the National Research Council (1992) called for the creation and restoration of 4 million ha of wetlands in the United States by 2010. There are also dozens of constructed wetlands applied in polluted river or lake water purification. According to a survey made by Nakamura et al. (2007), one third of operating constructed wetlands have surface areas ranging between 1,000 m^2 and 10,000 m^2, and 20% of the wetlands have surface areas less than 100 m^2. All constructed wetlands are free water surface flow wetlands, except for one subsurface flow wetland. 60% of constructed wetlands are planted with *P. australis*, whereas others are planted with watercress, cattail, or water hyacinth. Most of these constructed wetlands received river or lake water with a hydraulic loading rate (HLR) over 0.45 m d^{-1} and hydraulic retention time (HRT) below 5 hours. Water depths are relatively shallow from 0.05 to 0.20 m. Concentration ranges of total nitrogen (T-N) and total phosphorus (T-P) are 0.43-4.04 mg L^{-1} and 0.02-0.70 mg L^{-1}, respectively. As a result, constructed wetlands in Japan receive influent containing relatively low concentrations of nutrients (N, P) and rather high concentrations of suspended solids (SS) compared with wastewater from a secondary or tertiary treatment. SS would be deposited in constructed wetland planted with macrophytes that also contribute to sediment accumulation. It has been reported that dense *Typha* has elicited greater sediment deposition and reduced re-suspension in open water zones (Anderson and Mitsch, 2006). However, in constructed estuarine wetlands, sediment and phosphorus accumulation tends to occur rapidly in the first few years and then decline after 10 years depending on the conceptual model of the wetland (Craft, 1997).

Lake Kasumigaura (220 km^2, Figure 1) , the second largest freshwater lake in Japan, is well known for its eutrophication, with mean concentrations of T-N and T-P of 1.10 mg l^{-1} and 0.10 mg l^{-1}, respectively. N loads to Lake Kasumigaura from domestic wastewater and non-point sources account for 35% and 40%, respectively, of the lake's entire N load (Japan Society on Water Environment, 1999). As shown in Figure 1, 56 rivers flow into this lake. Because most of the nutrient load is carried to the lake by rivers, to reduce the nutrient load to the lake, it is necessary to reduce the nutrient concentrations in the rivers. Several wetlands were constructed along Lake Kasumigaura to purify nutrient polluted rivers, most of which were based on the common reed. On the other hand, compared with the reed stand distribution around the entire littoral zone of Lake Kasumigaura about 100 years ago, most of the reed stands have decreased in size for various reasons relating to the past fifty years. For example, construction of bank and some facilities on the lakeside for fishery or yacht harbor. Besides these reasons, a rise in the water level of the lake and an increase in depth due to the removal of gravelwas also thought to eliminate reed stands (Onuma and Ikeda, 1999). It is drawing a lot of attention for reconstructing reed stand around this lake from nineties. Hence, besides purifying polluted river water, constructed wetlands with *P. australis* would contribute to the restoration of reed stands along the lake. However, different species of *P. australis* show enormous variations in morphology, ecology and cytogenetics (Haslam, 1972; Clevering and Lissner, 1999). Some studies have suggested that the high ecological plasticity of the species *P. australis* is often based on the high number of reed clones (Haslam, 1972; van der Toorn, 1972). Furthermore, Neuhaus et al. (1993) reported the stability of reed stands is related to genetic diversity, and monoclonal stands should have limited adaptive response to changing site conditions. In addition, the genetic variation and population genetic structure of aquatic plants are highly influenced by environmental conditions including water quality (Barrett et al., 1993; Laushman, 1993). Despite this, there are scarcely data for the genetic diversity evaluation of *P. australis* in constructed wetland. Hence, it is necessary to evaluate whether monoclonal reed stands form in constructed wetland and further compare the genetic diversity of *P. australis* in constructed wetland with that in the native wetland around the lake as well as evaluate the effect of constructed wetland on nutrient reduction, sediment accumulation, and plant transition during a relatively long term.

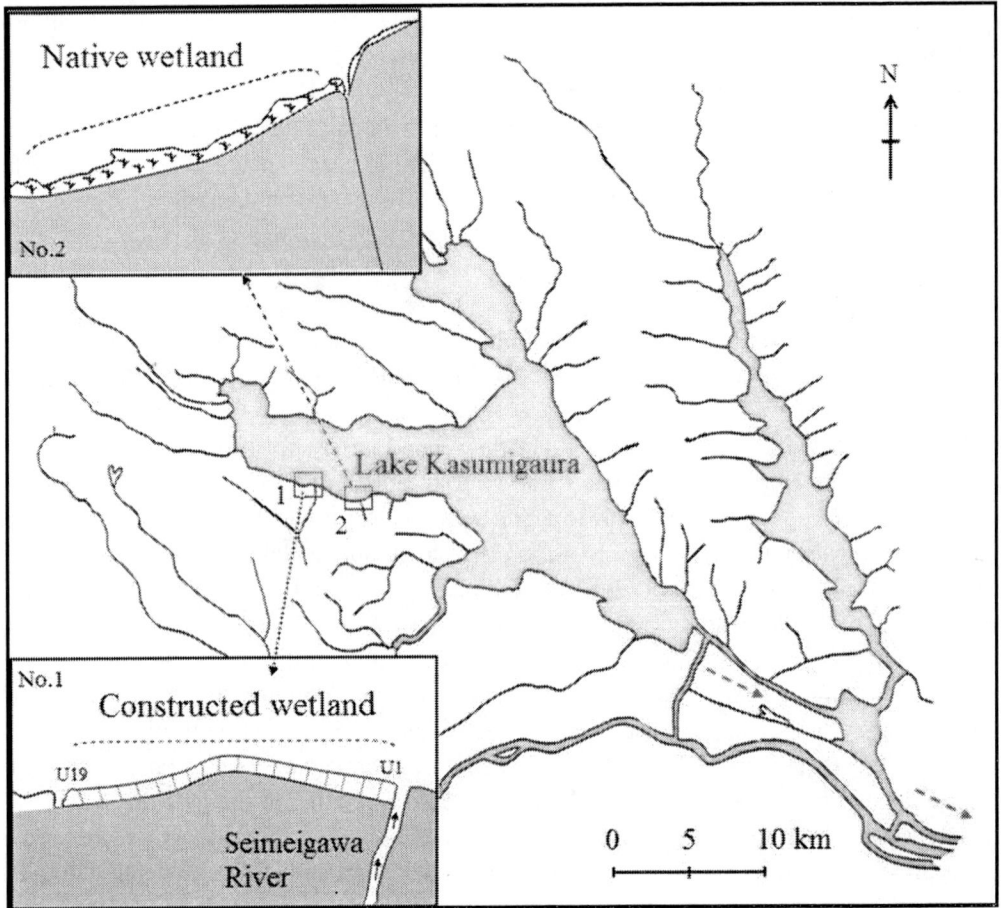

Figure 1. Location of the Seimeigawa River, constructed wetlands (No. 1) and native wetlands (No. 2). Outlines of constructed and native wetlands are shown in the insets.

2. METHOD AND MATERIALS

2.1. Site Description

The Seimeigawa River (36°01'52''N, 140°16'44''E; Figure 1) is one of several rivers that drain water from the nearby basin and transport domestic wastewater and agricultural land runoff to Lake Kasumigaura. The average flow rate of Seimeigawa River was 0.52 m^3 s^{-1} and T-N and T-P concentrations were 3.05 mg L^{-1} and 0.28 mg L^{-1}, respectively.

To reduce nutrient loading from the river to Lake Kasumigaura using constructed wetland, a series of surface-flow wetlands with *P. australis* that taken from littoral zones of Lake Kasumigaura were constructed near the estuary of the Seimeigawa River (No. 1 in Figure 1) by the Kasumigaura River Office, Japan. There are 19 units (U1 to U19, with one unit being 50 m × 40 m; the total area of the 19 units is 38,000 m^2) at the lakeside were constructed from 1991. By 1995-1996, all 19 units had been constructed and applied for purifying river water from the Seimeigawa River. The designed flow rate into the entire constructed wetland is 0.21 m^3 s^{-1} (hydraulic loading rate: 0.48 m^3 m^{-2} d^{-1}), which accounted for about 40% of river flow. As a comparative analysis of the genetic diversity of *P. australis* between the constructed and native wetlands, another site of native wetland (No. 2 in Figure 1) was chosen for native *P. australis* sampling. The native wetland is one of the largest natural reed-beds around Lake Kasumigaura (36°01′46″N, 140°16′59″E). The native site is 2.0 ha in area, 20–50 m in width and 600 m in length, and situated to the east of the river mouth of the Seimeigawa River. The site comprises mainly reeds (*P. australis*), a *Carex–Phragmites* community, cattail (*Typha latifolia*) or wild rice (*Zizania latifolia*), and scattered willows (*Salix* spp.) or rose (*Rosa multifilia*) bushes from lakeside to bankside.

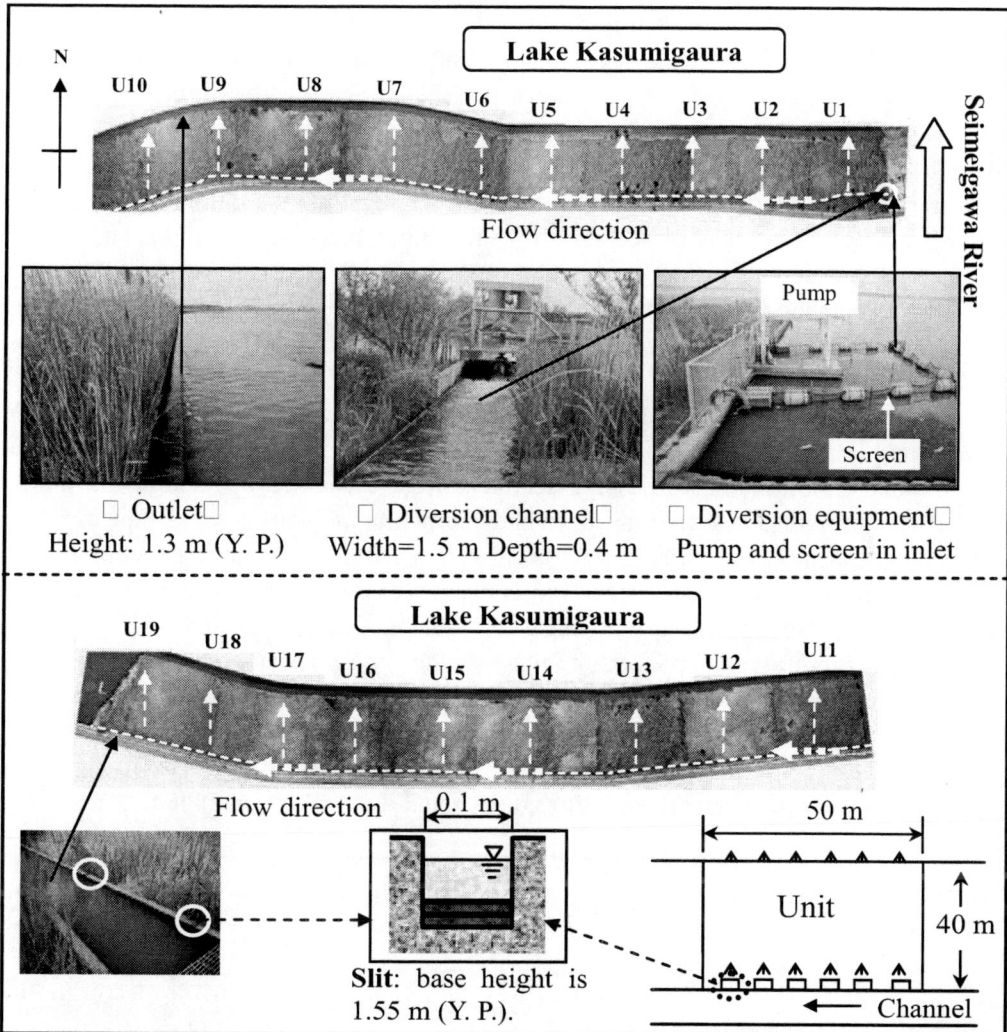

Figure 2. Details of constructed wetlands (No.1). Pictures of each unit (top: U1 - U10; bottom: U11 - U19) and their equipment including pump, diversion channel, slits, and outlet are presented. The outline of each unit is shown at the bottom. There are six slits of equal width for each unit and the design of the slit is shown (picture provided by the Kasumigaura River Office, Japan).

Details of the constructed wetlands including information on the diversion equipment, channel, outlet, and slits for each unit are presented in Figure 2. To separate river water from the channel into each unit equally, there are 6 slits (Y.P. 1.55 m; Y. P.: Yedokawa Peil, a kind of base level in this basin) of equal width (7 m) for influent at the inlet of each unit. The sides facing the lake have outlets with height of 1.3 m (Y. P.). The width of the channel is 1.5 m upstream and decreases to 0.6 m at the end of channel with flow diversion for each unit.

2.2. Sampling and Measurement

Most units were constructed by 1995. Around the completion of the wetlands construction, investigations of the water quality and sediment component were conducted during 1995-1996. Then, after 1999, as shown in Table 1, investigation of the water quality was conducted every year (U2, U5, U11, and U18). Measurements of sediment accumulation, hydraulic retention time, and vegetation were conducted for some years.

There are nine sampling points for sediment sampling for each unit, which consist of three points at three lines that located at 5 m, 15 m, and 25 m of the flow direction. Samples from each point were taken using cores and grouped by depth (0-5 cm; 5-10 cm; 10-20 cm; 20-30 cm) with other points. Sediment samples were analyzed for water content, organic content, nitrogen content, and phosphorus content as described by JSWA (1984). To evaluate the sediment accumulation, ground levels of the sediment surface were measured in 1999, 2002, and 2005 based on a 5 m grid system. The sediment surface being higher than 1.2 m (Y. P.) was considered to be net accumulation during the experimental period. Nitrogen and phosphorus contents accumulated in the sediment were calculated by multiplying the net sediment accumulation and average nitrogen and phosphorus contents in the sediment.

Samples of influent and effluent were taken periodically. For effluent, six samples were taken along the outlet side of each unit and combined as one sample. Water samples were analyzed for SS, T-N, T-P, and particulate and soluble phosphorus. T-N and T-P were analyzed by absorption spectrophotometry after decomposition with potassium peroxodisulfate ($K_2S_2O_8$) (JSWA, 1984). The ion concentration was measured using an ion chromatograph (ICS-90, Dionex, USA).

Table 1. Investigation schedule for constructed wetlands from 1996 to 2005

Year	Water quality	Sediment accumulation	Vegetation	Genetic analyses
1996	○	○	○	×
1999	○	○	×	×
2000	○	×	○	×
2001	○	×	×	×
2002	○	○	×	×
2003	○	×	×	×
2004	○	×	○	×
2005	○	○	×	○

a

U5   U4   U3   U2   U1

40 m

Site D   Site C   Site B   Site A

b

40 m

Site O   Site P   Site Q   Site R

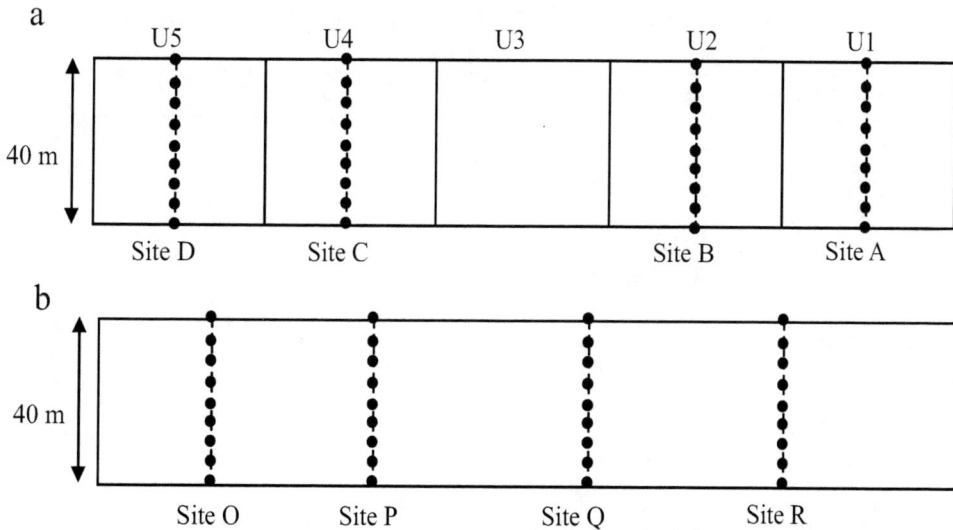

Figure 3. Details of *P. australis* sampling sites for genetic diversity investigation in constructed wetland (a: sites A-D) and native wetland (b: sites O-R). At each site, nine points (●) were sampled at 5 meter intervals.

2.3. Community Diversity and Genetic Analyses of *P. Australis*

2.3.1. Plant Community Diversity

·Macrophytes coverage by the dominant community was estimated at the period of peak biomass (July or August) by ground surveys. A 2 m × 2 m grid system was used to identify the location of plant communities in U2, U5, U11, and U18. In some areas, different plant species mix and they are difficult to separate as different communities. In this chapter, these areas are referred to as a mixed community and half the area was used to calculate the community diversity index.

2.3.2. Plant material and Total DNA Extraction

Samples of *P. australis* used in this study were obtained from constructed wetland (No. 1: U1, U2, U4, U5) and native wetland (No. 2) around Lake Kasumigaura in 2005. About 5-10 g of fresh leaves per plant was collected. A total of 72 individuals from these populations were included in an inter-simple sequence repeat (ISSR) study. Frozen leaves were powdered in liquid nitrogen and genomic DNA was extracted. Approximately 0.8 g of leaf material was used for DNA extraction using the ISOPLANT II kit (NIPPON GENE CO., LTD, Tokyo, Japan) according to manual. The cell wall, cell membrane, and nucleus membrane of *P. australis* leaves are destroyed by benzyl chloride as the main ingredient in solution, and DNA melts to an aqueous phase. Extracted DNA was ethanol precipitated and then pelleted at $6,000 \times g$ for 5 min. The pellet was washed in 300 μL of 70% ethanol. The dried pellet was resuspended in 100 μL of TE (Tris-EDTA buffer) at pH 8.0. The DNA quality and quantity were estimated by separating 1 μL on 1.5% agarose gels.

Table 2. Name and sequence of primers used in the present study

Primer name	Sequence (5'-3')	Annealing temperature (°C)
AM-1	(GGC)₅AT	58
AM-2	(AAG)₅GC	38
AM-3	(AAG)₅TG	38
AM-4	(AAG)₅CC	40
AM-5	(AGC)₅CA	57
AM-6	(AGC)₅GG	51
AM-7	(GGC)₅TA	60
AM-8	(AGC)₅GA	53
AM-9	(AAG)₅CG	40

2.3.3. ISSR Polymerase Chain Reaction (PCR) Amplification

Primers shown in Table 2 (Parsons et al., 1997) were selected for the preliminary ISSR analysis of *P. australis*. PCRs were carried out on a thermalcycler (Biometra, Inc., Japan) and programmed as shown in Table 3. Reactions were carried out in a volume of 25 μl, containing 0.2 mM of each dNTP (deoxyribonucleotide triphosphate), 2 mM $MgCl_2$, 0.2 mM primer, 0.625 U Taq DNA polymerase and 2 μl (approximately 120 ng) of DNA template. Amplification products were electrophoresed on 1.5% agarose gels stained with ethidium bromide, visualized under ultraviolet light, and photographed. Sizes of amplification products were estimated using a 100 bp DNA ladder (Takara Bio, Inc., Japan). Basing on the result of the preliminary experiment, the primers of AM-3, AM-7, and AM-8 are suitable and used for band generation of *P. australis* from constructed and native wetlands.

Table 3. PCR program in thermocycler

Step	Temperature (°C)	Time (min)	
1	95	3	
2	95	0.5	
3	T(°C) in Table 2+ 2	1	2 cycles
4	72	2	
5	95	0.5	
6	T(°C) in Table 2	1	2 cycles
7	72	2	
8	94	0.5	
9	T(°C) in Table 2	1	40 cycles
10	72	2	
11	72	5	

2.3.4. Data Analysis

PCR-fingerprints of all reed samples at each sample point were made. Band positions on the gels were determined visually and the fingerprint patterns were transformed into a binary character matrix with 1 indicating the presence and 0 the absence of a band at a particular position in a lane. The percentages of polymorphic bands in constructed and native wetlands for genetic diversity comparison were also calculated. The genetic similarity coefficients were used to cluster samples based on their degree of genetic similarity, which calculated by the unweighted pair group method using arithmetic averages (UPGMA), and finally for computing the dendrograms.

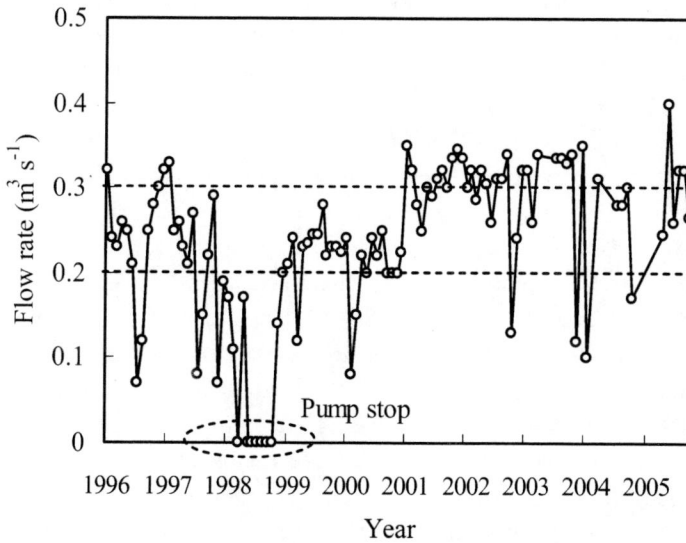

Figure 4. Flow rates entering the entire constructed wetland from 1996 to 2005 (data from the Kasumigaura River Office, Japan).

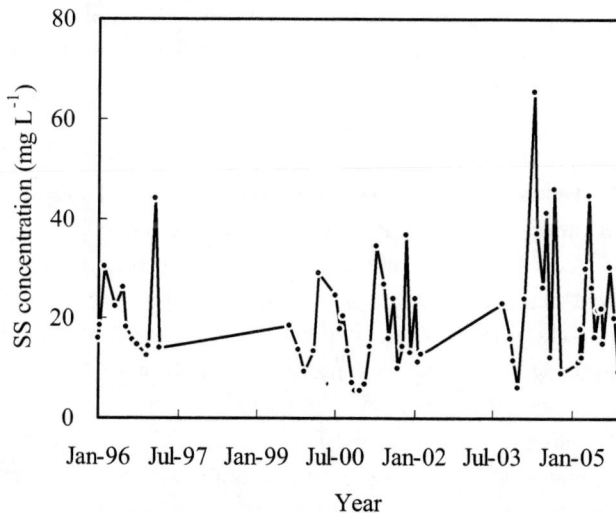

Figure 5. Monthly SS concentrations in influent from 1996 to 2005 (data from the Kasumigaura River Office, Japan).

3. RESULTS AND DISCUSSION

3.1. Hydraulic Load and Nutrient Reduction

The variation in the rate of flow into the entire constructed wetland from 1996 to 2005 is shown in Figure 4. The average flow rate was 0.25 m^3 s^{-1}. The flow can be separated into two stages in that the flow rate for the first 5 years was relatively low at 0.21 m^3 s^{-1} (HLR: 0.48 m^3 m^{-2} d^{-1}) while that for the second 5 years increased to 0.29 m^3 s^{-1} (HLR: 0.66 m^3 m^{-2} d^{-1}). However, the hydraulic retention time measured by the tracer was 84 minutes in 2005, which was significantly shorter than the designed value (5 hours) owing to the accumulated sediment reducing the volume of the unit and forming water ways. As shown in Figure 5, the SS concentration varied monthly with the average concentration being 20.2±11.5 mg L^{-1}, which resulted in about 160 ton SS entering the entire constructed wetland every year.

The average monthly T-N and T-P concentrations of influent varied in the range of 0.77 – 3.60 mg L^{-1} and 0.05 – 0.24 mg L^{-1}, respectively (Figure 6). For T-N, the major form of nitrogen flowing into the wetland was NO_3^-, accounting for 52% of T-N, whereas NH_4^+ accounted for 5% of T-N, which is similar to what is found for other rivers around Lake Kasumigaura (Zhou and Hosomi, 2008). Furthermore, it is obvious the concentrations of influent from spring to summer (April – September) are mostly in the low range (data encircled by the broken line in Figure 6), while those from autumn to winter (October – March) have higher values. The average concentrations of T-N over these two yearly periods from 1996 to 2005 were 1.45±0.40 mg L^{-1} and 2.42±0.53 mg L^{-1}, respectively. The difference is probably owing to the average temperature decreasing from October to March, which would influence denitrification in the river water. On the other hand, the concentrations of T-P during different seasons were similar (April – September: 0.15±0.04 mg L^{-1}; October – March: 0.13±0.04 mg L^{-1}). The T-P in influent was approximately 60% particulate phosphorus and 40% soluble phosphorus.

To evaluate nutrient removal efficiencies, the average concentration reductions in U2, U5, and U11 for samples collected periodically from 1996 to 1997 and every year from 1999 were calculated and are shown in Figure 7. During the initial stage from 1996 to 1999, the annual average removal efficiencies of T-N and T-P remained approximately 13% and 14%, respectively. However, after about 5 years running, both the concentration removal efficiencies of T-N and T-P varied wildly. By 2005, the annual removal efficiency of T-N was 2.0% while that of T-P was 1.7% after 10 years running. Moreover, as shown in Figure 8, T-N and T-P removal from spring to summer (April – September) appeared to significantly vary from 2000 after 5 years running while T-N and T-P removal from autumn to winter (October – March) almost had a positive nutrient removal effect through the 10 years. T-N removal efficiencies from autumn to winter were higher probably owing to a higher T-N (NO_3^-) concentration in the influent during this period compared with the concentrations from spring to summer as mentioned above, which would enhance the denitrification rate according to Michaelis-Menten kinetics (Grant, 1991; Zhou and Hosomi, 2008). In contrast, from spring to summer, along with the temperature rising (average water temperature from spring to summer: 25°C; average water temperature from autumn to winter: 11°C), microbial decomposition of residue in sediment would be enhanced and release nutrients into water

(Sartoris et al., 2000; Lan et al., 2006), which probably decreased nutrient removal efficiencies.

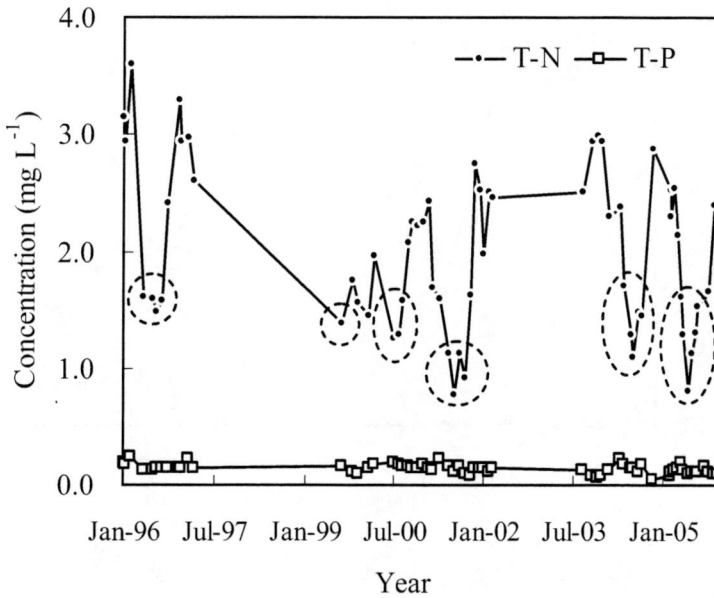

Figure 6. Monthly T-N and T-P concentrations in influent from 1996 to 2005 (data from the Kasumigaura River Office, Japan).

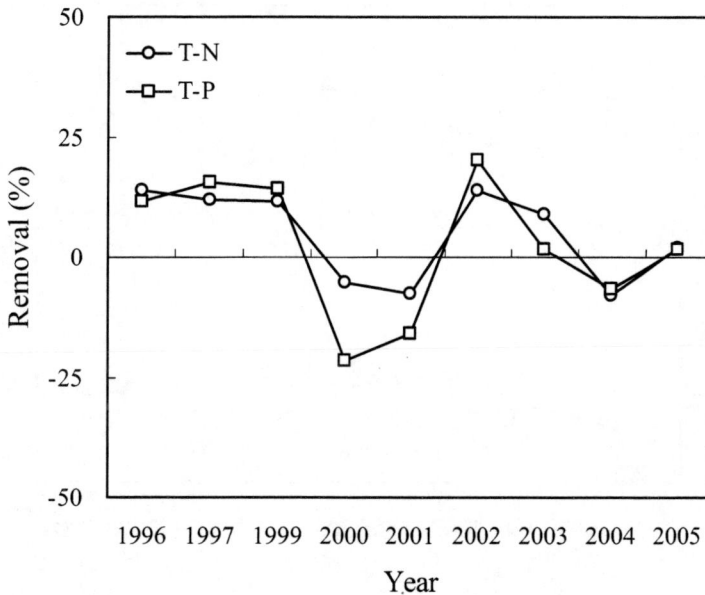

Figure 7. Average T-N and T-P removal from 1996 to 2005 (data from the Kasumigaura River Office, Japan).

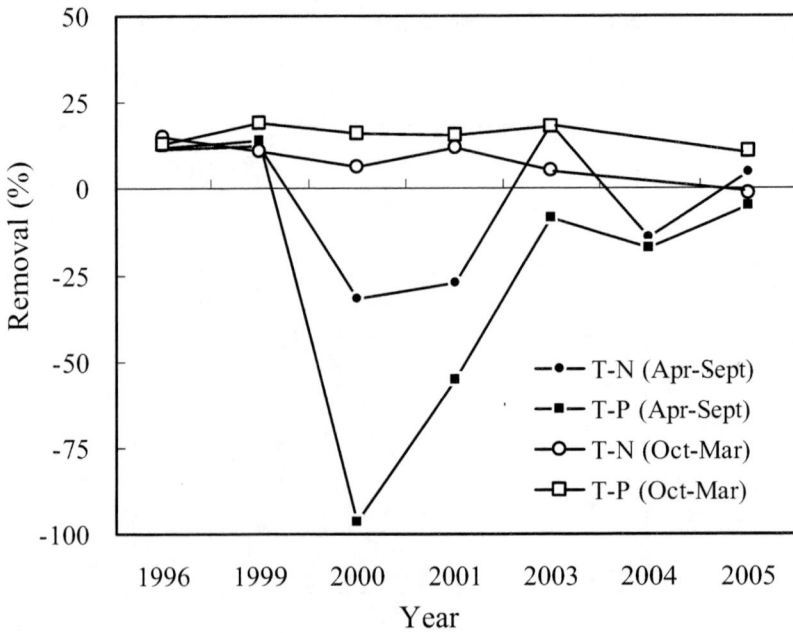

Figure 8. Comparison of T-N and T-P removal between April-September and October-March from 1996 to 2005 (data from the Kasumigaura River Office, Japan).

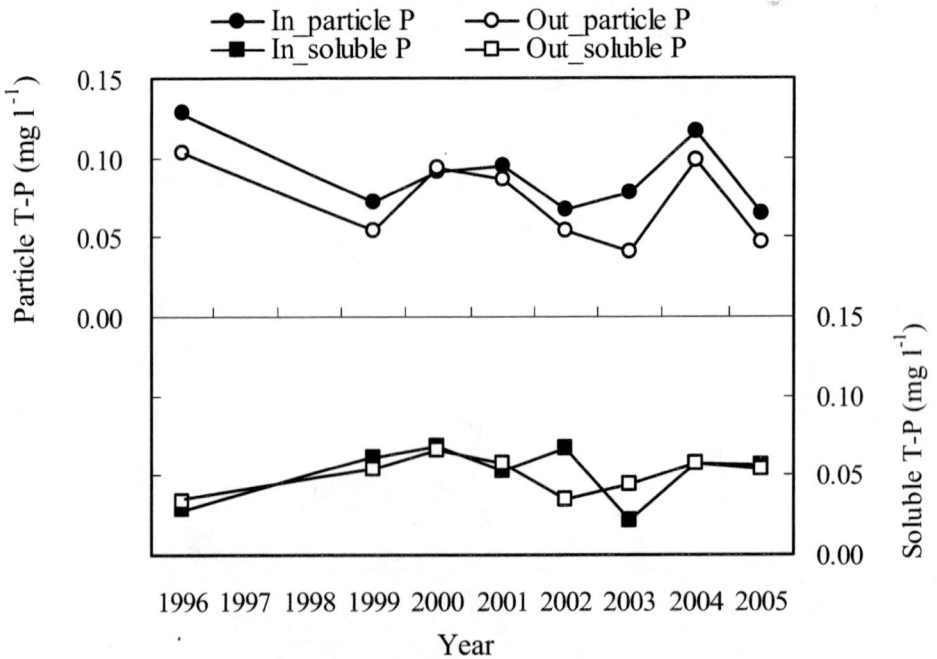

Figure 9. Average concentrations of particle phosphorus and soluble phosphorus in influent and effluent from 1996 to 2005 (data from the Kasumigaura River Office, Japan).

Table 4. Comparison of SS, T-N, and T-P removal in one unit between 1996 and 2005 (kg yr^{-1})

	U2		U5		U11		Average	
	1996	2005	1996	2005	1996	2005	1996	2005
SS	1388	1222	1535	185	2632	720	1852±680	709±519
T-N	116	85	133	43	117	73	122±10	67±22
T-P	8.1	5.2	11.2	1.3	4.3	2.3	7.9±3.5	2.9±2.0

* Data from the Kasumigaura River Office, Japan.

On the other hand, phosphorus readily adsorbs to soil in rivers and this pathway is the most prevalent mode of transport. (Fennessy et al., 1994) determined sedimentation to be an important phosphorus retention mechanism in wetland systems. Similarly, T-P concentration decreases are probably due to sedimentation in this study. As shown in Figure 9, particulate phosphorus was removed continuously through the 10 years, while the decrease in soluble phosphorus was not significant and sometimes the phosphorus concentration in the effluent was higher than that in the influent. The variation in soluble concentrations in influent and effluent is considered to be associated with uptake by plants and algae as well as release by residue decomposition.

To evaluate the SS and nutrient removal loads, data for U2, U5, and U11 collected every month in 1996 and 2005 were calculated and are shown in Table 4. The average removal of SS, T-N, and T-P decreased in 2005 compared with 1996 and accounted for 38%, 55%, and 37% of removal loads in 1996, which was probably due to the reduced HRT as described below. It is interesting that the decrease in SS and T-P removal loads were almost same, suggested again that sedimentation is the major mechanism for phosphorus removal in this study. In contrast, the constructed wetland retained a relatively high T-N removal capacity probably owing to the denitrification potential of mature wetland (Bachand and Horne, 2000b). However, despite the total amount of SS, T-N, and T-P removed decreasing to some degree compared with removal loads in 1996, the constructed wetland has proven to remove pollutants in river water purification even after 10 years running. In particular, the T-N removal rate in 2005 was calculated as 34 g-N m^{-2} yr^{-1}, which is similar to the result (average NO$_3^-$ retention of 39 g-N m^{-2} yr^{-1}) for the long term running of the created wetland (Mitsch et al., 2005a).

Table 5. Amounts of nutrients accumulated in sediment by 2005 and total nutrient loadings and removal during 10 years (kg per unit)

Unit	Sediment accumulation		Accumulated removal		Nutrient loading	
	T-N	T-P	T-N	T-P	T-N	T-P
U2	207	67	825	62	9892	698
U5	199	66	489	29	8390	542
U11	137	40	556	14	5626	374
U18	228	47	571	37	3186	184
Average	193	55	610	36	6774	450

* Data from the Kasumigaura River Office, Japan.

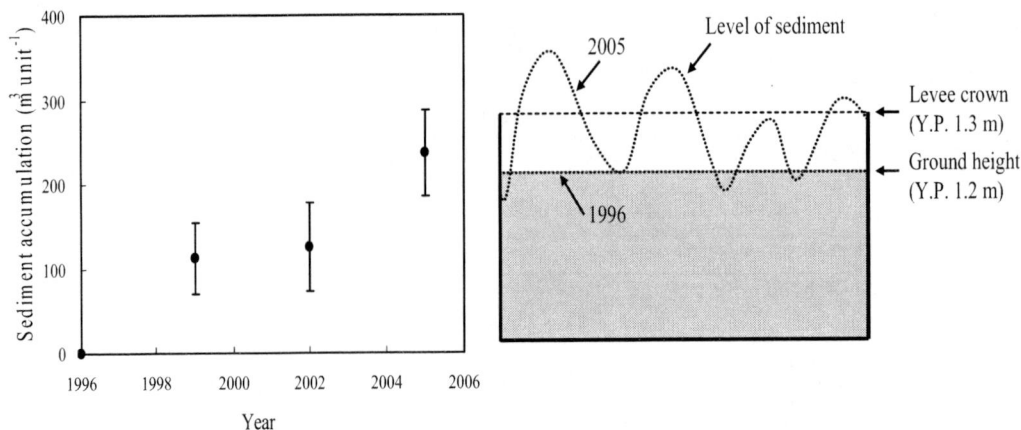

Figure 10. Average sediment accumulation from 1996 to 2005 (left) and a description of the shape of the unit sediment surface in 2005 (right) (data from the Kasumigaura River Office, Japan).

3.2. Sediment Accumulation

Figure 10 shows the average increase in sediment accumulation in U2, U5, U11 and U18. By 2005, the average sediment accumulation (180-296 m^3) already exceeded the designed volume (200 m^3) after 10 years running. Because the sediment surface in a unit is convexo-concave, some channelized flow formed even though the height at some places was higher than the designed levee crown (Figure 10, right). The HRT was measured as 30 minutes on September 2004, one tenth the retention time designed in 1996. Sediment accumulation occurred mostly owing to the prevailing influence of SS sedimentation and dead plant accumulation. Sedimentation can be estimated by calculating the total removed suspended solid from the influent. However, the amount of net solid removal ranged from 4.14 to 22.5 ton for each unit over the 10 years. The volume of removed solid modified by moisture content and bulk density ranged from 11 to 62 m^3, which was significantly lower than the accumulated sediment volume, suggesting that most sediment accumulated from dead plant. Supposing sediment accumulation only consisted of removed suspended solid and dead plant in each unit, the volume of dead plant accumulation can be calculated by subtracting the removed suspended solid volume from the accumulated sediment volume. The accumulation of dead plant in each unit was calculated as 3.4 kg m^{-2} yr^{-1}, which is within the P. australis production range of $0.2 - 6.0$ kg m^{-2} yr^{-1} (Shimatani et al., 2003). The average total sediment accumulation including removed SS and dead plant was $3.61 - 4.43$ kg m^{-2} yr^{-1}, which is similar to the sediment accumulation (3.9 ± 0.3 kg m^{-2} yr^{-1}) reported by Anderson and Mitsch, 2006) in investigating created riverine marshes over 10 years.

The changes in average organic matter, T-N, and T-P contents in soil at 0-5 cm, 5-10 cm, 10-20 cm, and 20-30 cm in U2, U5, U11, and U18 from 2000 to 2004 are shown in Figure 11. Compared with the stable values below 5 cm depth, contents of organic matter, T-N, and T-P in surface sediment increased from 2000 to 2004 by 1.6, 3.7, and 3.6 times, suggesting the nutrient sediment and residue accumulated with time. The amounts of nutrients accumulated in sediment by 2005 and total nutrient loadings and removal during the 10 years are shown in

Table 5. The T-P content retained in sediment was calculated as having an average rate of 2.8 g P m^{-2} yr^{-1}, which is within the range of 0.3–3.6 g P m^{-2} yr^{-1} for T-P deposition reported for natural wetlands (Mitsch et al., 1979; Mitsch and Reeder, 1991). However, compared with accumulated T-P removed from the inflow, there was more T-P accumulated in sediment, which was probably owing to macrophytes pumping phosphorus from deep sediments, simulated at a rate of 0.31–1.66 g P m^{-2} yr^{-1} (Wang and Mitsch, 2000). In contrast, comparing the amounts of nutrients accumulated in sediment with those removed during the 10 years, unlike the T-P content in sediment being similar or slight higher than the amount removed, the T-N content in sediment is significantly less than the amount removed. The average T-N removal rate was calculated as 31 g N m^{-2} yr^{-1} while only 9.6 g N m^{-2} yr^{-1} was stored in sediment. The loss of nitrogen was probably due to denitrification in the wetland (Bachand and Horne, 2000b; Spieles and Mitsch, 2000) since NO$_3^-$ is a major form of nitrogen in influent.

3.3. Transition of Plant Species

P. australis was planted in each unit and covered 100% of the area of each unit at the beginning of the experiment after construction was completed in 1996. By 2000, the coverage of the uniform P. australis community decreased to 77% owing to other plant invasion (e.g. Zizania latifolia, Carex dispalata, Lycopus lucidus). Most of the invading species are indigenous plants around the basin (Kusumoto et al., 2007), which are probably in the soil or came from river water. Furthermore, although the P. australis community still dominated in each unit, different spatial community diversity developed in 2004. In particular, the mixing of P. australis and Zizania latifolia, and P. australis and Carex dispalata expanded. As a result, the average uniform P. australis community coverage was 53% in 2004, a further decrease from 2000. In contrast, although the distributions in each unit differed, the average coverage of Zizania latifolia and Carex dispalata increased significantly and finally reached 13% and 15%, respectively. On the other hand, not only macrophytes but also woody terrestrial plants such as Salix spp. were observed after several years running.

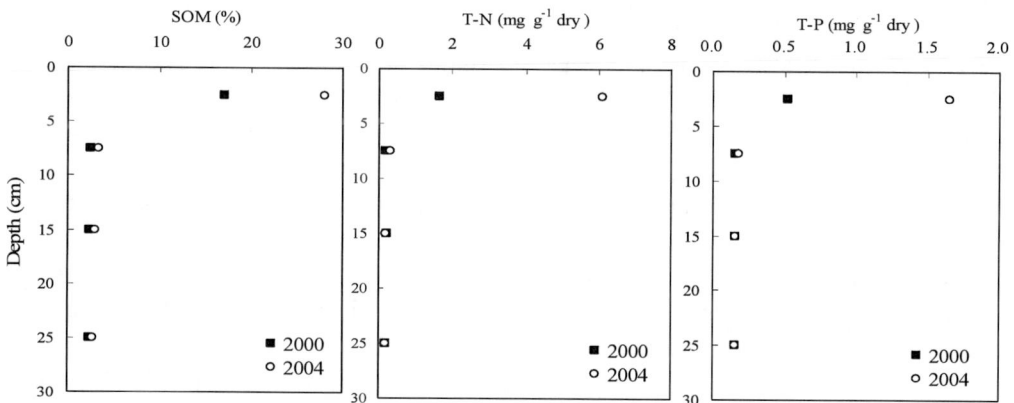

Figure 11. Organic matter, nitrogen, and phosphorus contents in sediment at different depths (data from the Kasumigaura River Office, Japan).

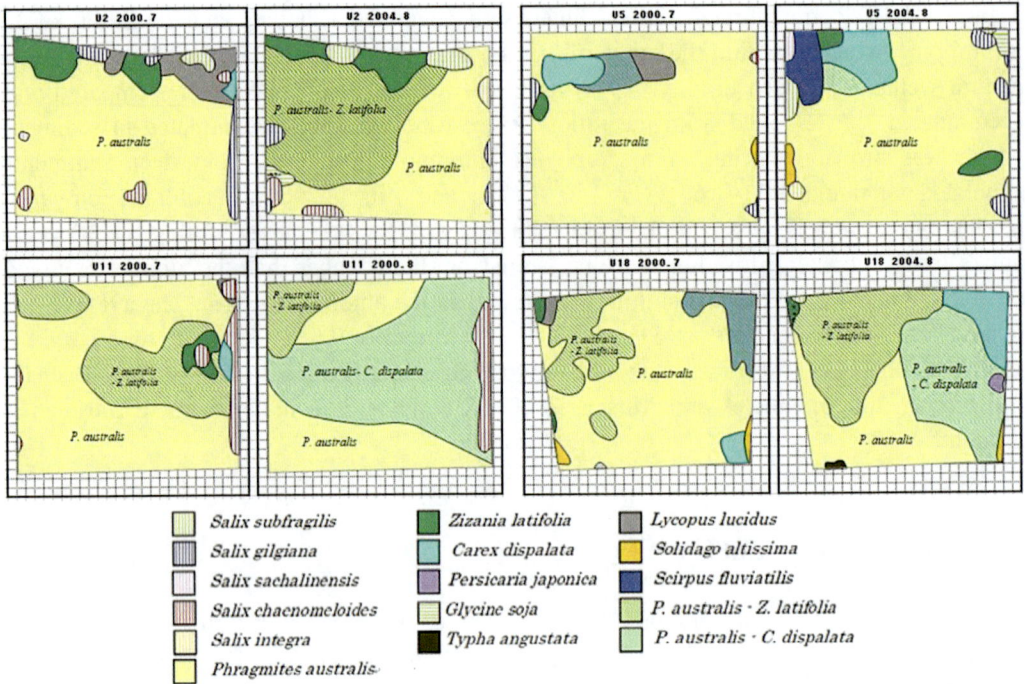

Figure 12. Vegetation community coverage in U2, U5, U11, and U18 in July 2000 and August 2004 (data from the Kasumigaura River Office, Japan).

The macrophyte community diversity index (CDI) reported by Mitsch et al. (2005b) was used to quantify the spatial diversity in the constructed wetland. The index is expressed as

$$CDI = \sum_{i=1}^{N}(C_i \ln(C_i))$$

where C_i is the percentage coverage of community "i" (0–1) and N is the number of plant/aquatic communities. In 1996, the CDI was zero for U2, U5, U11, and U18 since all of units were covered only by $P.$ $australis$. However, after 4 years and 8 years running, as shown in Figure 11, the CDI of U2, U5, U11, and U18 significantly increased. By 2004, the average CDI of these four units was 0.88±0.04, which was also significantly higher than that in 2000 (0.64±0.14).

The hydrology of wetland plays an important role in directing vegetation change by affecting the germination potential, growth capability and mortality of plants (Ellison and Bedford, 1995). Although macrophyte vegetation affects and is affected by water quality in riparian wetlands (Fink and Mitsch, 2007), the vegetation transition in this study probably occurred due to hydrology changes along with the ground level being elevated by sediment accumulation. Particularly after several years running, some areas rose above the flood water level to form land inside the constructed wetland (Figure 11, right), which is suitable for other vegetation invasion such as that by $Salix$ spp. The phenomenon agrees with the result reported by Grace and Wetzel (1981) that less flood-tolerant species are superior competitors in drier locations, allowing invasion of occupied space when water tables are low, because more energy is needed for tolerance to flood conditions.

3.4. Genetic Analyses of *P. Australis*

Using AM-3, AM-7, and AM-8 primers, 39-40 bands were identified of which 64-73% were polymorphic at each site (A-D) of constructed wetland. In contrast, 28-38 bands were identified of which 43-67% were polymorphic at each site (O-R) of native wetland (Table 6). Although the average percentages of polymorphic bands of constructed wetland (68±5%) were slightly higher than those of native wetland (57±11%), there was no significant difference ($p = 0.142$) between these two groups basing on statistic analysis.

Figure 14 shows the UPGMA-based dendrogram of reed samples from constructed and native wetlands basing on the AM-7 primer, which has the similar patterns with the results basing on the other two primers. The genetic similarity coefficients within the reed stand in the constructed wetland ranged from 0.62 to 1.00 while those in native wetland ranged from 0.53 to 1.00. However, as shown in Figure 14, more gene types appeared in the constructed wetland than in the native wetland. In the constructed wetland, samples had significant variations among sampling points even though the sampling intervals were only 5 m, except at site D. Compared with other sites of the constructed wetland, reed from most sampling points at site D had a relatively high similarity coefficient (>0.94). On the other hand, the reed from the native wetland can be divided into several main clusters according to sampling sites. One group consisted of site Q and site R, which had a similarity coefficient between them of approximately 0.95. Reed from all sampling points at site P also had a high similarity coefficient (>0.90) while reed from the two halves of site O can be divided into two groups.

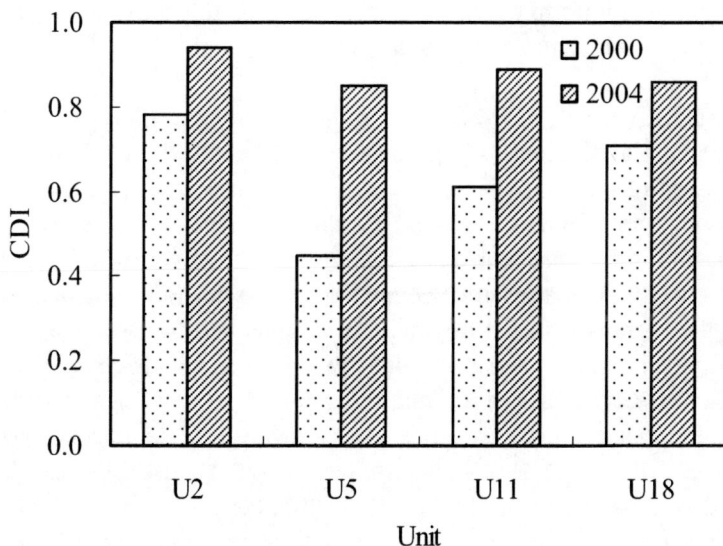

Figure 13. Vegetation diversities in U2, U5, U11, and U18 in July 2000 and August 2004. CDI is the community diversity index, which is defined in the text.

Table 6. Numbers of polymorphic and monomorphic bands determined by AM-3, AM-7, and AM-8 ISSR-PCR primers. Total polymorphic bands, total monomorphic bands, and percentage of polymorphic bands of *P. australis* at each site are indicated

Site	AM-3		AM-7		AM-8		Total		
	PB	MB	PB	MB	PB	MB	PB	MB	PPB %
A	3	5	9	6	13	3	25	14	64
B	6	3	9	5	13	3	28	11	72
C	3	6	10	5	12	3	25	14	64
D	6	4	9	5	14	2	29	11	73
O	6	5	10	4	8	3	24	12	67
P	2	6	9	6	10	5	21	17	55
Q	1	5	5	6	6	5	12	16	43
R	3	4	9	7	12	2	24	13	65

PB: polymorphic bands; MB: monomorphic bands; PPB: percentage of polymorphic bands.

Because seedlings of *P. australis* were taken from littoral zones around Lake Kasumigaura and planted in the constructed wetland at the beginning of construction, it can be considered that various genetic types of *P. australis* were introduced at the same time, which probably resulted in a relatively high number of gene types (Figure 14 (a)). Moreover, although the investigation was conducted 10 years after construction, the constructed wetland was considered to be in an establishment stage compared with the mature native wetland. At this stage, *P. australis* not only competes with itself to fill the spaces between plants, but competes with other invading species while coping with ground level elevation as mentioned above, which results in a higher genetic diversity owing to different clones intermingling. Furthermore, as shown in Figure 12, although the CDI was similar between U2 (Site B) and U5 (Site D) in 2004, the uniform distribution of *P. australis* in U5 (80%) was approximately twice that in U2 (42%), which probably results in a higher similarity coefficient at site D than at site B.

In contrast, the native wetland (No. 2) along Lake Kasumigaura has existed for 100 years as reported by Onuma and Ikeda (1999), and is thus considered well adapted to the local environmental conditions and to have entered a stationary stage. In addition, seedling establishment is not very frequent in mature *P. australis* populations (Barrett et al., 1993). Populations initiated by seeds are probably initially genetically diverse and over time become dominated by a few clones as a result of competition and selection. Genetic diversity would decrease during the stationary stage, in which a small number of clones well adapted to the local environmental conditions prevail. Relatively low genetic diversity and monoclonal populations at each site of the native wetland could be the result of such a selection process and be an indication that *P. australis* stands have grown under stable conditions for a long time (Watkinson and Powell, 1993). Lambertini et al. (2008) also reported that one large monoclonal stand was present in an old wetland with rather stable environmental conditions over a long time period whereas polyclonal stands were younger and characterized by disturbance.

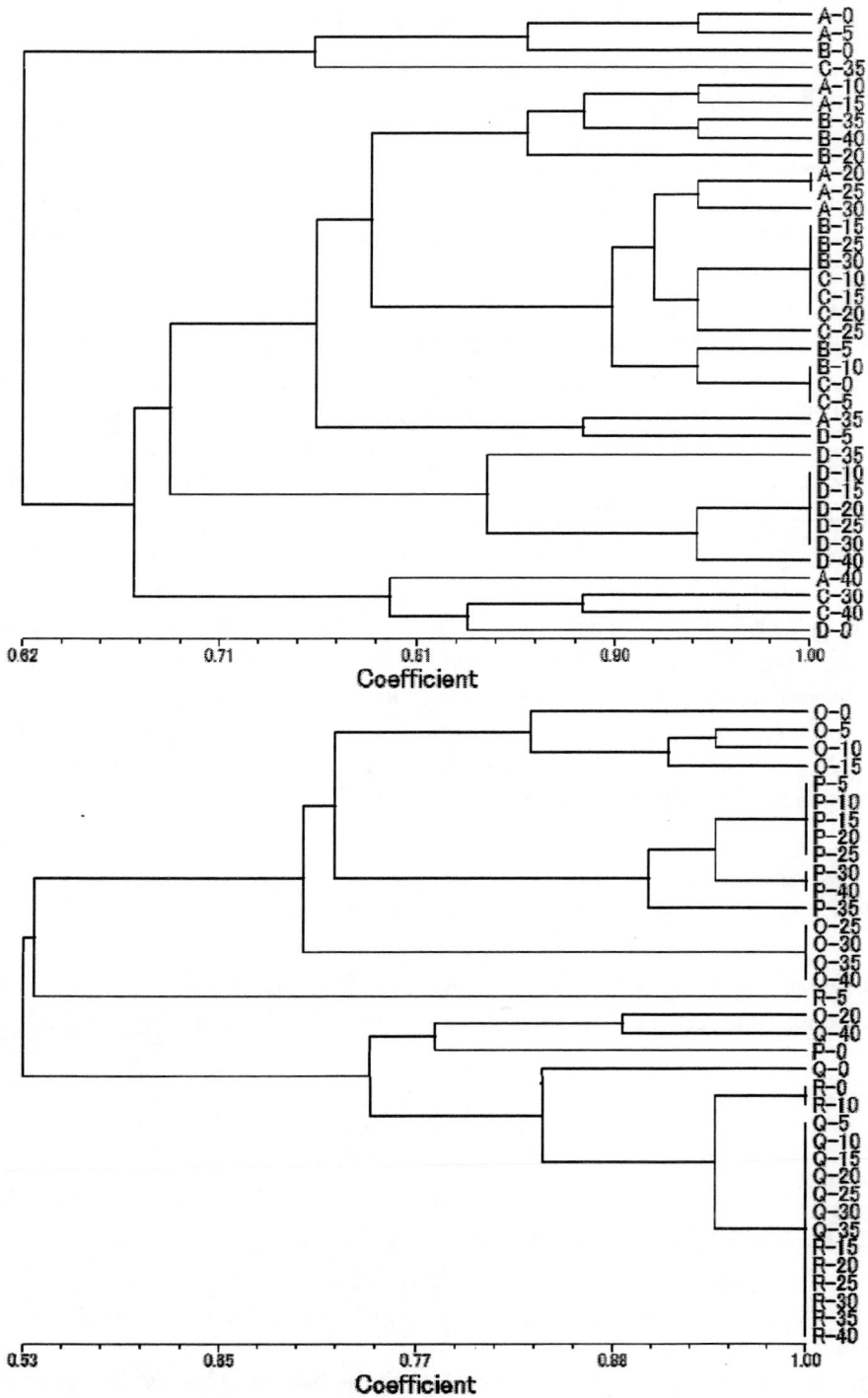

Figure 14. Dendrograms of Phragmites australis samples from constructed wetland (a) and native wetland (b) revealed by UPGMA cluster analysis based on genetic similarity coefficients. Results are based on PCR fingerprinting analysis with primer AM-7.

4. CONCLUSION

After 10 years running, sediment accumulated in the constructed wetland and reduced the HRT by forming channelized waterways between elevated lands, which would influence nutrient removal efficiencies. However, although nutrient removal efficiencies varied after several years running, particularly from spring to summer, considerable amounts of SS, nitrogen, and phosphorus were retained by the constructed wetland in 2005 after 10 years running. The average total sediment accumulation was $3.61 - 4.43$ kg m^{-2} yr^{-1}, which resulted in organic matter, nitrogen and phosphorus contents in the surface sediment being significantly higher than those in deeper soil. The average rate of T-P retained in sediment was calculated as 2.8 g P m^{-2} yr^{-1}. However, unlike the amount of T-P in sediment being similar or slight higher than the amount removed from the inflow river water, the amount of T-N in sediment is significantly less than that removed, suggesting that phosphorus was removed along with particle sedimentation while most nitrogen was removed by denitrification.

On the other hand, with the ground level being elevated by sediment deposition, woody terrestrial plants and other aquatic plants invaded the constructed wetland probably owing to hydrology changes, which developed population diversity in the constructed wetland. By 2004, the average CDI of the four units was 0.88 ± 0.04, significantly more than that at the initial stage of running the constructed wetland. In addition, compared with the genetic diversity of *P. australis* in the native wetland, the constructed wetland showed more gene types probably owing to the constructed wetland being in an establishment stage while the native wetland had already entered a stationary stage.

ACKNOWLEDGEMENT

We express our great thanks to the Kasumigaura River Office (Ministry of Land, Infrastructure, Transport and Tourism, Japan) for providing all the data in section 3.1-3.3.

REFERENCES

Anderson, C.J. and W.J. Mitsch, 2006. Sediment, carbon, and nutrient accumulation at two 10-year-old created riverine marshes. *Wetlands,* 26: 779-792.

Bachand, P.A.M. and A.J. Horne, 2000a. Denitrification in constructed free-water surface wetlands: I. Very high nitrate removal rates in a macrocosm study. *Ecological Engineering,* 14: 9-15.

Bachand, P.A.M. and A.J. Horne, 2000b. Denitrification in constructed free-water surface wetlands: II. Effects of vegetation and temperature. *Ecological Engineering,* 14: 17-32.

Barrett, S.C.H., C.G. Eckert and B.C. Husband, 1993. Evolutionary Processes in Aquatic Plant-Populations. *Aquatic Botany,* 44: 105-145.

Brix, H., 1994. Functions of Macrophytes in Constructed Wetlands. *Water Science and Technology,* 29: 71-78.

Clevering, O.A. and J. Lissner, 1999. Taxonomy, chromosome numbers, clonal diversity and population dynamics of *Phragmites australis. Aquatic Botany,* 64: 185-208.

Comin, F.A., J.A. Romero, V. Astorga and C. Garcia, 1997. Nitrogen removal and cycling in restored wetlands used as filters of nutrients for agricultural runoff. *Water Science and Technology,* 35: 255-261.

Coops, H., N. Geilen, H.J. Verheij, R. Boeters and G. vanderVelde, 1996. Interactions between waves, bank erosion and emergent vegetation: An experimental study in a wave tank. *Aquatic Botany,* 53: 187-198.

Craft, C.B., 1997. Dynamics of nitrogen and phosphorus retention during wetland ecosystem succession. *Wetlands Ecology and Management,* 4: 177-187.

Ellison, A.M. and B.L. Bedford, 1995. Response of a Wetland Vascular Plant Community to Disturbance - a Simulation Study. *Ecological Applications,* 5: 109-123.

Fennessy, M.S., C.C. Brueske and W.J. Mitsch, 1994. Sediment Deposition Patterns in Restored Fresh-Water Wetlands Using Sediment Traps. *Ecological Engineering,* 3: 409-428.

Fink, D.F. and W.J. Mitsch, 2004. Seasonal and storm event nutrient removal by a created wetland in an agricultural watershed. *Ecological Engineering,* 23: 313-325.

Fink, D.F. and W.J. Mitsch, 2007. Hydrology and nutrient biogeochemistry in a created river diversion oxbow wetland. *Ecological Engineering,* 30: 93-102.

Grace, J.B. and R.G. Wetzel, 1981. Habitat Partitioning and Competitive Displacement in Cattails (Typha) - Experimental Field Studies. *American Naturalist,* 118: 463-474.

Grant, R.F., 1991. A Technique for Estimating Denitrification Rates at Different Soil Temperatures, Water Contents, and Nitrate Concentrations. *Soil Science,* 152: 41-52.

Haslam, S.M., 1972. Biological flora of the British Isles. *Phragmites communis Trin. J. Ecol.,* 60: 585-610.

Jing, S.R., Y.F. Lin, D.Y. Lee and T.W. Wang, 2001. Nutrient removal from polluted river water by using constructed wetlands. *Bioresource Technology,* 76: 131-135.

JSWA, J.S.W.A., 1984. *Analysis Methods of Wastewater.* Tokyo, 649 pp.

Kadlec, R.H. and R.L. Knight, 1996. *Treatment wetlands.* Lewis Publishers, Florida.

Karunaratne, S., T. Asaeda and K. Yutani, 2003. Growth performance of Phragmites australis in Japan: influence of geographic gradient. *Environmental and Experimental Botany,* 50: 51-66.

Kira, T., 1991. *Brief review on ecology of reed.* Report of Lake Biwa Research Institute, 9: 29-37.

Kovacs, M., I. Precsenyi and J. Podani, 1978. Accumulation of Elements in Reeds of Lake Balaton (Phragmites Communis). *Acta Botanica Academiae Scientiarum Hungaricae,* 24: 99-111.

Kusumoto, Y., S. Yamamoto, T. Ohkuro and M. Ide, 2007. The relationship between the paddy landscape structure and biodiversity of plant community in the Tone river basin. *Jornal of The Japanses Institute of Landscape Architecture* 70: 445-448.

Lambertini, C., M.H.G. Gustafsson, J. Frydenberg, M. Speranza and H. Brix, 2008. Genetic diversity patterns in Phragmites australis at the population, regional and continental scales. *Aquatic Botany,* 88: 160-170.

Lan, N.K., T. Asaeda and J. Manatunge, 2006. Decomposition of aboveground and belowground organs of wild rice (Zizania latifolia): mass loss and nutrient changes. *Aquatic Ecology,* 40: 13-21.

Laushman, R.H., 1993. Population-Genetics of Hydrophilous Angiosperms. *Aquatic Botany,* 44: 147-158.

Mitsch, W.J., J.W. Day, L. Zhang and R.R. Lane, 2005a. Nitrate-nitrogen retention in wetlands in the Mississippi river basin. *Ecological Engineering,* 24: 267-278.

Mitsch, W.J., C.L. Dorge and J.R. Wiemhoff, 1979. Ecosystem Dynamics and a Phosphorus Budget of an Alluvial Cypress Swamp in Southern Illinois. *Ecology,* 60: 1116-1124.

Mitsch, W.J., J.C. Lefeuvre and V. Bouchard, 2002. Ecological engineering applied to river and wetland restoration. *Ecological Engineering,* 18: 529-541.

Mitsch, W.J. and B.C. Reeder, 1991. Modelling nutrient retention of a freshwater coastal wetland: estimating the roles of primary productivity, sedimentation, resuspension and hydrology. *Ecological Modelling* 54: 151-187.

Mitsch, W.J., L. Zhang, C.J. Anderson, A.E. Altor and M.E. Hernandez, 2005b. Creating riverine wetlands: Ecological succession, nutrient retention, and pulsing effects. *Ecological Engineering,* 25: 510-527.

Nakamura, K., T. Chiba, K. Sato, Y. Morita, M. Hosomi and s. Tanaka, 2007. A survey of constructed wetlands in Japan. *Report of Research Institute of River & Watershed Environment Management* (RIREM), 19: 41-46.

National-Research-Council, 1992. *Restoration of aquatic ecosystems.* National Academy Press, Washington, DC, 522 pp.

Neuhaus, D., H. Kuhl, J.G. Kohl, P. Dorfel and T. Borner, 1993. Investigation on the Genetic Diversity of Phragmites Stands Using Genomic Fingerprinting. *Aquatic Botany,* 45: 357-364.

Onuma, H. and H. Ikeda, 1999. Reduction processes of reed community in the Lake Kasumigaura. *Bulletin of Environmental Research Center, the University of Tsukuba,* 24: 49-58.

Parsons, B.J., H.J. Newbury, M.T. Jackson and B.V. FordLloyd, 1997. Contrasting genetic diversity relationships are revealed in rice (Oryza sativa L) using different marker types. *Molecular Breeding,* 3: 115-125.

Reilly, J.F., A.J. Horne and C.D. Miller, 2000. Nitrate removal from a drinking water supply with large free-surface constructed wetlands prior to groundwater recharge. *Ecological Engineering,* 14: 33-47.

Sartoris, J.J., J.S. Thullen, L.B. Barber and D.E. Salas, 2000. Investigation of nitrogen transformations in a southern California constructed wastewater treatment wetland. *Ecological Engineering,* 14: 49-65.

Shimatani, Y., M. Hosomi and K. Nakamura, 2003. *Water Quality Improvement by ecotechnology.* Soft Science, Inc., Tokyo.

Spieles, D.J. and W.J. Mitsch, 2000. The effects of season and hydrologic and chemical loading on nitrate retention in constructed wetlands: a comparison of low- and high-nutrient riverine systems. *Ecological Engineering,* 14: 77-91.

vanderToorn, J., 1972. Variability of Phragmites australis (Cav.) Trin ex Steudel in relation to the environment. *Van Zee tot Land,* 48: 1-122.

Vymazal, J., 2005. Constructed wetlands for wastewater treatment. *Ecological Engineering,* 25: 475-477.

Wang, N.M. and W.J. Mitsch, 2000. A detailed ecosystem model of phosphorus dynamics in created riparian wetlands. *Ecological Modelling,* 126: 101-130.

Watkinson, A.R. and J.C. Powell, 1993. Seedling Recruitment and the Maintenance of Clonal Diversity in Plant-Populations - a Computer-Simulation of Ranunculus-Repens. *Journal of Ecology*, 81: 707-717.

Yuji, S. and M. Hosomi, 2002. *Killifish and Reed.* IWANAMI SHOTEN, Tokyo, 186 pp.

Zhou, S. and M. Hosomi, 2008. Nitrogen transformations and balance in a constructed wetland for nutrient-polluted river water treatment using forage rice in Japan. *Ecological Engineering*, 32: 147-155.

INDEX

D

E

N

T